S0-BUA-574

ELECTRONIC CONCEPTS, PRINCIPLES, AND CIRCUITS

ELECTRONIC CONCEPTS, PRINCIPLES, AND CIRCUITS

Charles F. Wojslaw
San Jose City College

Reston Publishing Company, Inc.
A Prentice-Hall Company
Reston, Virginia

Library of Congress Cataloging in Publication Data
Wojslaw, Charles F
   Electronic concepts, principles, and circuits.

   Includes bibliographical references and index.
   1. Electronic circuits.   2. Electronics. I. Title.
TK7867.W58   621.3815   80-143
ISBN 0-8359-1660-X

© 1980 by Reston Publishing Company, Inc.
A Prentice-Hall Company
Reston, Virginia 22090

All rights reserved. No part of this book may be reproduced, in any way or by any means, without permission in writing from the publisher.

10 9 8 7 6 5 4 3 2 1

Printed in the United States

*To my beautiful wife, Toni
and our children, Christopher and Nicole*

CONTENTS

PREFACE xi

Part I ELECTRONIC CONCEPTS, PRINCIPLES, AND CIRCUITS 1

1 INTRODUCTION 3
1.1 The Electronics Industry, 3
1.2 The Electronics Profession, 5
1.3 What Is a Concept?, 6
1.4 What Is a Principle?, 6
1.5 Basic Principles, 7
1.6 Applications of the Concepts and Principles, 15

2 AMPLIFICATION CONCEPTS AND CHARACTERISTICS 17
2.1 The Concept of Amplification, 17
2.2 Amplifier Characteristics, 18
2.3 Amplifier Components, 22
Problems, 30

3 AMPLIFIER DC PRINCIPLES AND BIAS CIRCUITS 33
3.1 Introduction, 33
3.2 Dc Equivalent Circuit, 34
3.3 Bias Circuit Models, 35
3.4 Voltage-Divider Bias Circuit, 35
3.5 Voltage-Divider Bias Circuit With Two Dc Supplies, 38
3.6 Collector Feedback Bias Circuit, 39
3.7 Modified Collector Feedback Bias Circuit, 41
3.8 Emitter-Bias Circuit, 42
3.9 Dc Biasing of AM Radio Circuits, 43
3.10 Bias Circuit Design, 45
3.11 An Introduction to Stability Factors, 47
3.12 JFET Bias Circuits, 49
Problems, 51

4 AMPLIFIER AC PRINCIPLES 53
4.1 Introduction, 53
4.2 Amplifier Ac Equivalent Circuit, 54
4.3 Transistor Ac Model or Equivalent Circuit, 58
4.4 Small-Signal Voltage Gain, 60
4.5 Input Resistance, 67
4.6 Output Resistance, 72
4.7 Common-Base (CB) Amplifier, 75
4.8 Interfacing, 76
4.9 Effect of Source Resistance on Voltage Gain, 76
4.10 Current Gain, 78
4.11 Power Gain, 82

4.12　Power and Voltage Gain of Cascaded Stages, 84
4.13　Differential Amplifier, 85
4.14　JFET Common Source Amplifier, 87
Problems, 88

5　AMPLIFIER FREQUENCY RESPONSE　91
5.1　Upper and Lower Cutoff Frequencies, 91
5.2　Lag Network, 92
5.3　Amplifier Lag Networks, 97
5.4　Lead Network, 102
5.5　Amplifier Lead Networks, 106
5.6　Bandwidth, 109
5.7　Sinusoidal and Pulsed Responses, 109
Problems, 110

6　POWER AMPLIFIERS　113
6.1　Definition, 113
6.2　Dc Load Line, 113
6.3　Ac Load Line, 114
6.4　Classes of Operation, 116
6.5　Class A Power Amplifiers, 118
6.6　Class B and Class AB Power Amplifiers, 124
6.7　Power Amplifier ICs and Applications, 130
6.8　Class C Operation, 133
Problems, 134

7　AMPLIFIER INTEGRATED CIRCUITS AND APPLICATIONS　137
7.1　Operational Amplifier Fundamentals, 138
7.2　Integrated Operational Amplifier, 141
7.3　Feedback and the Amplifier, 142
7.4　Operational Amplifier Applications, 146
7.5　Advanced Amplifier Applications, 156

7.6　Data Sheet Parameters, 161
7.7　Practical Considerations in an Amplifier Circuit, 167
7.8　Inverting Amplifier Circuit Design, 169
7.9　Current-Mode Operational Amplifier, 170
7.10　Comparators, 176
Problems, 182

8　REGULATION CONCEPTS AND PRINCIPLES　185
8.1　The Concept of Regulation, 185
8.2　Power Supply Fundamentals, 186
8.3　Voltage Regulator Characteristics, 188
8.4　Basic Regulator Circuits, 189
Problems, 194

9　REGULATOR INTEGRATED CIRCUITS AND APPLICATIONS　197
9.1　History of Integrated Regulators, 197
9.2　Integrated Positive Regulator, 198
9.3　Positive Regulator Applications, 202
9.4　Negative Regulator Applications, 204
Problems, 207

10　OSCILLATION CONCEPTS AND PRINCIPLES　209
10.1　The Concept of Oscillation, 209
10.2　Oscillator Characteristics, 210
10.3　Theory of Oscillation, 212
10.4　Basic Oscillator Circuits, 215
Problems, 222

11　OSCILLATORS, INTEGRATED CIRCUITS, AND APPLICATIONS　225
11.1　Other Sinusoidal Oscillators, 225
11.2　Oscillator Integrated Circuits, 230

11.3 Phase Locked Loop, 236
11.4 Local Oscillator of an AM Radio, 238
Problems, 241

12 FILTER CONCEPTS AND PRINCIPLES 243
12.1 The Concept of Filtering, 244
12.2 Filter Types, 244
12.3 Filter Characteristics, 244
12.4 Basic Filters, 250
Problems, 254

13 ACTIVE FILTERS AND APPLICATIONS 257
13.1 Active Filter Characteristics, 257
13.2 First-Order Active Filters, 258
13.3 Second-Order Active Filters, 260
13.4 Filter Applications, 263
Problems, 270

Part II SEMICONDUCTOR, TRANSISTOR, AND INTEGRATED CIRCUIT THEORY 271

14 SEMICONDUCTOR, PN JUNCTION, AND TRANSISTOR THEORY 273
14.1 Semiconductor Theory, 273
14.2 *pn* Junction Theory, 281
14.3 Transistor Theory: Bipolar Transistors, 288
14.4 Transistor ac Equivalent Circuits, 291

15 INTEGRATED CIRCUIT THEORY 293
15.1 Definitions, 293
15.2 Classifications, 294
15.3 Features, 295
15.4 IC Fabrication, 296
15.5 Planar Process, 298
15.6 Semiconductor Material, 299
15.7 Bipolar Processing, 300
15.8 IC Components: Bipolar, 306
15.9 References for Topics in Chapters 14 and 15, 314

APPENDIX A DATA SHEETS OF REPRESENTATIVE ICs & TRANSISTORS USED IN TEXT 315
B SELECTED PROOFS & & DEVELOPMENTS 330

INDEX 335

PREFACE

A limited number of basic circuits are used as building blocks to form analog systems. Each of these basic circuits illustrates a fundamental concept. While the circuit technology implementing the concepts has changed over the past few years, the concepts or ideas behind the circuits have remained the same. The interrelationships within the circuit are still established by universal circuit principles and are only slightly modified by the characteristics of the particular active device used.

A few electronics books have been used for decades and have withstood the test of time, quite an accomplishment in the rapidly advancing electronics field. It is my belief that the reason for the sustained success of these books has been the ability of the author to clearly articulate the concepts and principles of the topics presented. In other words, the emphasis and focus have been on the ideas behind the electronic circuits, rather than on the circuits themselves. In this manner, the author was able to transcend the current circuit design technique. With this thought in mind, the framework of this book was woven around the ideas and laws governing the behavior of a number of classes of circuits. This technique, hopefully, will carry the student through subsequent generations of circuits without rendering him obsolescent.

The key analog concepts presented in this book are amplification, regulation, oscillation, and filtering. The concepts are illustrated using two generations of circuits: the discrete and the integrated. The theory of operation of the active devices found in the circuits is provided in a supplemental section and within the text material.

This book is written for electronic technology students, for use in core courses in the second year of a two-year technology program. Prerequisites include dc and ac theory and some schooling in complex numbers and circuit theorems. The value and importance of complex numbers and circuit theorems is clearly demonstrated in their application in this text.

My sincere appreciation goes to my colleagues from industry and the college. Their willingness to share their knowledge and skills has allowed me to grow as a technologist and an educator. I offer a special thank-you to my students who used the original manuscript; to Bill Nash, a distinguished colleague; and to Toni, my loving wife.

ELECTRONIC CONCEPTS, PRINCIPLES, AND CIRCUITS

The most complex electronic entity is called a *system*. Each electronic system may be broken down into subunits, each comprising a number of circuits. Each circuit represents a unique organization of electronic components, and it performs a high-level function. The purpose of this section is to define the conceptual basis of these circuits and to analyze and mathematically describe the various interrelationships within the circuit. A comprehensive coverage of all circuits is not possible in a single book; thus, the coverage is limited to the key concepts and circuits found in the analog domain. These concepts include amplification, regulation, oscillation, and filtering. The corresponding circuits are the amplifier, regulator, oscillator, and filter. The concepts are illustrated using simple discrete component circuits and integrated circuits, or ICs.

in the beginning.....

INTRODUCTION

Electronics is a branch of physics, a technical discipline that has spawned an industry whose growth curve is now beginning to accelerate. This acceleration is evidenced by the many products currently impacting the consumer, and by the great need for trained personnel. The training of electronics people requires the concurrent development of high-level conceptual and manipulative skills, and these skills must be sound and fundamental to overcome obsolescence as the technology of the discipline advances. A sound foundation is laid through the development of the concepts and principles behind electronic circuits. This type of foundation transcends any type of technology used to build the circuits, be it the assembly of discrete conponents or the equivalent integrated versions. Certain principles have universal application, and although they are covered in more elementary texts, they are reexamined here in light of their application to the circuits presented in this book.

1.1 THE ELECTRONICS INDUSTRY

The electronics industry is becoming an important segment of our national economy. The widespread movement of electronic products into the consumer, business, and industrial markets has dramatically increased the growth of the industry and the sale of its equipment. This growth has been possible because of technological developments in the production of electronic products. These developments have resulted in high-performance, low-cost, reliable, lightweight, small, and energy-conserving devices and systems. Many industries now look to the electronics industry for solutions. The future for the electronics field is indeed bright, and its growth should continue as its products reach further and further into other industries.

Electronic products can be seen all around us. They are found where we live, work, and play. The market area where electronic products have had the most dramatic impact is the consumer area. The low-cost items sold in this marketplace sell in large quantities and represent gross sales in the millions and millions of dollars. If one translates these sales to jobs, it is understandable why the need for skilled electronic technologists exists.

Mechanical timepieces, such as clocks and watches, are proven products. They are economical and reliable. Yet the competitive price, the numerical readout, the additional reliability, and the attractiveness of luminescent displays have prompted people to buy digital clocks and watches. The watch and clock industry, once centered abroad, is now moving into the silicon valleys of the major electronic centers.

The electronic calculator has displaced the slide rule and other mechanical computing machines and is now used by the housewife, student, and professional. The numerical data and calculations housed in books are now readily available at out fingertips. Programmable versions of the calculators adapt this tool to problem-solving tasks in various other industries.

Attachments to our television sets provide us with new educational and entertainment dimensions during our leisure hours. These attachments, with interchangeable cartridges, have further promoted television as the major form of home entertainment. Lower power consumption, solid-state reliability, electronic tuning, built-in clocks, and larger viewing screens are a few of the recent features added to enhance the value and pleasure of television itself. Video tape recorders and closed-circuit television systems allow us to record our favorite shows and to produce our own.

The electronic industry's fight for a greater portion of the entertainment dollar is further carried by high-fidelity and quadrasonic stereo systems, AM and FM receivers, electronic organs, cameras, jukeboxes, pinball machines, and tape recorders. Of course, the list will not end with these products.

The need for greater economy, reliability, efficiency, and control has prompted automobile manufacturers and other transportation companies to look to electronics for solutions. Digital clocks, seat-belt alarms, electronic tachometers and ignitions, and controls for pollution, fuel, speed, and the drive train are being added to most new cars. Computer-controlled mass-transit systems with their electrically operated cars offer an alternative to the automobile in urban areas. Airplanes replaced their hydraulic controls with electrically operated controls and now use computers to monitor the myriad of airborne electronic subsystems. The avionics systems of airplanes represent a significant portion of the plane's cost.

The effect of computers on our day-to-day living is profound. Computers come in three sizes: small, medium, and large; but it has been the small computer that has created the uproar. The power of the one-chip computer called a microprocessor has prompted business and industry to adapt it to their applications. Production is increased and routine, boring human tasks are eliminated through the microprocessor-based process control equipment found in mills, refineries, and print shops.

It would be economically impractical to maintain the money transactions at banks, the sale of stocks at the exchanges, and the goods inventory at large stores, or to generate payrolls for businesses without the computer. Electronic cash registers and point-of-sales (POS) systems are rapidly appearing at the check-out stands of grocery stores and small retail stores. Economy is not the only benefit reaped from the use of a computer. Accuracy, speed, and efficiency make the computer an invaluable business and industrial tool.

Computers, more than any other electronic product, have brought out the need for interdisciplinary people. The extent of the application of the computer in the business, industrial, transportation, and medical fields would be much greater if skilled electronic technologists possessed a high degree of acumen in the field of the computer's application. Alternately, skilled people in other disciplines with a knowledge of electronics are difficult to find.

The heartbeat of the world is continuously monitored by the communications media. No longer are countries and their peoples just dots on a map. They are as close as our living rooms. Low-power, lightweight electronic satellites relay the events of the world to us as they happen. Visual communications are not limited to television. Sections of the country are already using video telephones, and only the cost of replacing the older systems is hindering their widespread usage. The popularity of citizen's band (CB) transceivers began with ham operators. Truck drivers adapted it for truck-to-truck communication, and now large-scale car-to-car and car-to-home CB transmission is imminent. Children mimic their parents using walkie-talkies. Answering service monitors, dictaphones, typewriters, and reproduction machines add to the list. The electronic black boxes associated with lasers and fiber-optics further push progress in the communications field.

People, as they live better, want to live longer. The medical profession removes cataracts, seals wounds, and burns cancerous tissues using lasers. Microminiature pacemakers help people's hearts beat. Critically ill patients are monitored in intensive-care units by sophisticated electronic equipment. Fluid analyzers, fluid-processing machines, X-ray scanners, and prosthetic devices are added to the number of products where electronics have helped to aid the medical profession.

The center of the educational system is still the student and the teacher. Learning begins with teaching and continues forever through studying. However, this procedure today is supplemented with closed-circuit TV (CCTV), video-tape players, interactive person–machine interfaces, film strips, and other audiovisual aids.

The hush-hush military and aerospace industries are at the forefront in using the sophistication and power of electronics. The communications, navigation, radar, sonar, and fire-control systems in military hardware represent the ultimate in electronic sophistication.

1.2 THE ELECTRONICS PROFESSION

The highlighting of the electronic products found in other fields has revealed the significant and increasing influence of the electronics industry. This influence can be translated to gross sales, which in turn can be expressed as job opportunities, both now and in the future. As the industry grows, opportunities and positions will increase for skilled people. Not only will the demand for engineers, technicians, assemblers, programmers, and technical supervisory personnel increase, but so will the need for the large number of people in other disciplines whose skills are necessary to operate a successful company.

The technological advancements made by the industry must be credited to the knowledge, skill, and creativity of its personnel. These achievements are considerable when one looks back at the level of technology just ten years ago. By today's standards that level appears archaic, and probably what we do today will appear archaic ten years from now.

The technological developments and the growth of the companies within the industry have placed a demand on its employees. More than ever, it is becoming necessary that employees update their skill level to the point where their capabilities match the demands of the job. It is now common to see practicing technologists returning to the classroom to update or increase their knowledge in new areas. This is occurring at all levels, and the demand is reaching back to the educational institutions.

Colleges are required to provide the student with the knowledge and skill necessary for employment. Implied in this charter is the fact that the student's skill level must be near that practiced by industry today. With electronics technology rapidly advancing, *educational systems must provide the all-important fundamental concepts AND illustrate them with current devices and circuits.*

1.3 WHAT IS A CONCEPT?

A concept is a general notion or idea. It is an idea of something formed by mentally combining all the idea's characteristics or particulars. A concept is a *word or group of words that conveys a mental image.* It is not something physical but may be *associated* with an object or a person. Beauty is a concept. It is a word that conveys a mental image of a thing or person that delights the senses or the mind. Beauty is the concept, and an art object or an attractive person are physical items that are described as beautiful.

To a large extent, *electronics is based on concepts.* The concept of *energy*, paramount in the discipline, is the ability to do work. This far-reaching concept is the basis for many types of physical systems, including mechanical, light, thermal, nuclear, chemical, radiant, and electrical. Energy exists in two forms, kinetic and potential. Kinetic electrical energy, or the energy of motion, is called *current.* Potential electrical energy, or the energy of position, is called *potential*, *potential difference* (pd), or *voltage*. The electronic technologist must be skilled in the measurements of these energy types and their effect, control, and application.

At the component level, *resistance* is the concept of converting energy, and *capacitance and inductance* are the energy-storing concepts. The components whose behavior, in general, is associated with these concepts are called *resistors*, *capacitors*, and *inductors*.

The *advanced electronic concepts* of amplification, regulation, filtering, oscillation, and so on, describe the abilities of reasonably complex circuits. These circuits are used as functional blocks to form electrical systems that perform sophisticated tasks that benefit people.

1.4 WHAT IS A PRINCIPLE?

A principle is a fundamental, primary, or general law or truth from which others are derived.

In any given circuit, there will be a number of voltages and currents. The objective of analyzing a circuit is to find their values and to express their interrelationships. Often, the input and output relationship will identify and define the circuit's function. The derivation of these relationships will be through the employment or application of fundamental laws.

The understanding of how a circuit functions, the interrelationships of the potentials and currents, the determination of numerical values, and the input–output relationship all depend upon the thorough understanding of a small number of *basic principles*. For as complicated as electronics seems, the mastery of this discipline hinges on the continual application of these fundamental laws. When these basic principles are applied to a unique circuit or system, new principles or laws will be developed that are more specific. They will be more closely tied to describing a given relationship in the circuit they are associated with.

It is impossible to remember the large number of principles and equations that describe electronic circuits, even a particular one. However, if the basic laws are thoroughly understood and one understands how that particular circuit functions, then the specific relationships can easily be developed. *The key is the thorough understanding of the basic principles and how a circuit functions.*

Because of the importance of these basic laws to the circuits we will be discussing, they will be briefly reexamined. These laws should have

BASIC PRINCIPLES

been previously studied in depth, and they are thoroughly covered in dc and ac circuits books and courses. The *objective of this coverage is to*

1. Rekindle the reader's interest in these basic principles.
2. Highlight their meaning.
3. Illustrate them with circuits appropriate to the material in the book.

1.5 BASIC PRINCIPLES

Kirchhoff's Voltage or Potential Law (KVL)

This law states that the algebraic sum of the potential differences or voltages around any closed path of a circuit is zero.

The circuit of Figure 1-1a contains one three-terminal and three two-terminal unidentified components. Applying KVL to the three closed paths or loops, we obtain

$V_1 + V_2 + V_{BE} = 0$ (loop 1)

$V_3 + V_{CB} + V_2 = 0$ (loop 2)

$V_1 + V_3 + V_{CE} = 0$ (loop 3)

V_{BE}, V_{CB}, and V_{CE} represent the three potential differences associated with the three-terminal device. V_1, V_2, and V_3 represent the potential drops or rises associated with the two-terminal components.

The polarities of the component voltages are usually known or can be inferred. For sources, the + and − terminals are labeled. For a specific transistor type, the directions of the currents and the polarities of the terminal voltages are specified if the device is to operate in a particular mode. If a component is in series with a transistor's lead, the direction of its current and hence its voltage polarities are also established. The direction of the loops is arbitrarily chosen and basically represents the flow of the loop currents. In writing a loop equation, if the current direction matches the polarity of the potential difference, a + sign is assigned to the variable. Conventional current entering a terminal will establish that terminal as the more positive end. It is important to remember that the signs associated with a component's potential difference identify *only* the more positive or more negative side. These signs *do not* identify a terminal's polarity with respect to ground or the circuit reference. A negative sign is assigned to the variable if the loop current is opposite to that of the polarity of the potential

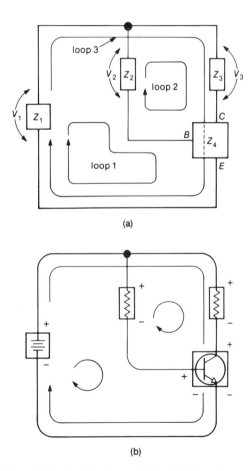

Figure 1-1. Application of Kirchhoff's voltage law: (a) loops; (b) voltage polarities.

drop it would establish. For the polarities and directions assigned in Figure 1-1b, we obtain

$$-V_1 + V_2 + V_{BE} = 0 \quad \text{(loop 1)}$$
$$V_3 + V_{CB} - V_2 = 0 \quad \text{(loop 2)}$$
$$V_1 + V_3 + V_{CE} = 0 \quad \text{(loop 3)}$$

There are six variables or unknowns in the above equations. At least three of them must be known to solve the remaining variables for their numerical values. A set of equations can only be solved if there are as many unknowns (or less) as there are independent equations. The equations can be solved using any of the substitution or simultaneous equations techniques. Typically, several of the quantities will either be a known constant or will be specified due to some condition.

For loops that have voltage sources, the sum of the voltage drops equals the sum of the voltage rises. The voltage rises are the source voltages. If component 1 is a voltage source in our example, then

$$V_2 + V_{BE} = V_1 \quad \text{(loop 1)}$$

where V_1 is the voltage rise and V_2 and V_{BE} are the voltage drops.

Kirchhoff's Current Law (KCL)

This law states that the algebraic sum of the currents entering and leaving a node is zero. Simply, whatever goes in must come out. A point at which two or more components have a common connection is called a *node*.

The circuit of Figure 1-2a contains one three-terminal and four two-terminal components. The components are not identified. Applying KCL to the four nodes, we obtain

$$I_1 + I_4 = 0 \quad \text{(node 1)}$$
$$I_4 + I_C + I_2 = 0 \quad \text{(node 2)}$$
$$I_2 + I_B + I_3 = 0 \quad \text{(node 3)}$$
$$I_E + I_3 + I_1 = 0 \quad \text{(node 4)}$$

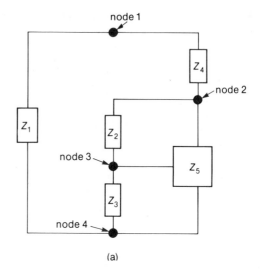

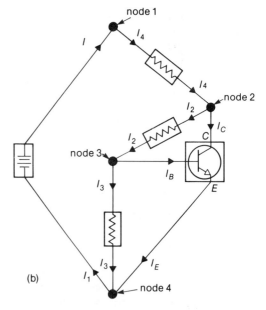

Figure 1-2. Application of Kirchhoff's current law: (a) nodes; (b) current polarities.

I_E, I_B, and I_C represent the three-terminal component currents, and I_1, I_2, and I_3 represent the two-terminal component currents.

The assignment of a polarity sign to the current entering a node is arbitrary, but the

BASIC PRINCIPLES

opposite sign must be used to designate currents leaving it. A plus sign is used to designate currents entering a node in Figure 1-2b. For the indicated current directions, we obtain

$I_1 - I_4 = 0$ (node 1)
$I_4 - I_C - I_2 = 0$ (node 2)
$I_2 - I_B - I_3 = 0$ (node 3)
$I_E + I_3 - I_1 = 0$ (node 4)

Circuit currents, typically, vary from 10^{-7} ampere (A) to 10^{-2} A. If a current(s) is smaller than the others by a factor of 100, it can be neglected or assumed to be zero for all practical purposes. This, often, will greatly simplify a relationship or equation. The same rationale can be used for voltages; however, the typical range of potential values is not as great. In the above example, if I_B is significantly smaller ($\frac{1}{100}$) than I_2 or I_3, then we can say

$I_2 \cong I_3$ (node 3)

When an equation has been solved, the answer will contain a + or − sign. If the answer is positive, then the assumed direction of the current (or the polarity of the voltage drop) is correct. If the answer is negative, then the actual current direction (or voltage polarity) is opposite to that of the assumed direction.

Kirchhoff's voltage and current laws are derived from the universal conservation of energy principle.

Thevenin's Theorem

Electronic components arranged in a specific manner make a *circuit*. Circuits arranged in a specific manner make an *electronic system*. The relationships and effects of circuits on each other are very important. For most cases, the interest in circuits is not in the detailed behavior of the components inside the circuit, but in the terminal behavior at the circuit's output terminals. This type of approach is often referred to as the *black-box concept*. Thevenin's theorem is an analysis tool that allows a complex circuit with two output terminals to be replaced by a single voltage source in series with a resistor. *This voltage source*, called a Thevenin or open-circuit voltage, *and* the resistor, called *the Thevenin resistance, will generate at the output terminals the same potential and current as the circuit it replaces*. The voltage is designated as V_{TH} and the resistance is designated as R_{TH}. The complete circuit shown in Figure 1-3 can be replaced by a Thevenin equivalent circuit that will generate the same potential and current at the load as the circuit it replaced. If various loads were to be connected to the output of this circuit, the effect of each load could be easily determined by considering its effect on the Thevenin circuit, rather than the complex original circuit.

The values for V_{TH} and R_{TH} are obtained by the following procedure or recipe. *To obtain the value of V_{TH}:*

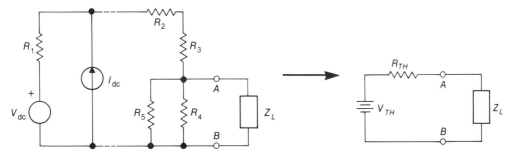

Figure 1-3. Simplifying a circuit using Thevenin's theorem.

1. Disconnect, between the two terminals of interest, the circuit or component that is *not* going to be Theveninized; that is, open circuit the load.

2. Determine (by measurement or calculation) the voltage between these two terminals. This open-circuit voltage is V_{TH}.

To obtain the value of R_{TH}:

1. Open circuit or disconnect the load.

2. Set all sources to zero. Ideal voltage sources become short circuits and ideal current sources become open circuits.

3. Determine (by measurement or calculation) the resistance between the two terminals of interest. This resistance is the Thevenin resistance R_{TH}.

An application of Thevenin's theorem is shown in Figure 1-4a. The components Z_3, Z_4, Z_5, and Z_6 represent a load between terminals A and B. The circuit to the left of these terminals is to be Theveninized. With the load disconnected, V_{TH} or the open-circuit voltage (V_{OC}) will be the voltage drop across R_2. Resistors R_1 and R_2 form a series circuit with the source V_{CC}, and the voltage V_2 will be some proportion of V_{CC}. According to voltage-divider action,

$$V_2 = V_{TH} = V_{OC} = \frac{R_2}{R_1 + R_2} V_{CC}$$

To determine R_{TH}, V_{CC} is set to zero volts or is replaced with a short circuit. The resistance between terminals A and B is the parallel combination of R_1 and R_2. Thus R_{TH} is

$$R_{TH} = R_1 \| R_2 = \frac{R_1 R_2}{R_1 + R_2}$$

The simplified circuit is shown in Figure 1-4b. The potential and current associated with terminals A and B are the same for both circuit versions. Thevenin's theorem applies to ac and dc linear circuits.

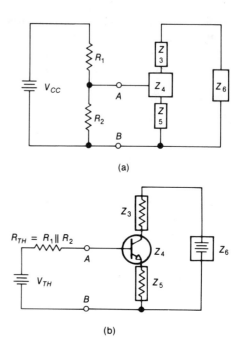

Figure 1-4. Application of Thevenin's theorem: (a) complete circuit; (b) simplified, equivalent circuit.

Norton's Theorem

Norton's theorem is the dual of Thevenin's theorem. Thevenin's theorem allowed us to replace a circuit, connected between two terminals, with a single voltage source and a series resistor. *Norton's theorem allows us to replace a circuit (or a portion of it) with a single current source and a parallel resistor.* Norton's theorem is an analysis tool that also simplifies circuits or converts them to an equivalent form. The current source in the Norton equivalent circuit is called I_N, or *the Norton current*, and the parallel resistor is called R_N, or *the Norton resistance*. An example of this type of circuit reduction is depicted in Figure 1-5.

The values for I_N and R_N are obtained by the following procedures or recipes. *To obtain the value of I_N:*

BASIC PRINCIPLES

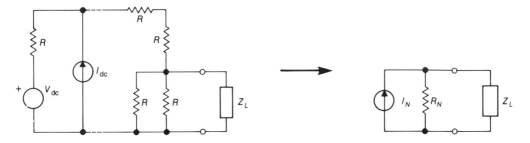

Figure 1-5. Simplifying a circuit using Norton's theorem.

1 Disconnect, between the two terminals of interest, the circuit or component that is *not* going to be Nortonized; that is, open the load.

2. Connect a short between the two terminals and determine the current (by measurement or calculation) through the short. This short-circuit current is I_N.

To obtain the value of R_N:

1. Open or disconnect the load.

2. Set all sources to zero. Ideal voltage sources become shorts (0 V), and ideal current sources become opens (0 A).

3. Determine the resistance between the two terminals of interest. This resistance is R_N.

The components $Z_3, Z_4, Z_5,$ and Z_6 in Figure 1-6a represent a load between the terminals a and b. The circuit to the left of these terminals is to be Nortonized. The load is disconnected and terminals a and b are shorted. The current through this short is the Norton current, I_N, and is also referred to as the short-circuit current, I_{SC}. The short shunts R_2, and the circuit that remains is a single voltage source, V_{CC}, and a resistor, R_1. According to Ohm's law,

$$I_{SC} = I_N = \frac{V_{CC}}{R_1}$$

To determine R_N, V_{CC} is set to zero volts or is replaced with a short circuit. The resistance between terminals a and b is the parallel combination of R_1 and R_2. Thus,

$$R_N = R_1 \| R_2 = \frac{R_1 R_2}{R_1 + R_2}$$

The simplified circuit is shown in Figure 1-6b. The potential and current associated with ter-

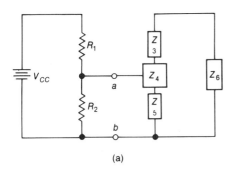
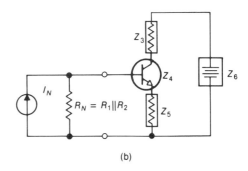

Figure 1-6. Application of Norton's theorem: (a) complete circuit; (b) simplified, equivalent circuit.

minals *a* and *b* are the same for both circuit versions.

The circuits chosen to illustrate the basic principles are amplifier dc bias circuits. They represent practical examples of the application of network theorems, and we will see them again. The identification of the components was obscured to allow us to focus on the universality of these fundamental laws. In general, Thevenin's theorem is more frequently employed because most signal sources are potential sources. A circuit can either be Theveninized or Nortonized.

Miller's Theorem

Perhaps a more appropriate place for this theorem would be with the discussion of amplifiers in feedback circuits. However, in reality all circuits with active devices contain some degree of feedback, and thus Miller's theorem has widespread application. The feedback in circuits may be introduced intentionally, or it may be present owing to the device and distributed components. The analysis and understanding of these circuits are greatly eased through the use of this theorem, and hence it deserves sharing the limelight with the other powerful basic principles or analytical tools.

Miller's theorem applies to amplifying devices or circuits (transistors and operational amplifiers) with components or networks connected from the output back to the input (feedback). We will confine our coverage of Miller's theorem to negative feedback circuits. Negative feedback implies the connection of a component from the output back to the (inverting) input such that the effect of the input signal is diminished. The interlead or can capacitance from the collector to base of a transistor is an example of an amplifying device with negative feedback.

Miller's theorem states that an impedance connected from the output back to the input (inverting) of an amplifying device or circuit can be replaced by an

(a) *equivalent impedance from the input to ground* AND an

(b) *equivalent impedance from the output to ground.*

If Z_F is the feedback impedance (Figure 1-7), then the equivalent input impedance is

$$Z_{in} \text{ (Miller)} = \frac{Z_F}{A+1}$$

and the equivalent output impedance to ground is

$$Z_{out} \text{ (Miller)} = Z_F \frac{A}{A+1}$$

A is the voltage gain of the amplifying device or circuit. Figure 1-7b illustrates the feedback impedance transformed into the Miller equivalent input and output impedances.

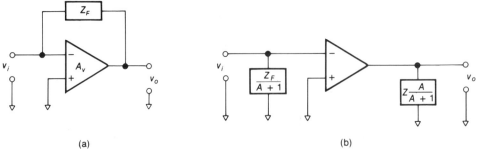

Figure 1-7. Miller's theorem: (a) amplifier circuit with feedback; (b) millerized circuit.

Z_F can represent any two-terminal component or circuit, but the resistor and the capacitor are the most frequently used cases. The capacitor C_F in the amplifier circuit of Figure 1-8b is connected from the transistor's collector back to the base. The base and collector signals are 180° out of phase with respect to each other, and the capacitor represents a negative feedback connection. This feedback capacitance can be translated to a capacitance from the base to ground and from the collector to ground by employing Miller's theorem. The voltage gain of the circuit is the ratio of the resistances of the collector and emitter.

$$A_v \cong \frac{r_C}{r_E}, \quad r_E \gg r'_e$$

The input Miller capacitance is

$$C_{in} \text{ (Miller)} = C_F(A_v + 1) = C_F\left(\frac{r_C}{r_E} + 1\right)$$

The output Miller capacitance is

$$C_{out} \text{ (Miller)} = C_F\left(\frac{A_v + 1}{A_v}\right) = C_F\left\{\frac{\left(\frac{r_C}{r_E}\right) + 1}{\frac{r_C}{r_E}}\right\}$$

The *Miller input impedance* equals the feedback impedance *divided by (A + 1)*. The *Miller input capacitance* equals the feedback capacitance *multiplied by (A + 1)*. The Miller input capacitance is larger than the feedback capacitance because impedance and capacitance are inversely proportional; that is, $Z_C = 1/(2\pi f C)$ in the steady state.

For the circuit in Figure 1-8b, $R_C = 5.1$ kilohms (kΩ), $R_E = 470$ ohms (Ω), and $C_F = 25$ picofarads (pF). The Miller input capacitance is

$$C_{in} \text{ (Miller)} = C_F(A_v + 1)$$
$$= 25 \text{ pF} \left(\frac{5100}{470} + 1\right) = 296 \text{ pF}$$

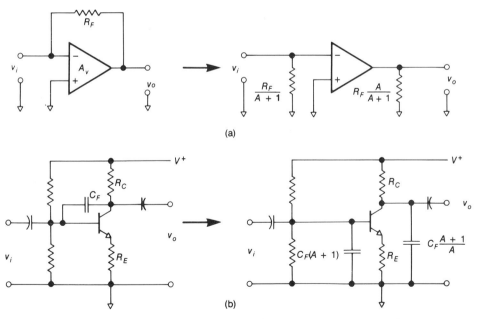

Figure 1-8. Application of Miller's theorem: (a) feedback resistor case; (b) feedback capacitor case.

The output Miller capacitance is

$$C_{out} \text{ (Miller)} = C_F \frac{A_v + 1}{A_v}$$

$$= 25 \text{ pF} \left\{ \frac{(5100/470) + 1}{5100/470} \right\}$$

$$= 27.3 \text{ pF}$$

For reasonably high values of voltage gain, the value of the Miller output capacitance is about the same as the value of the feedback capacitance.

The resistor R_F in the amplifier circuit of Figure 1-8a is connected from the output of the amplifier back to the inverting (−) input. This feedback resistance can be translated to a resistance from the input to ground and from the output to ground using Miller's theorem. If the voltage gain of the amplifier is A_v, the Miller input resistance is

$$R_{in} \text{ (Miller)} = \frac{R_F}{A_v + 1}$$

and the output resistance is

$$R_{out} \text{ (Miller)} = R_F \left(\frac{A_v}{A_v + 1} \right)$$

For the circuit in Figure 1-8a, the voltage gain is 100,001 and the feedback resistor is 100 kΩ.

$$R_{in} = \frac{R_F}{A_v + 1} = \frac{100 \text{ k}\Omega}{100{,}001} \cong 1 \text{ }\Omega$$

$$R_{out} = R_F \frac{A_v}{A_v + 1} \cong R_F, \quad \text{since } A_v \gg 1$$

Superposition Theorem

The *superposition theorem* states that *in a linear network with two or more sources the current or voltage for any terminal or component is the algebraic sum of the effects produced by each source acting separately.*

The application of this theorem is effective in circuits with multiple sources. In a circuit of this type, a particular voltage or a particular current may be determined by algebraically summing the voltages or currents from each source acting alone. To see the effect of only one source, all other sources are turned off or set to zero volts or zero amperes. This is accomplished by shorting ideal voltage sources (0 V) and opening ideal current sources (0 A).

The superposition theorem can be derived from Kirchhoff's current and voltage laws. It, however, provides a greater insight into the circuit when compared to the nodal and mesh analysis techniques associated with KCL and KVL.

An application of the superposition theorem is illustrated in Figure 1-9a. The circuit consists of two voltage sources, V_{s1} and V_{s2}. The objective of the analysis is to determine V_B (with respect to ground). The circuit is designed such that the current entering the B terminal of Z_2 is extremely small compared to I_1 and can be assumed zero. Thus, $I_1 \cong I_2$, and R_1 and R_2 form a voltage divider with the two sources.

With V_{s1} on and V_{s2} off (shorted), the circuit reduces to a voltage divider with a single source and is shown in Figure 1-9b. The voltage at B due to this source is

$$V_{B1} = \frac{R_2}{R_1 + R_2} (V_{s1})$$

With V_{s1} off (shorted) and V_{s2} on, the circuit reduces to a voltage divider with a single source (Figure 1-9c). The voltage at B due to this source is

$$V_{B2} = \frac{R_1}{R_1 + R_2} (V_{s2})$$

With both sources on,

$$V_B = V_{B1} + V_{B2}$$
$$= \frac{R_2}{R_1 + R_2} (V_{s1}) + \frac{R_1}{R_1 + R_2} (V_{s2})$$

The source V_{s1} will have a positive (+) polarity or sign attached to it and V_{s2} a negative (−) sign. The sign of V_B can be either depending on the resistor and source values.

APPLICATIONS OF THE CONCEPTS AND PRINCIPLES

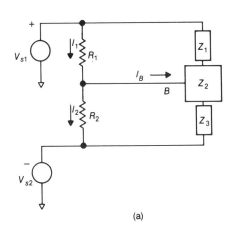

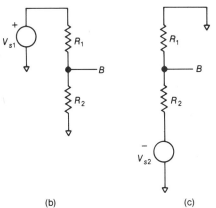

Figure 1-9. Application of the superposition theorem: (a) complete circuit; (b) V_{s1} on and V_{s2} off; (c) V_{s1} off and V_{s2} on.

1.6 APPLICATIONS OF THE CONCEPTS AND PRINCIPLES

The objective of analyzing a circuit is to *theoretically define* the values or evaluate the variables associated with a circuit, and to mathematically model the interrelationships. The values obtained, as a result of the analysis, provide a *basis for comparison with those* values determined by constructing the actual circuit and making *measurements*. If the measured and calculated values correlate within a percentage of difference that can be justified, the theoretical performance and the actual performance of a circuit are confirmed and joined. The successful union of *theory and practice advances the skill level* of the technologist and is not necessarily confined to any particular technical level. A skill level that is salable is highly unlikely without one aspect or the other. While the emphasis for different technologists may shift to some degree from one to the other, the electronics technician is required to display a level of competence in both areas. He must be able to conceptually understand the function and operation of a circuit and to build, troubleshoot, and test it. The culmination of his skill lies in his ability to confirm the theoretical operation of a circuit by testing the physical model and determining if the theory accurately matches the actual circuit.

Electronics, like any physical science, is based upon certain fundamental laws and truths. These laws form the basis of the discipline, and a working knowledge of their meaning is an essential requirement to achieve a reasonable skill level.

The origin of the medium of the discipline is at the atomic level or is microscopic in nature. Since people work at the macroscopic level, one has to rely upon concepts, principles, and models to understand the macroscopic effects that result from the atomic behavior. As long as the macroscopic results confirm, or verify, the sets of words that describe what is happening at the atomic level, those words are accepted as truths.

The concepts, principles, and models represent the working knowledge of the trade. All technical personnel within the discipline communicate by these truths and utilize them as the foundation to establish higher-level concepts and skills. Without them we would be like blind men trying to perform a sophisticated manual art. Possible, but very, very difficult.

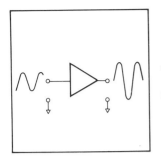

AMPLIFICATION CONCEPTS AND CHARACTERISTICS 2

To amplify means to increase the magnitude. This chapter discusses the concept of amplification, amplifier characteristics, and the devices or components that make up an amplifier.

Amplification *is the idea or concept, and the circuit whose behavior follows the concept is the* amplifier. *The amplifier is described by its characteristics. The characteristics represent the big picture of the amplifier, because they describe the amplifier as a functional block.*

Inside, the amplifier is made up of components, both active and passive. The passive elements arise from three sources; discrete, device, and distributed. The active devices provide the gain in the circuit, and hence they play a key role in amplifiers. The bipolar transistor is the active device emphasized because of its widespread usage in discrete form and its dominant role in integrated circuits. An introduction to the unipolar (JFET) transistor is presented.

2.1 THE CONCEPT OF AMPLIFICATION

Amplification is the act or state of amplifying, or the state of being amplified. To amplify means to make larger or greater; to enlarge; to extend. These definitions apply to the literal interpretation of the word amplification and are obtained from a dictionary. Electrically, to amplify means *to increase the magnitude*, but this definition requires a closer look.

Of all the advanced electronic concepts, amplification is the most important. The requirement to amplify is present in almost all electronic systems. From communications and control to generation and processing, amplifying devices, circuits, and systems abound. The ac voltage amplifier is the most frequently used circuit to employ this concept.

It is common and convenient to view an ac voltage amplifier as a circuit that makes a larger version of its input. However, amplification affects only the magnitude of the input signal. Time is *not* magnified. For sinusoidal, time-varying voltages, the frequency or the time period of one cycle remains constant within the constraints of the amplifier's performance. Using an oscilloscope, we trace, or follow, what appears to be the same signal from its input, through the circuit, to its output. However, the output signal is not really part of the input signal.

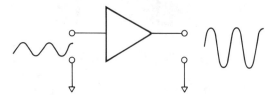

Figure 2-1. Concept of amplification.

The amplifier circuit is a functional block that is powered by a dc voltage source. Its input is usually an ac voltage, and as a result of its input *and* the features of the circuit, the amplifier creates a "replica" of the input signal. The output of the block is then *an image* of the input, but with a greater magnitude in amplitude but of the same frequency.

Amplification is an idea or notion. It reflects the ability of an electronic circuit to develop a replica of the input at the output. The output replica will have a greater amplitude, but both input and output signals will be cyclic with the same time period. Figure 2-1 illustrates the concept.

Most amplifiers are viewed as voltage-gain circuits. However, these same circuits can also produce current and power gain. The labeling of a particular circuit as a voltage amplifier, power amplifier, or current amplifier is dependent on its ultimate application, the emphasis of the circuit design, and the gain values of each variable.

Amplification is not restricted to voltage, power, or current. A circuit that increases the magnitude of any electronic quantity would literally qualify that circuit as an amplifier. Capacitance and resistance are excellent examples of other quantities whose magnitude is increased by amplifier circuits. Commonly, these particular circuits are referred to as capacitance and resistance multipliers. While amplifiers can be a broad category of circuits, emphasis will be directed toward the ac voltage and ac power amplifiers.

2.2 AMPLIFIER CHARACTERISTICS

Symbology

The amplifier, either discrete or integrated, is symbolically represented in block-diagram form by a triangle. This symbol can represent a single- or multiple-circuit amplifier. The amplifier is considered a two-terminal pair network; that is, it has two input terminals and two output terminals. The input and output signals of most discrete amplifiers are referenced to ground or the circuit's reference. The output of most integrated amplifiers is ground referenced, but the input is a difference of potential. Discrete amplifiers are usually powered with a single dc potential source, but the typical IC amplifier uses two. Figure 2-2 shows the schematic representations of the *typical* discrete and integrated amplifiers.

Gain

In an amplifier the input and output voltages are related by the circuit's voltage gain. Mathematically, this is expressed as

$$A_v = \frac{V_{\text{out}}}{V_{\text{in}}}$$

The voltage gain, A_v, is then the ratio of the output voltage to the input voltage. It is a dimensionless number; that is, it does not have a unit of measurement. This number remains constant within the operating limits (usually defined by frequency) of the amplifier and can

AMPLIFIER CHARACTERISTICS

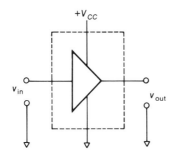
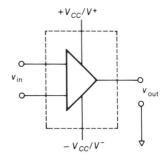

Figure 2-2. Typical amplifier schematic representations.

be any value from less than one to hundreds of thousands. The letter A designates amplification, and its subscript identifies voltage gain, power gain A_p, or current gain A_i.

Primary Characteristics

Gain is certainly the *most important* amplifier *characteristic*. It is, however, only one of five important primary characteristics. They are as follows:

1. *Voltage gain*, A_v
2. *Input resistance*, r_i
3. *Output resistance*, r_o
4. *Upper cutoff frequency*, f_{cu}
5. *Lower cutoff frequency*, f_{cl}

The resistances, r_i and r_o, apply to low- and moderate-frequency amplifiers. They are determined from ac equivalent circuits and are represented using lowercase letters. Uppercase letters designate resistances determined from dc equivalent circuits. For high-frequency amplifiers, input impedance, Z_i, and output impedance, Z_o, would be more appropriate. For dc coupled amplifiers, that is, those amplifiers that have no coupling capacitors, the lower cutoff frequency f_{cl} will be at dc or 0 hertz (Hz). The difference between the upper and lower cutoff frequencies defines the amplifier's bandwidth.

Definitions

Gain (A): the ratio of the output over the input. Normally, the output and input variables are the same and gain is a dimensionless number.

Input resistance (r_i): the equivalent resistance between a circuit's input terminals under ac conditions.

Output resistance (r_o): the equivalent resistance between a circuit's output terminals under ac conditions. This resistance does not include the load.

Upper cutoff frequency (f_{cu}): the higher frequency at which the amplifier's gain is down 3 decibels (dB) (or 0.707) of its midfrequency value.

Lower cutoff frequency (f_{cl}): the lower frequency at which the gain is down 3 dB (or 0.707) of its midfrequency value.

Bandwidth (BW): the frequency range or passband where an amplifier's gain is ideally constant. The bandwidth, Figure 2-3, is defined as $BW = f_{cu} - f_{cl}$.

Secondary Characteristics

The primary characteristics listed for the amplifier pertain to a general-purpose circuit or device. They represent a *generalization*. There will be specialized cases where other charac-

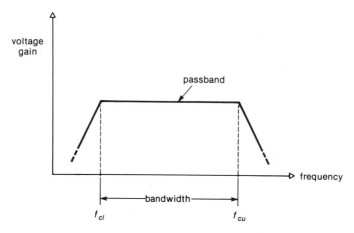

Figure 2-3. An amplifier's bandwidth.

teristics will be paramount because of the application and design of the amplifier.

There are a large number of *secondary characteristics* associated with discrete and monolithic amplifiers. Their importance is a function of the particular amplifier design and its intended application. Many of the secondary characteristics are related and can be categorized. Some of the categories and specific characteristics are as follows:

1. *Noise:* input noise voltage, input noise current, and total input noise. These characteristics are especially important in the preamplifier section of an amplifier system. The sources or types of noise are thermal, shot, 1/f, partition, and popcorn.

2. *Transient response:* rise time, settling time, recovery time, overload recovery, and propagation delay. The performance of high-speed, fast-response amplifiers under pulsed conditions can not be adequately described unless the transient response characteristics are known.

3. *Distortion:* harmonic and total harmonic distortion (THD). This characteristic reflects the purity of an amplifier's ability to reproduce a sinusoidal signal.

In addition to the primary and secondary characteristics, an amplifier's performance is also described with data sheet parameters. These are discussed as an integrated-circuit amplifier topic.

Inputs and Outputs

Amplifiers are used as functional blocks in electronic systems. In a system, the amplifier inputs and outputs will be connected to other circuits. The circuit driving the amplifier's inputs will be a transducer, a source, or another circuit. For any of these cases, *the driving circuit (Figure 2-4) can be modeled as a voltage source with a series resistance (impedance) or a current source with a parallel resistance (impedance). The load can be a transducer or another circuit.* The driving circuit can be a Thevenin or a Norton equivalent circuit, and the load can be the input resistance of the next stage. If an actual voltage source drives the amplifier, then R_{TH} (Thevenin resistance) is R_S (source resistance). The load is usually modeled as a resistor.

The input and output of most electronic *systems* are sensors or transducers. Transducers basically convert energy of one type to another.

AMPLIFIER CHARACTERISTICS

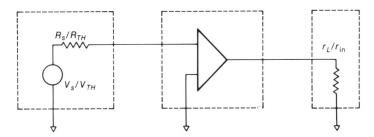

Figure 2-4. Input and output circuits associated with an amplifier.

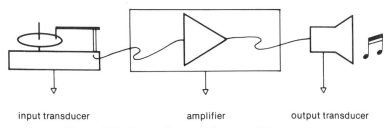

Figure 2-5. Block diagram of an amplifier system.

The microphone and its associated circuit converts the mechanical energy of sound to an analogous electrical-energy signal. One common microphone is a sound-to-resistance converter. The resistance of the device varies with the vibrating columns of air, generated by voice or an instrument. The speaker is a transducer that converts an electrical-energy signal to the mechanical energy of sound. It is typically the load of an amplifier circuit and is modeled as a resistor (8 or 16 Ω). A photocell is a light-to-electrical energy transducer and is modeled as a current source with a high shunt resistance. A simplified amplifier system is illustrated in Figure 2-5.

Input and Output Resistance

The performance of the amplifier in a system can be different than when it is acting alone. This is undesirable and can be minimized by interfacing the output resistance of one stage with the input resistance of the next stage. The interfacing of the two resistances is usually made by making the output resistance very low and the input resistance very high. However, in communications systems, the two values are made equal to ensure maximum power transfer from source to load.

The input and output resistances of an amplifier circuit (Figure 2-6) are not only important when interfacing circuits, but these characteristics are also important in maximizing the circuit's voltage, current, and power gain. Interfacing

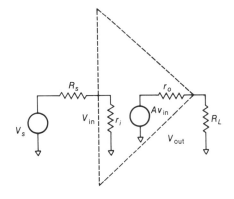

Figure 2-6. Amplifier equivalent circuit.

refers to the union of two circuits whose individual characteristics are used to establish the overall characteristic.

Cutoff Frequencies

The gain of an amplifier circuit remains constant. That is, it remains constant for a certain range of frequency. *The upper and lower limits of this frequency range are called the upper and lower cutoff frequencies, and they define the bandwidth of the amplifier.* Within the circuit are *RL* and *RC* networks. The reactance of the inductances and capacitances varies with frequency, and these networks define the cutoff frequencies. The reactive components form ac voltage dividers with associated resistances. For *RC* lag networks, the capacitances are in shunt with the signal path, and the signal diminishes as the frequency increases. For *RC* lead networks, the capacitances are in series with the signal path, and the signal diminishes as the frequency decreases.

2.3 AMPLIFIER COMPONENTS

Passive Elements

The resistance, capacitance, and inductance of an electronic circuit exist in three forms:

1. *Discrete*
2. *Device*
3. *Distributed*

Discrete components are the individual elements that are in a circuit. They come in all sizes, shapes, values, and ratings.

Distributed resistance, capacitance, and inductance are associated with the physical interconnections, devices, and assemblies of a circuit. An example of distributed resistance is the resistance of a solder connection or that of a copper trace on a printed circuit board. Normally, these resistances are very low compared to discrete resistors and are typically in the milliohm region. An example of distributed capacitance is the capacitance between a terminal and an adjacent ground plane. Typical values are in the low picofarad region. An example of distributed inductance is the inductance of a piece of wire whose value is in the low nanohenry range. The values of distributed resistance, capacitance, and inductance are generally low compared to discrete component values but may *or may not* be neglected depending upon the frequency of the signals in the circuit, current levels, and other factors.

Device resistances, capacitances, and inductances usually model a device mechanism. These components are used in an equivalent circuit to simulate a physical effect. An example is the emitter diffusion resistance, r'_e, of a transistor whose nominal value is 25 Ω. Their values are low but typically greater than distributed values.

Discrete resistors in amplifier circuits are used *primarily* (1) to establish the circuit's dc currents and voltages, and (2) to establish the circuit's gain.

The resistors and the transistor are configured to establish the transistor's dc current (I_C) and dc voltage (V_{CE}). *This static or dc operating point is called the quiescent or Q point.* Some resistances are not discrete and will be associated with the transistor. They are reflected in the transistor's equivalent circuit or model.

Discrete resistors used in small-signal amplifiers are typically in the kilohm region. For power amplifiers, the resistors are in the hundreds of ohms region, and for low-power amplifiers, the resistor values are in the tens and hundreds of kilohms regions.

Discrete capacitors in amplifier circuits are used primarily as follows:

1. To couple the ac signal in and out of the circuit.

2. To provide dc isolation when interfacing to other circuits.

3. To control the frequency response and gain of the amplifier.

4. To decouple or bypass the circuit's dc supplies.

For most of the above cases, the capacitors are of such a value that, within the frequency limitations of the amplifier, they can be viewed as equivalent to a short circuit or an open circuit. The exception is the capacitors chosen to control the frequency response of the amplifier. Some capacitances are not discrete. There are distributed circuit capacitances and transistor or device capacitances, and these are reflected in the circuit and transistor models. Frequently, they dominate in establishing the frequency limitations of the amplifier.

The typical value of coupling and bypass capacitors is in the microfarad region. Capacitances used in establishing the upper cutoff frequency are in the picofarad region.

Diodes and zener diodes are occasionally used as voltage-control devices. Inductors are seldom used in low-frequency amplifiers, but they abound in high-frequency communication circuits. Integrated-circuit amplifiers primarily use (internally) transistors, resistors, and diodes. Capacitors are normally connected external of the IC, and inductors are realized through active circuit techniques, that is, a circuit that simulates the effect of an inductor.

Transformers are valuable in output stages where they are used to buffer or interface the output impedance of the amplifier circuit to that of the load. They are also used as part of frequency-selective, tuned LC circuits. The IF transformer in a communication's receiver is an example of the latter case.

Active Devices

Bipolar Transistor

Every gain-producing circuit must have at least one gain-producing element or component. This component is referred to as an active device, and the transistor is the prime example. Inductors, capacitors, and resistors are passive devices.

An active element, or device, is an element that is theoretically capable of delivering an infinite amount of energy. A passive element is an element that is *not* capable of delivering an infinite amount of energy. Resistors, capacitors, and inductors are passive elements; however, the resistor converts energy, whereas the ideal capacitor and inductor store energy. Capacitors and inductors are also called reactive elements.

The most prominent active device is the *npn* (*pnp*) bipolar transistor, but the special characteristics of the MOS and JFET unipolar transistors make them invaluable in certain applications.

In the simplest and fewest words possible, the bipolar *npn* or *pnp* transistor is a three-terminal active device capable of current gain. The device leads are called base (B), collector (C), and emitter (E). The normal input is the base current (I_B), and the normal output is the collector current (I_C). The collector current is greater than the base current by the factor of beta (β). β, or the ratio of I_C to I_B, is a dimensionless number and reflects the gain of the device. Typical values for β_{dc} are from 35 to 300, with 100 a nominal value.

Three voltages and three currents are associated with the transistor; they are shown in Figure 2-7. I_C and V_{CE} are generally established by the circuit that surrounds the transistor. V_{BE} is a constant of approximately 0.6 Vdc. The B-E junction of the transistor will be forward biased, and 0.6 V represents this junction's barrier potential. V_{CB} is seldom used and can be determined (by KVL) if V_{CE} and V_{BE} are known. I_B and I_C are related by β. If β is high, I_E is approximately equal to I_C. Alpha, α, relates the collector and emitter currents, and its value is very close to 1. The power dissipated by the transistor is equal to the product of I_C and V_{CE}. The relationships of the

AMPLIFICATION CONCEPTS AND CHARACTERISTICS

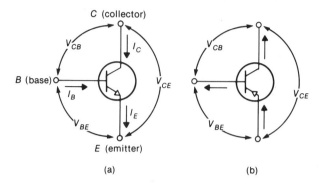

Figure 2-7. Bipolar transistor voltages and currents: (a) *npn*; (b) *pnp*.

$$\beta = \frac{I_C}{I_B}$$
$$\alpha = \frac{I_C}{I_E}$$
$$I_C = \beta I_B$$
$$I_C = \alpha I_E$$
$$I_E = I_B + I_C$$
$$\alpha = \frac{\beta}{\beta + 1}$$
$$I_E = (\beta + 1)I_B$$
$$I_E = \frac{\beta + 1}{\beta} I_C$$
$$V_{CE} = V_{CB} + V_{BE}$$
$$P_D = I_C \cdot V_{CE}$$

Figure 2-8. Bipolar transistor dc relationships.

voltages and currents are listed in Fig. 2-8. It is assumed that the reader is aware of the internal operations of the transistor, and the terminal relationships are presented here as a review. Section II provides a summary of the physics necessary to understand the internal mechanisms of the transistor. The transistor will be modeled for those cases where the device's internal mechanisms are directly related to circuit performance.

npn and *pnp* Transistors

The *pnp* transistor is the *complement* of the *npn*. Its mode of operation is the same as the *npn* with one exception. The directions of the *pnp*'s currents and the polarities of the *pnp*'s voltage are opposite to those of the *npn*.

Every *npn* circuit has a complementary *pnp* circuit. The polarities of the supplies or the location of the supplies and circuit common are the only differences. The analysis in both cases is the same. The *pnp* circuit can be analyzed directly, or it can be converted to its complementary *npn* circuit by changing the *pnp* to an *npn* and voltage translating the supplies and ground.

The *npn* circuit is the more common of the two. However, one must be equally adept in analyzing both types. In integrated circuits, the *npn* is also more common. It requires one less manufacturing step to make the *npn*, and it has slightly better performance ratings.

Unipolar (JFET) Transistor

Simply, the *n*- or *p*-channel, junction, field-effect, unipolar transistor (JFET) is a three-terminal active device capable of gain. The device (Figure 2-9) leads or terminals are called gate (G), source (S), and drain (D). The normal signal input is the gate-to-source voltage (V_{GS}), and the normal output is the drain current (I_D). Transconductance g_{fs} is the ratio of the change in I_D (ΔI_D) to the change in V_{GS} (ΔV_{GS}), and it reflects the gain of the device. It is commonly called a figure of merit, and it has the dimension of Siemens (S) (formerly mhos). Transconductance typically varies from 0.1 to 1000 μmhos or μS.

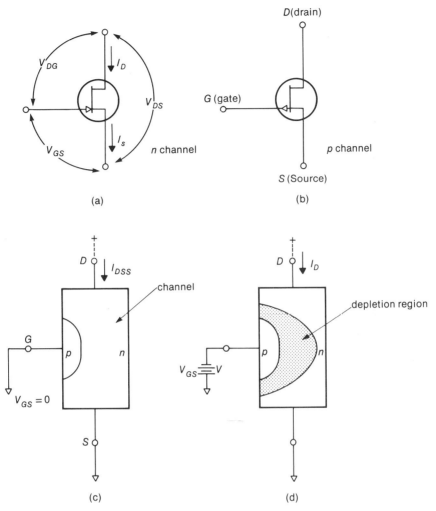

Figure 2-9. Junction field effect transistor: (a) n channel; (b) p channel; (c) $V_{GS} = 0$ V; (d) $V_{GS} = -V$.

If the drain-to-source region (the channel) is made of n-type semiconductor material, the device is an n-channel JFET. If the channel is made of p-type material, it is called a p-channel device. In the transistor symbol, the gate's arrow points in for an n-channel device and points out for a p-channel device.

The structure of an n-channel JFET is shown in Figure 2-9c. The pn junction formed between the gate and the channel is never forward biased. It is either biased at zero volts (V_{GS}), or it is reverse biased. The JFET operates by means of the creation and control of the depletion region that exists in all reverse-biased pn junctions. The controlling effect is a field produced by the input voltage, which is why JFETs are called field-effect devices.

When V_{GS} is zero volts (Figure 2-9c), the pn junction is unbiased, and the channel appears as a piece of semiconductor material between the source and drain. The drain-to-source current (I_{DS}) is established by the external circuit, but it is limited (I_{DSS}) by the resistance of the channel. When V_{GS} is a negative voltage (*n* channel), the *pn* junction is reverse biased, and a depletion region (Figure 2-9d) is formed. The depletion region increases the channel's resistance because of the reduced number of mobile carriers, and it decreases the drain-to-source current. Increasing the magnitude of the gate-to-source voltage will ultimately fully deplete the *n* region or channel and cut off the drain current. The value of V_{GS} that pinches the channel off is called the pinch-off voltage V_p.

The junction-field-effect transistor, or JFET, is strictly a depletion-mode device; that is, its drain current decreases for an increasing gate-to-source voltage (V_{GS}). The maximum I_D for the transistor occurs when V_{GS} is zero, and it is called I_{DSS}. As the magnitude of V_{GS} increases, the drain-to-source region or channel is depleted of carriers, resulting in a lower value of I_D. When V_{GS} equals V_p, the drain current is negligible or zero. I_D, I_{DSS}, V_{GS}, and V_p are mathematically related by the following relationship:

$$I_D = I_{DSS}\left(1 - \frac{V_{GS}}{V_p}\right)^2$$

The *pn* junction of the JFET is typically reverse biased, and it provides an extremely high input impedance. This device characteristic makes it very easy to design *circuits* with a high input impedance. This is a desirable circuit characteristic, and the input stage of stereo amplifiers is an example where this type of circuit is frequently found.

The n/p enhancement- or depletion-mode, metal-oxide semiconductor (MOS) transistor is another member of the unipolar class of transistors. In MOSFET devices, a glasslike dielectric (S_iO_2) separates the *p* and *n* regions. This type of device is heavily used in digital integrated circuits, but it is seldom encountered in discrete or integrated amplifiers.

The Collector Family of Curves

An electronic test instrument that measures the performance of the transistor is the curve tracer. The output of this instrument is a two-dimensional cathode-ray tube (CRT) display of I_C (vertical) versus V_{CE} (horizontal). Superimposed on the CRT is a grid. The I_C and V_{CE} values for the divisions on the grid are established by the sensitivity controls of the curve tracer. For a given value of I_B, a line is traced showing the possible combinations of values for I_C and V_{CE}. In actuality, many lines are traced, and each line represents a different value of I_B. The picture we see is referred to as the *collector family of curves*. A typical collector family of curves is shown in Figure 2-10a.

A transistor in an amplifier circuit must have its base-to-emitter junction forward biased and the base-to-collector junction reverse biased. This condition is not met until V_{CE} is approximately 1 V. On the collector family of curves, I_C rises sharply until V_{CE} reaches approximately 1 V. I_C then remains relatively constant (for a given value of I_B) for various values of V_{CE} above 1 V until the breakdown voltage is reached. This flat region reflects the nondependence of I_C on V_{CE} within the transistor; however, circuit conditions will relate the two.

Bipolar Transistor's Operating Point

The transistor in a bias circuit will be operating at a particular value of I_C and V_{CE}. These two pieces of information define a point on the family of curves called the quiescent or Q point. In an amplifier, this point will move (in real time) when ac signals are introduced into the amplifier circuit. This point moves along a straight line called the *load line*, and it is determined by the resistances in the ac equivalent circuit.

Amplifying circuits are classed according to the extent of the movement of the operating point. If the transistor's operating point moves approximately 10% from its dc value along the load line, the amplifying circuit is called a *small-signal amplifier*. If the operating point moves about 90% over the load line, the circuit is called a *large-signal amplifier*.

Unipolar (JFET) Transistor's Operating Point

The curve tracer can also measure the performance of unipolar transistors. For these transistors, the output of the instrument is a two-dimensional display (CRT) of drain current, I_D, versus drain-to-source voltage, V_{DS}. The I_D and V_{DS} values for the divisions of the display's grid are established by the sensitivity controls. For a given value of V_{GS}, a line is traced showing the combinations of values of I_D and V_{DS}. Several lines are traced on the screen, and each line represents a different value of V_{GS}. The maximum I_D occurs at V_{GS} equals zero volts. For low values of V_{DS}, I_D increases linearly and behaves like a voltage-controlled resistor. The value of V_{DS} where I_D begins to remain constant is called the *threshold voltage*, V_T. I_D remains constant for various values of V_{DS} until V_{DS} reaches the breakdown region.

The picture of I_D versus V_{DS} (Figure 2-10b) for many values of V_{GS} is called the *drain family of curves*. The JFET in a bias circuit will be operating at a particular value of I_D and V_{DS}. These two pieces of information, I_{DQ} and V_{DSQ}, define the quiescent or Q point on the family of curves. In an amplifier, this point will move in real time when ac signals are introduced into the amplifier circuit. This point moves along a straight line called the load line, which is determined by the resistances in the ac equivalent circuit.

Current Gain, β, Variations

In many applications, electronics equipment must operate over a wide temperature range and various operating conditions. The performance of the circuits must be maintained, and the sensitivity of its characteristics must be minimized. This is usually accomplished through circuit techniques.

The current gain of a bipolar transistor varies to some degree. It is sensitive to collector current, temperature, and frequency. The variation of β, or h_{fe}, versus I_C, T, and f is illustrated by the graphs in Figure 2-11.

The temperature dependence is minimized by designing circuits whose operating points (I_C and V_{CE}) are not directly proportional to β. The sensitivity of β to collector current is minimized by designing or using circuits whose operating points depend on a high value of β

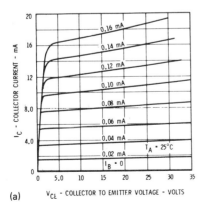
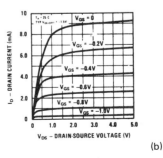

Figure 2-10. Collector and drain family of curves: (a) collector family; (b) drain family.

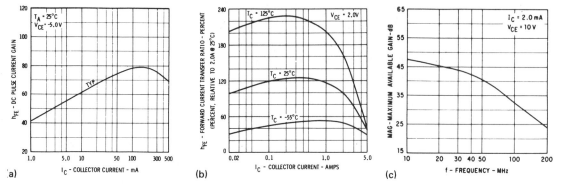

Figure 2-11. β or h_{fe} variations: (a) with I_C or collector current; (b) with T or temperature; (c) with f or frequency.

but not its specific value. For small-signal amplifiers, I_C is chosen near where β peaks on the graph.

An amplifier cannot have a bandwidth greater than that of the transistor. Transistors in amplifiers are selected with gain-frequency capabilities in excess of that of the circuit. Then the reactive components of the device in conjunction with discrete components establish the circuit's bandwidth.

There are two betas. β_{dc} is the dc value of beta and is used in bias circuit and low-frequency applications calculations. β_{ac} decreases as frequency increases. The frequency where β_{ac} is 1 is called f_t and is the highest possible frequency at which a transistor can be used.

The exact value of β for the transistor at its operating point can be determined using the curve tracer. Select the nearest convenient values of I_B above and below the operating point. Find the corresponding values of I_C and compute β.

$$\beta = \frac{I_{C1} - I_{C2}}{I_{B1} - I_{B2}} = \frac{\Delta I_C}{\Delta I_B}$$

Base Curve, I_E Versus V_{BE}

For silicon transistors in amplifier circuits, V_{BE} is generally assumed to be a constant of 0.6 V. However, V_{BE} slightly deviates from this nominal value and is a function of I_E, T (temperature), and V_{CE}. A graph of I_E versus V_{BE} is shown in Figure 2-12 and is called the *base curve*. This diode-type curve displays the exponential relationship of I_E and V_{BE} and is typical of a bipolar *pn* junction. The equation that models this behavior is given as

$$I_E \cong I_S e^{V_{BE}/V_t} \qquad \text{for } I_E \gg I_S$$

I_S and V_t are constants at 25°C. I_S is called the *reverse saturation current* and is a function of the physical characteristics of the junction. V_t, at room temperature, is approximately 0.026 V.

V_{BE} remains small for very low values of I_E until the knee portion of the curve is reached. After the knee, I_E increases dramatically for slight increases in V_{BE}. Beyond the knee, the I_E and V_{BE} relationship is nearly linear, indicating a resistive-type effect. This effect reflects the bulk resistance of the *B-E* junction, and its value is equal to the slope of this line (in siemens).

The exact value of V_{BE} is often required. For circuits that operate over temperature extremes, for class AB and Class B circuits, and for precision amplifiers, the exact value of V_{BE} for a specific condition must be known. Graphs of I_E versus V_{BE}, T, and V_{CE} are usually given in the specifications data for most transistors.

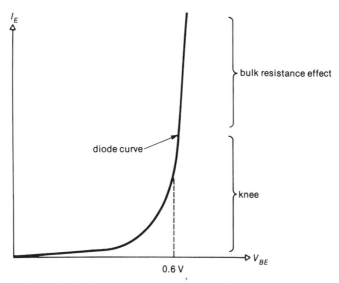

Figure 2-12. The base curve: I_E versus V_{BE}.

From these graphs, one can extract the necessary information, although these curves represent typical or average data. The I_E-V_{BE} relationship for any specific transistor may be measured by using the curve tracer.

Just as I_C and V_{CE} define the operating point of the transistor, I_E and V_{BE} define the operating point of the transistor's B-E junction. The specific value of I_B and V_{BE} that this junction is operating at in a circuit locates a point on the base curve. The identification of this point is necessary to determine the value of the emitter diode ac resistance, r'_e, in the ac equivalent circuit of the transistor. In small-signal amplifiers, the junction's operating point will only move about 10% from its dc value. In large-signal amplifiers, the operating point will move nearly over the entire base curve.

The transistor is a three-terminal component, and the base-to-emitter junction is one of two *pn* junctions that make up the device. The emitter current is not normally expressed in terms of V_{BE} but in terms of other currents or circuit relationships. The graph will be used to relate to and describe components in the transistor's model or equivalent circuit.

Dc Transistor Model or Equivalent Circuit

In analyzing an amplifier circuit, it is often necessary to consider the internal operation of the transistor. *The physical mechanisms within the device can be modeled by an equivalent electronic circuit. An equivalent circuit, or model, is an arrangement of resistances, capacitances, inductances, and sources whose terminal behavior simulates the effect of the device it represents.* Usually, an equivalent circuit is valid only for a given set of conditions. The components of the transistor equivalent circuit, and their values, play a key role in identifying and evaluating many amplifier characteristics.

The equivalent circuits of Figure 2-13 are the first- and higher-order models of the bipolar transistor. While there are other versions and more detailed models, these are sufficiently

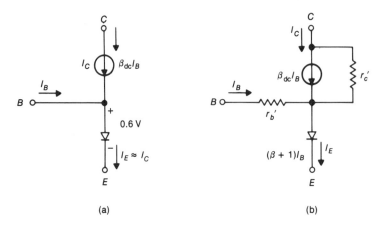

Figure 2-13. Transistor dc equivalent circuits: (a) first order model; (b) higher order model.

accurate for our purposes and adequately illustrate the transistor's role in the amplifier. They are valid for bias-circuit (dc conditions) calculations.

The circuit of Figure 2-13a consists of a diode and a dependent current source. The diode models the forward-biased B-E pn junction. The value of the collector current source depends on I_B, and if β is high, it is approximately equal to I_E. Usually, circuit conditions will establish one of the three transistor currents. The remaining two can be calculated once the first is known. The transistor's current gain β is usually assumed or can be measured using the curve tracer. Once β is known, α can be calculated.

Figure 2-13b includes the effect of the base spreading or ohmic resistance, r_b', and r_c'. The typical value of r_b' is in the hundreds of ohms region and is usually small compared to discrete resistors in series with the base. The typical value of r_c' is near 100 kΩ and is usually large compared to discrete resistors in parallel with it. The value of r_b' is usually given in the transistor data sheet, or it can be calculated from h-parameter data. The value of r_c' can be determined from the collector family of curves.

For a given value of I_B, it is the reciprocal of the slope of the I_C-V_{CE} curve in the flat region, or

$$r_c' = \frac{\Delta V_{CE}}{\Delta I_C}$$

I_C is smaller than I_E by the factor α_{dc}, which is a function of β_{dc}'.

These equivalent circuits model the operation of the transistor in dc circuits. The second-order model is the more accurate version, although both are adequate to achieve typical laboratory accuracies. A different model is used for ac circuits, and it is presented under the amplifier ac principles topic.

Problems

1. The input voltage of an amplifier is 10 mV (rms). The peak-to-peak value of the output voltage is 4 V. Calculate the amplifier's voltage gain.

2. The upper cutoff frequency of an audio amplifier is 18 kHz. Estimate the amplifier's bandwidth if its lower cutoff frequency is 100 Hz.

3 The voltage measured at the input of an amplifier is 50 mV peak. The amplifier's input current is 2 µA peak. What is the amplifier's input resistance?

4 What is a 10 µF coupling capacitor equivalent to in an ac amplifier whose mid-frequency is 20 kHz?

5 What is a transistor capacitance of 30 pF equivalent to in an ac amplifier whose mid-frequency is 20 kHz?

6 What is the percent of difference between a transistor's emitter current I_E and its collector current I_C if β is equal to 100?

7 If the emitter current I_E of a transistor is 10 mA and its current gain is 100, what is the exact value of I_C and I_B?

8 Determine the power dissipated by a transistor if the dc values of I_C and V_{CE} are 50 mA and 10 V.

9 Using Figure 2-10a, estimate the transistor's current gain if its collector current is 10 mA and its collector-to-emitter voltage is 10 V.

10 Using Figure 2-10a, estimate r'_c if I_B is 0.10 mA.

11 Using Figure 2-11a, determine the percent of increase in β if I_C is increased from 10 to 50 mA.

12 What is the gate-to-source voltage of a JFET if it is operated at a drain current of 4 mA and its drain-to-source voltage is 2 V? Use Figure 2-10b.

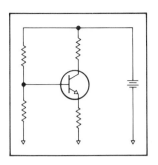

AMPLIFIER DC PRINCIPLES AND BIAS CIRCUITS

3

The ac amplifier is a circuit that creates a replica of its input at the output. The output replica will have a greater amplitude but will be of the same frequency as the input. Superposition allows us to examine the dc and ac behavior of the amplifier circuit independently. This chapter discusses the dc principles of the amplifier and bias circuits.

The dc equivalent circuit of an amplifier is called a dc bias circuit. Bias circuits establish the dc voltages and currents to which the ac signal will be algebraically added. I_C and V_{CE} establish the circuit's dc operating or Q point. Several common bias circuits are (1) voltage divider, (2) collector feedback, (3) modified collector feedback, and (4) emitter bias. Stability factors reflect the degree of variation of the Q point when other circuit or device quantities change.

3.1 INTRODUCTION

The primary characteristics of an amplifier are gain, input and output resistance, and the lower and upper cutoff frequencies. To determine and evaluate these characteristics, the complete amplifier circuit must be analyzed. This analysis is quite detailed and involves the dc and ac responses of the circuit. Fortunately, through the superposition theorem, the dc and ac responses of the amplifier can be separated and looked at individually with the actual response being the sum of the two. The complete amplifier circuit will be reduced to a dc equivalent circuit and analyzed. From this analysis, the dc response of the amplifier will be determined and information extracted that is necessary to describe the amplifier's overall response. This procedure is repeated by looking at the ac equivalent amplifier circuit. By using the information determined from the two, the overall amplifier response can be determined.

A complete, discrete-component, amplifier circuit is shown in Figure 3-1. Our ultimate objective will be to evaluate this circuit's primary characteristics. This particular circuit is but one of many circuit configurations used to amplify. It, however, is one of the more common ones and will serve as a vehicle to develop a quantitative understanding of the circuit's characteristics. Other circuit variations will be examined as well. There is no need to examine all possible

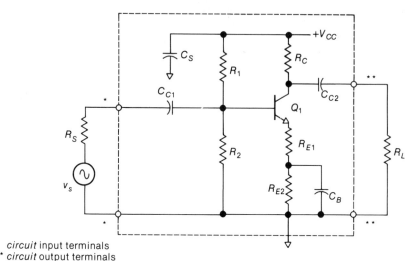

* *circuit* input terminals
** *circuit* output terminals

Figure 3-1. Amplifier circuit.

circuit configurations, because most amplifier circuits used today are monolithic or integrated. For most cases, the understanding of discrete amplifier circuits serves as a foundation for the understanding of the IC versions. The presentation of the discrete amplifier is slanted toward that knowledge needed to understand the integrated amplifier.

3.2 DC EQUIVALENT CIRCUIT

The *dc equivalent circuit of an amplifier is called a dc bias circuit, or simply a bias circuit.*

In this circuit, the dc supply(ies) establishes the dc voltages and dc currents of the passive and active components. When an ac signal is introduced into the circuit, these voltages and currents will vary in a manner similar to that of the signal. This signal will be added (algebraically in real time), and the composite will be averaged about the quiescent (no-signal) value.

A complete analysis will determine all the dc voltages and currents; however, the transistor collector current (I_C) and collector-to-emitter voltage (V_{CE}) are the most important. I_C and V_{CE} establish the dc operating point of the transistor and circuit. This operating point is called the quiescent or Q point.

The complete amplifier is reduced to its *dc equivalent circuit* by replacing all capacitors and inductors with their low-frequency or dc equivalents, that is, opens and shorts. Inductors are rarely encountered in low-

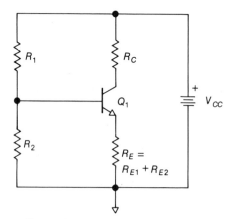

Figure 3-2. Dc equivalent circuit.

and moderate-frequency amplifiers, but most amplifiers contain a minimum of two discrete capacitors. The dc equivalent circuit of the amplifier is shown in Figure 3-2.

3.3 BIAS CIRCUIT MODELS

Most bias circuits can be reduced, through analysis techniques, to one of the *two bias circuit models or prototypes* shown in Figure 3-3. These models represent a *simplified bias circuit* and are called base-bias or emitter-bias models. In the base-bias circuit, the base is biased by the source, V_{BB}, and the resistor R_B. The source, V_{BB}, is usually not a separate voltage supply. It is typically derived, by a resistor-divider circuit, from the V_{CC} supply. Both have the same polarity. V_{BB} and R_B are often V_{TH} and R_{TH} derived from a more complex circuit.

In the emitter-bias circuit, the emitter is biased by the source, V_{EE}, and the resistor R_E. The V_{CC} and V_{EE} supplies are of the opposite polarity. Since circuits reduced to this model require two supplies, they are less frequently encountered than the base-bias circuits. In both circuits, the collector is biased by the source V_{CC} and the resistor R_C.

Most discrete amplifiers use a single supply, and low-frequency integrated amplifiers usually use two. For single-supply circuits, the input and output signals are ground referenced, and all points within the amplifier are electrically biased *between* ground and the supply. Integrated amplifiers can operate from a single supply, but additional components must be used to bias the inputs and outputs above ground. For these circuits, the signals are ac coupled in and out of the amplifier circuit. Dual-supply circuits can provide both voltage and both current polarities, and the input and output signals can vary about zero volts.

3.4 VOLTAGE-DIVIDER BIAS CIRCUIT

Determining I_C and V_{CE}

The dc equivalent circuit in Figure 3-2 is called a voltage-divider bias circuit, and it is the most important member of the base-bias class. This circuit contains four resistors and one transistor. Resistors R_1 and R_2 form a voltage divider and establish the base voltage V_B. Resistor R_C is in series with the transistor's collector lead, and R_E is in series with the emitter. The circuit can be reduced to the model by Theveninizing the R_1, R_2, and V_{CC} circuit associated with the base. Figure 3-4 shows the reduction of the circuit to its model form.

The circuit, also called an emitter feedback

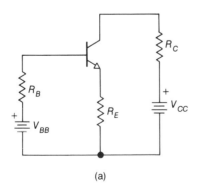
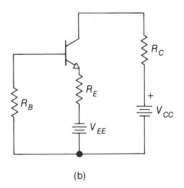

Figure 3-3. Bias circuit models: (a) base bias; (b) emitter bias.

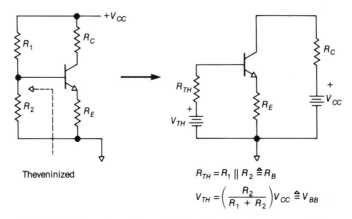

Figure 3-4. Reducing the voltage-divider bias circuit to its model.

bias circuit, derives its name from the voltage-divider action developed by R_1 and R_2. The *key to the analysis of this circuit* is the node or junction of R_1, R_2, and the base of Q_1. If we apply Kirchhoff's current law to this node, then

$$I_1 = I_2 + I_B$$

For a well-designed bias circuit, the base current I_B will be *significantly* smaller than I_2; that is,

$$I_B \ll I_2$$

Hence,

$$I_1 \cong I_2$$

If the currents I_1 and I_2 are approximately equal to each other, then

$$V_B \cong \frac{R_2}{R_1 + R_2} V_{CC}$$

or the base voltage is established by the voltage-divider action of R_1 and R_2.

To guarantee that

$$I_B \ll I_2$$

the currents I_1 and I_C must be within an order of magnitude (factor of 10) of each other, *and* the dc beta (β) of Q_1 must be high, that is, $\gtrsim 100$. Ordinarily, these conditions are not difficult to achieve. I_B will be $1/\beta$ of I_C, and if I_C and I_1 are approximately the same, then I_B will be $1/\beta$ of I_1 also. Hence, I_B will be small compared to I_1.

The base voltage is established by the R_1-R_2 voltage divider, and the emitter voltage is related to the base voltage by the device's barrier potential, V_{BE}. Thus,

$$V_E = V_B - V_{BE} = \left(\frac{R_2}{R_1 + R_2}\right) V_{CC} - V_{BE}$$

For an *npn* transistor, the base voltage must be more positive than the emitter voltage by V_{BE}. One end of R_E is at 0 V (the circuit reference) and the other end is at V_E. Hence,

$$I_E = \frac{V_E}{R_E} = \frac{\left(\frac{R_2}{R_1 + R_2}\right) V_{CC} - V_{BE}}{R_E}$$

If β_{dc} of Q_1 is high, then

$$I_C \cong I_E$$

and $\quad I_C \cong \dfrac{\left(\dfrac{R_2}{R_1 + R_2}\right) V_{CC} - V_{BE}}{R_E}$

By applying Kirchhoff's voltage law to the V_{CC}, R_C, V_{CE}, R_E loop,

$$V_{CC} = I_C R_C + V_{CE} + I_E R_E$$

VOLTAGE-DIVIDER BIAS CIRCUIT

Since I_C, I_E, R_C, R_E, and V_{CC} are known,

$$V_{CE} = V_{CC} - I_C(R_E + R_C)$$

Example

$R_1 = 10$ kΩ, $R_2 = 3.3$ kΩ, $V_{CC} = 20$ Vdc, $R_E = 4.3$ kΩ, $R_C = 5.6$ kΩ, $V_{BE} = 0.6$ V, $\beta = 100$.

Step 1. $V_B = \left(\dfrac{R_2}{R_1 + R_2}\right) V_{CC}$

$= (3.3$ k$\Omega/13.3$ k$\Omega)20$ V

$= 4.96$ V.

Step 2. $V_E = V_B - V_{BE} = 4.96$ V $- 0.6$ V

$= +4.36$ V.

Step 3. $I_C \cong I_E = V_E/R_E = 4.36$ V$/4.3$ kΩ

$= 1.01$ mA.

Step 4. $V_{CE} \cong V_{CC} - I_C(R_E + R_C)$

$= 20$ V $- 1.01$ mA$(9.9$ k$\Omega)$

$= 9.96$ Vdc.

To check the validity that $I_B \ll I_1$, we will calculate I_1 and compare its value with I_C.

$$I_1 = \frac{15.04 \text{ V}}{10 \text{ k}\Omega} = 1.50 \text{ mA}$$

If $I_C = 1.01$ mA and the β of the transistor is 100, then

$$I_B = \frac{I_C}{\beta} = \frac{1.01 \text{ mA}}{100} = 0.010 \text{ mA}$$

which is indeed small compared to the 1.50 mA value of I_1.

A Second Technique for Determining I_C and V_{CE}

The value of I_C and V_{CE} may also be determined from the base-driven bias model of the voltage-divider bias circuit. After Theveninizing

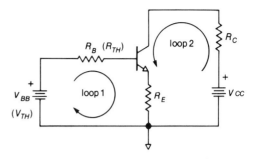

$R_B = R_1 \| R_2$

$V_{BB} = \left(\dfrac{R_2}{R_1 + R_2}\right) V_{CC}$

Figure 3-5. Theveninized voltage-divider bias circuit.

the base circuit (Figure 3-5) an equation employing KVL is written for loop 1.

$$V_{BB} = I_B R_B + V_{BE} + I_E R_E$$

The emitter and base currents of the transistor are related by

$$I_E = (\beta + 1)I_B$$

Substituting for I_B,

$$V_{BB} = \left(\frac{I_E}{\beta + 1}\right) R_B + V_{BE} + I_E R_E$$

Solving for I_E yields

$$I_E = \frac{V_{BB} - V_{BE}}{R_E + R_B/(\beta + 1)}$$

Since

$$I_C \cong I_E$$

then $I_C \cong \dfrac{V_{BB} - V_{BE}}{R_E + R_B/(\beta + 1)}$

and $V_{CE} \cong V_{CC} - I_C(R_E + R_C)$

Approximations

If β_{dc} is high, R_B is low, and V_{BB} or V_{TH} is large compared to V_{BE}, then I_C and V_{CE} may be approximated by the following equations:

$$I_C \approx \frac{V_{BB}}{R_E}$$

$V_{CE} \approx V_{CC} - I_C(R_E + R_C)$ (first-level approximation)

If a greater degree of accuracy is desired, or if V_{BB} is not significantly greater than V_{BE}, then

$$I_C \cong \frac{V_{BB} - V_{BE}}{R_E}$$

$V_{CE} \cong V_{CC} - I_C(R_E + R_C)$ (second-level approximation)

Normally, a third-level approximation is the highest degree of accuracy one has to use. At this level the effect of R_B is included.

$$I_C = \frac{V_{BB} - V_{BE}}{R_E + R_B/(\beta + 1)}$$

$$= \frac{\left(\frac{R_2}{R_1 + R_2}\right)V_{CC} - V_{BE}}{R_E + R_B/(\beta + 1)}$$

$V_{CE} = V_{CC} - I_C(R_E + R_C)$ (third-level approximation)

3.5 VOLTAGE-DIVIDER BIAS CIRCUIT WITH TWO DC SUPPLIES

The circuit in Figure 3-6 is a voltage-divider bias circuit with a *pnp* transistor and two dc supplies. A complete analysis will be made to determine all voltage drops, all voltages with respect to ground, and all currents. The analysis is similar to that used in the single-supply bias circuit; however, ground-referenced voltages can be positive or negative.

The potential difference across R_1 and R_2 is 20 V.

$+V_{EE} - (-V_{CC}) = 20$ V

I_B is again assumed small, and the sum of the voltage drops across R_1 and R_2 must equal this potential difference.

$V_{R1} + V_{R2} = 20$ V

The amount of each drop is determined by voltage-divider action and KVL.

$$V_{R1} = \left(\frac{R_1}{R_1 + R_2}\right) 20\text{ V} = 13.6\text{ V}$$

$$V_{R2} = \left(\frac{R_2}{R_1 + R_2}\right) 20\text{ V} = 6.4\text{ V}$$

V_{R1} and V_{R2} are the voltage drops across R_1 and R_2. They are not voltages with respect to ground. Conventional current flows from the $+V_{EE}$ supply, through R_1 and R_2, to the $-V_{CC}$ supply. This current establishes the polarity of the voltage drops shown in Figure 3-6.

The base voltage V_B, with respect to ground, is less positive than $+V_{EE}$ (a ground-referenced voltage) by 6.4 V or the voltage drop across R_2.

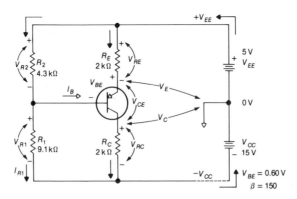

Figure 3-6. Voltage-divider bias circuit with a *pnp* and two dc supplies.

$V_B = +V_{EE} - V_{R2} = -1.4$ V

For the *pnp* transistor, the emitter voltage, with respect to ground, is more positive than the base by V_{BE}.

$V_E = V_B + V_{BE} = -0.8$ V

The voltage drop across R_E is the difference in potential between the $+V_{EE}$ supply and V_E. Hence,

$V_{RE} = +V_{EE} - V_E = 5.8$ V

The emitter current is determined from Ohm's law.

$I_E = I_{RE} = \dfrac{V_{RE}}{R_E} = 2.9$ mA

Since β for the transistor is high,

$I_C \cong I_E = 2.9$ mA

Since I_C is now known, the voltage drop across R_C can be calculated.

$V_{RC} = I_C R_C = 5.8$ V

The collector voltage V_C, with respect to ground, is more positive than $-V_{CC}$ by 5.8 V

$V_C = -V_{CC} + V_{RC} = -9.2$ V

V_{CE} is the difference in potential between V_C and V_E.

$V_{CE} = V_C - V_E = -8.4$ V

Since the voltage drops across R_1 and R_2 are known,

$I_{R1} = \dfrac{V_{R1}}{R_1} = 1.49$ mA

$I_{R2} = \dfrac{V_{R2}}{R_2} = 1.49$ mA

If the transistor's β is 100 and I_C is 2.9 mA,

$I_B = \dfrac{I_C}{\beta} = 0.029$ mA

The base current is indeed small compared to I_{R1} and I_{R2}.

Summary

$I_C = 2.9$ mA $V_C = -9.2$ V
$V_{CE} = -8.4$ V $V_{R1} = 13.6$ V
$V_E = -0.8$ V $V_{R2} = 6.4$ V
$V_B = -1.4$ V $I_{R1} \cong I_{R2} = 1.49$ mA

The circuit dc currents and voltages are all known, and they serve as a basis for comparison with measured values.

3.6 COLLECTOR FEEDBACK BIAS CIRCUIT

The *collector feedback bias circuit* in Figure 3-7 contains one transistor and two resistors. The advantage of this bias circuit is the minimal number of components it requires. A disadvantage to this circuit is the Q point's dependence on β.

The current through the collector resistor, R_C, is the sum of I_B and I_C. The collector current is determined by applying Kirchhoff's voltage law to loop 1.

$V_{CC} = (I_B + I_C)R_C + I_B R_B + V_{BE}$

I_B and I_C are related by

$I_B = \dfrac{I_C}{\beta}$

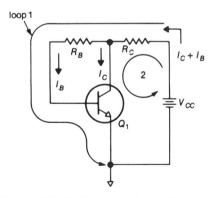

Figure 3-7. Collector feedback bias circuit.

Substituting for I_B,

$$V_{CC} = \left(\frac{I_C}{\beta} + I_C\right)R_C + \left(\frac{I_C}{\beta}\right)R_B + V_{BE}$$

Solving for I_C,

$$I_C = \frac{V_{CC} - V_{BE}}{R_C\left(\frac{\beta+1}{\beta}\right) + \frac{R_B}{\beta}} = \frac{V_{CC} - V_{BE}}{R_C/\alpha + R_B/\beta}$$

The collector-to-emitter voltage is determined by applying KVL to loop 2.

$$V_{CE} = V_{CC} - (I_B + I_C)R_C$$
$$= V_{CC} - \left(I_C + \frac{I_C}{\beta}\right)R_C$$
$$= V_{CC} - I_C R_C \left(\frac{\beta+1}{\beta}\right)$$

The formula for I_C illustrates the direct dependence of I_C on the transistor's current gain. If β increases (owing to temperature), I_C increases and the circuit's dc quiescent point changes. The economy, however, of the circuit's minimal number of parts makes it a frequently used circuit.

Collector-to-Base Feedback

Resistor R_B is connected from the collector of the transistor (the typical output) back to the base (the typical input). Because of this feedback connection, changes in the collector voltage affect the collector current. This helps reduce the effects of changing β. For the transistor,

$$I_C = \beta_{dc} \cdot I_B$$

If β increases (owing to temperature), the collector voltage decreases because of the increased collector current and the larger voltage drop across R_C. However, if the collector voltage decreases, the base current decreases, causing a decrease in collector current. The end result is that the collector current will change, but not nearly as much as the change in β.

Midpoint Bias

For many applications, *the optimum value of V_{CE} is when it is equal to one-half of V_{CC}*. For this case, we say the transistor is *midpoint biased*. When the transistor is midpoint biased, it will amplify the largest peak-to-peak signal possible for a given value of V_{CC}. For midpoint bias, $V_{CE} = V_C = 0.5 V_{CC}$; therefore,

$$V_{CE} = V_{CC} - I_C R_C \left(\frac{\beta+1}{\beta}\right) = 0.5 V_{CC}$$

$$I_C = \frac{0.5 V_{CC}}{R_C} \left(\frac{\beta}{\beta+1}\right)$$

The collector current is also related by

$$I_C = \frac{V_{CC} - V_{BE}}{R_C\left(\frac{\beta+1}{\beta}\right) + \frac{R_B}{\beta}}$$

Equating the two expressions,

$$\frac{0.5 V_{CC}}{R_C\left(\frac{\beta+1}{\beta}\right)} = \frac{V_{CC} - V_{BE}}{R_C\left(\frac{\beta+1}{\beta}\right) + \frac{R_B}{\beta}}$$

The above relationship can be simplified by the following approximations:

$$\frac{\beta+1}{\beta} \cong 1 \quad \text{and} \quad V_{BE} \approx 0 \text{ V}$$

This results in

$$\frac{0.5 V_{CC}}{R_C} \cong \frac{V_{CC}}{R_C + R_B/\beta}$$

or

$$\frac{1}{2R_C} \cong \frac{1}{R_C + R_B/\beta}$$

$$R_B = \beta_{dc} R_C$$

When R_B is β_{dc} times R_C, V_{CE} is approximately equal to one-half of V_{CC}.

Example

$V_{CC} = 15$ V, $R_C = 9.1$ kΩ, $R_B = 680$ kΩ, $\beta = 75$, $V_{BE} = 0.600$ V.

MODIFIED COLLECTOR FEEDBACK BIAS CIRCUIT

$$I_C = \frac{V_{CC} - V_{BE}}{R_C\left(\frac{\beta+1}{\beta}\right) + \frac{R_B}{\beta}} = 0.787 \text{ mA}$$

$$V_{CE} = V_{CC} - I_C R_C \left(\frac{\beta+1}{\beta}\right) = 7.74 \text{ V}$$

$$R_B = \beta_{dc} R_C = 683 \text{ k}\Omega \quad \text{(approximately midpoint biased)}$$

3.7 MODIFIED COLLECTOR FEEDBACK BIAS CIRCUIT

The collector feedback bias circuit can be modified to reduce the Q point's dependence on β by adding a resistor from the transistor's base to ground. *The modified collector feedback bias circuit (also called an nV_{BE} biasing circuit)* (Figure 3-8) consists of a transistor and the resistors R_C, R_2, and R_1.

The key to the analysis of this circuit is the barrier potential of Q_1. With Q_1 on and the emitter at 0 V, the base voltage will be a constant of approximately 0.6 V. This forces the voltage drop across R_2 to be of the same value.

$$V_{R2} = V_{BE}$$

I_{R2} is then established:

$$I_{R2} = \frac{V_{R2}}{R_2}$$

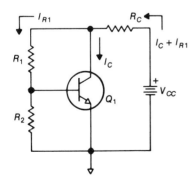

Figure 3-8. Modified collector feedback bias circuit (also called an nV_{BE} bias circuit).

For a well-designed circuit, the transistor's base current will be made much smaller than I_{R1} and I_{R2}.

$$I_B \ll I_{R1}, I_{R2}$$

Therefore,

$$I_{R1} \cong I_{R2}$$

V_{CE} is the sum of the voltage drops V_{R1} and V_{R2}. With I_{R1} known

$$V_{CE} = V_{R2} + V_{R1} = V_{BE} + \frac{V_{BE}}{R_2} R_1$$

$$= V_{BE}\left(1 + \frac{R_1}{R_2}\right) = nV_{BE} \quad (n \text{ is a constant})$$

V_{CE} is not directly related to β, as long as β is high enough to ensure a small base current.

The voltage drop across R_C is the difference in potential between V_{CC} and V_{CE}.

$$V_{RC} = V_{CC} - V_{CE} = V_{CC} - \left(1 + \frac{R_1}{R_2}\right)V_{BE}$$

The current through the collector resistor, R_C, is

$$I_{RC} = \frac{V_{RC}}{R_C} = \frac{V_{CC} - (1 + R_1/R_2)V_{BE}}{R_C}$$

This current consists of two parts, I_C and I_{R2}.

$$I_{RC} = I_C + I_{R1}$$

Solving for I_C yields

$$I_C = \frac{V_{CC} - (1 + R_1/R_2)V_{BE}}{R_C} - \frac{V_{BE}}{R_2}$$

Example

If $V_{CC} = 12$ V, $V_{BE} = 0.600$ V, $\beta = 100$, $R_C = 2.7$ kΩ, $R_1 = 10$ kΩ, $R_2 = 1$ kΩ, then

$$V_{R2} = V_{BE} = 0.600 \text{ V}$$

$$I_{R2} = \frac{V_{R2}}{R_2} = 0.600 \text{ mA}$$

$$I_{R1} \cong I_{R2} = 0.600 \text{ mA}$$

$V_{R1} = I_{R1}R_1 = 6.00 \text{ V}$

$V_{CE} = V_{R1} + V_{R2} = 6.60 \text{ V}$

$V_{RC} = V_{CC} - V_{CE} = 5.4 \text{ V}$

$I_{RC} = \dfrac{V_{RC}}{R_C} = 2.00 \text{ mA}$

$I_C = I_{RC} - I_{R1} = 1.40 \text{ mA}$

$I_B = \dfrac{I_C}{\beta} = 0.014 \text{ mA} \ll 0.600 \text{ mA}$

3.8 EMITTER-BIAS CIRCUIT

The *emitter-bias circuit* in Figure 3-9 uses two dc voltage supplies and is a member of the emitter-bias model class. The V_{EE} supply forward biases the transistor's *B-E* junction, and the V_{CC} supply reverse biases the *B-C* junction. This four-component bias circuit maintains a reasonably stable Q point (I_C and V_{CE}) but requires two supplies.

The *key to the analysis* of this circuit is to find I_E. The emitter current (and hence I_C and V_{CE}) can be found by applying KVL to the base-emitter loop in Figure 3-9b.

$I_B R_B + V_{BE} + I_E R_E - V_{EE} = 0$

I_B and I_E are related by α and β:

$I_B = \dfrac{\alpha}{\beta} I_E$

Substituting for I_B will reduce the equation to only the unknown I_E.

$\dfrac{\alpha}{\beta} R_B I_E + V_{BE} + R_E I_E - V_{EE} = 0$

Solving for I_E yields

$I_E = \dfrac{V_{EE} - V_{BE}}{R_E + (\alpha/\beta)R_B}$

I_E and I_C are related by α.

$I_C = \dfrac{\alpha(V_{EE} - V_{BE})}{R_E + (\alpha/\beta)R_B}$

The collector-to-emitter voltage V_{CE} is determined by applying KVL to the collector-emitter loop.

$-V_{CC} + I_C R_C + V_{CE} + I_E R_E - V_{EE} = 0$

I_E and I_C are related by α.

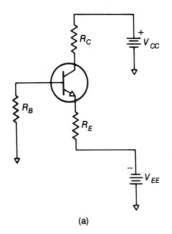
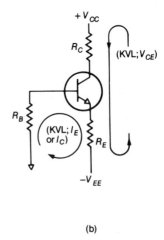

Figure 3-9. Emitter bias circuit: (a) circuit; (b) applying KVL to determine I_C and V_{CE}.

$$-V_{CC} + I_C R_C + V_{CE} + \frac{I_C}{\alpha} R_E - V_{EE} = 0$$

Since I_C is known at this point, the equation can be solved for V_{CE}.

$$V_{CE} = (V_{CC} + V_{EE}) - I_C \left(R_C + \frac{R_E}{\alpha} \right)$$

If β is high, I_C and V_{CE} may be approximated by the following equations:

$$I_C \cong \frac{V_{EE} - V_{BE}}{R_E + (R_B/\beta)}$$

$$V_{CE} = (V_{CC} + V_{EE}) - I_C(R_C + R_E)$$

If β is high, R_B and R_E are approximately equal, and V_{EE} is much greater than V_{BE}, a rough approximation can be made using the following equations:

$$I_C \cong \frac{V_{EE}}{R_E}$$

$$V_{CE} \cong (V_{CC} + V_{EE}) - I_C(R_C + R_E)$$

As long as R_B and R_E are reasonably close (a factor of 10) in value, the emitter of the transistor will be near zero volts. The base current will be one-hundredth ($\beta = 100$) of the emitter current, and the voltage drop across R_B will be small. The emitter voltage is the sum of the voltage drop across R_B plus V_{BE} of Q_1, and it is typically less than 1 V. With the emitter near zero volts, the V_{EE} voltage is dropped across R_E, which establishes I_E and hence I_C.

Example

Find the approximate and exact values of I_C and V_{CE} for the circuit in Figure 3-9 if V_{CC} = 18 V, V_{EE} = 8 V, $R_C = R_E = R_B$ = 12 kΩ, V_{BE} = 0.600 V, and β = 85 (α = 0.988).

Solution

The exact values of I_C and V_{CE} are

$$I_C = \frac{\alpha(V_{EE} - V_{BE})}{R_E + (\alpha/\beta) R_B}$$

$$= \frac{0.988(8 \text{ V} - 0.60 \text{ V})}{12 \text{ k}\Omega + \frac{0.988(12 \text{ k}\Omega)}{85}} = 0.602 \text{ mA}$$

$$V_{CE} = (V_{CC} + V_{EE}) - I_C(R_C + R_E/\alpha)$$
$$= 11.46 \text{ V}$$

The approximate values of I_C and V_{CE} are

$$I_C \cong \frac{V_{EE}}{R_E} = \frac{8 \text{ V}}{12 \text{ k}\Omega} = 0.667 \text{ mA}$$

$$V_{CE} \cong (V_{CC} + V_{EE}) - I_C(R_E + R_C)$$
$$= 26 \text{ V} - 0.667 \text{ mA}(24 \text{ k}\Omega) = 10 \text{ V}$$

The exact and approximate values are within 12% of each other.

3.9 DC BIASING OF AM RADIO CIRCUITS

The actual schematic of a popular, low-cost *AM radio* is shown in Figure 3-10. The high-frequency carrier, which is amplitude modulated, with the intelligence is received at the input transducer or antenna. The signal is processed or broken down until the intelligence or audio signal is recovered. That signal is then used to drive the output transducer or speaker.

This application of bias circuits also serves as an excellent example illustrating how circuits can be cascaded or put in series. The ac coupling capacitors C_3, C_{22}, C_{23}, and C_{26}, and the transformers T_1 through T_3 establish open circuits between the stages, isolating the dc currents and voltages of the individual circuits. The Q_1 through Q_5 stages or circuits are *collector-feedback biased*. Remember that inductors and transformer windings appear as short circuits to dc and capacitors appear as open circuits. The dc operating point of the audio amplifier or Q_5 circuit will be determined. The dc equivalent circuit of the Q_5 stage is shown in Figure 3-11. Resistor R_{14} is very small, and its junction with R_{11} will

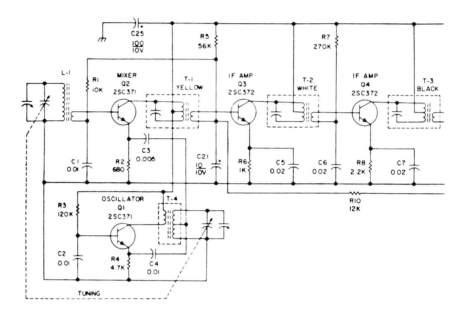

Figure 3-10. AM radio schematic.

be assumed to be at V_{CC} or 6 V. Resistor R_{20} is very large compared to R_{13}, and it will be assumed to be an open circuit. Resistor R_{13} and resistor R_{20} are components establishing the ac voltage gain of the Q_5 through Q_8 circuits. The collector current is determined by applying KVL to the V_{CC}-R_{11}-R_{12}-V_{BE}-R_{13} loop.

$$-V_{CC} + (I_B + I_C)R_{11} + I_B R_{12} + V_{BE} + I_E R_{13} = 0$$

The variables I_B and I_E are replaced with their equivalence in terms of I_C.

$$-V_{CC} + \left(\frac{I_C}{\beta} + I_C\right)R_{11} + \frac{I_C}{\beta}R_{12} + V_{BE} + \frac{I_C}{\alpha}R_{13} = 0$$

The equation is then solved for I_C.

$$I_C = \frac{V_{CC} - V_{BE}}{(R_{13}/\alpha) + (R_{12}/\beta) + [(\beta+1)/\beta]R_{11}}$$

The voltage V_{CE} is equal to V_{CC} less the voltage drops across R_{11} and R_{13} (apply KVL again).

$$-V_{CC} + (I_C + I_B)R_{11} + V_{CE} + I_E R_{13} = 0$$

$$-V_{CC} + \frac{\beta+1}{\beta}I_C R_{11} + V_{CE} + \frac{I_C}{\alpha}R_{13} = 0$$

Figure 3-11. Dc bias circuit—AM radio amplifier.

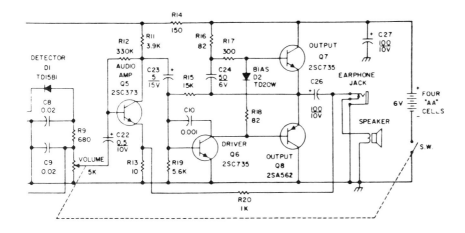

Figure 3-10. (*Continued*)

or $V_{CE} = V_{CC} - \dfrac{I_C}{\alpha}(R_{11} + R_{13})$

For the given circuit values on the schematic and $\beta_{dc} = 200$, $V_{BE} = 0.7$ V, the values of I_C and V_{CE} are

$I_C = 0.96$ mA

$V_{CE} = 2.3$ V

The value of the dc beta is very high, and alpha can be approximated as equal to 1 with minimal error. The value of R_{13} is very small, and the emitter is very near 0 Vdc.

3.10 BIAS CIRCUIT DESIGN

Philosophy

A design procedure is opposite to that of an analysis. In an analysis, the circuit is specified. The arrangement of the components, the type of components, and their values are given. The performance and capabilities of the circuit are determined by computing the values of the circuit's characteristics. For a given set of conditions, generally only one value or correct answer applies to any one characteristic.

In a design, the characteristics are specified, that is, the circuit's performance is established, but not the circuit itself. It is left to the designer to create a circuit such that its performance will meet the desired specifications. Generally, a large number of different circuits will meet the performance criteria. However, other practical considerations will significantly reduce the number of eligible circuits. In most designs, there is a degree of arbitrariness in the selection of the circuit and the number and type of components. However, the more detailed the specifications, the more restricted is the selection.

The presentation of design examples in this book serves three purposes:

1. Design examples provide the background material to help understand the purpose and in-

tent of each component in a circuit and how their values are derived.

2. An understanding of rudimentary design principles reinforces one's analytical ability. A person must be capable of analyzing a circuit before designing one.

3. In engineering and development environments, the technician must be aware of potential design problems and identify them from component malfunctions and/or assembly errors.

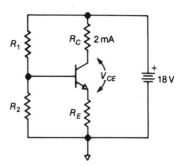

Figure 3-12. Bias circuit design.

Example

Requirements: Design a dc bias circuit whose Q point is 2 mA (I_C) and 8 V (V_{CE}). The measured Q point must be within 10% of the specified values. A single dc supply is available with a fixed value of 18 V. All standard values of 5%, carbon-composition resistors are in stock. The Q point should be relatively independent of device and temperature changes. The circuit will be used in a low-volume product and economy is not the overriding consideration. An abundance of low-cost, well-performing *npn* transistors is available.

Solution

The requirement of a relatively stable circuit suggests a voltage-divider bias circuit. Since economy is not dominant in the circuit selection, the extra components of this circuit do not restrict its use. The availability of *npn* transistors and a 18 V supply define the circuit shown in Figure 3-12.

V_{CE} is 8 V; hence, 10 V must be dropped across R_E and R_C. The emitter voltage V_E and R_E will establish the emitter current ($\cong I_C$). Initially, and somewhat arbitrarily, the 10 V is split between V_{RE} and V_{RC}. If V_E is 5 V, a 2.5 kΩ resistor will establish an emitter and collector current of approximately 2 mA. Since 2.5 kΩ is not a standard resistor value, other combinations of V_E and R_E are tried that provide a I_C of 2 mA. This combination must also satisfy the condition that

$$V_{CC} = 18\text{ V} = V_{RC} + V_{CE} + V_{RE}$$
$$= V_{RC} + 8\text{ V} + V_{RE}$$

After several iterations,

$R_E = 3$ kΩ
$V_E = 6$ V
$R_C = 2$ kΩ

satisfy the requirements.

The emitter voltage will be +6.00 V if the base voltage is more positive by 0.6 V or one V_{BE}. Hence, the resistor divider comprised of R_1 and R_2 must be such that the base voltage is +6.60 V. Mathematically,

$$\frac{R_2}{R_1 + R_2}(18\text{ V}) = 6.60\text{ V}$$

Arbitrarily, standard values are assigned to R_1, and the equation is solved for R_2 until R_2 is computed as a standard value. The values

$R_1 = 4.7$ kΩ
$R_2 = 2.7$ kΩ

will provide a base voltage of +6.57 V.

The circuit must now be checked to ensure that

$$I_B \ll I_1$$

The current through the voltage divider is

$$I_1 = \frac{18 \text{ V}}{7.4 \text{ k}\Omega} = 2.43 \text{ mA}$$

If $\beta = 75$, then

$$I_B = \frac{2 \text{ mA}}{75} = 0.0267 \text{ mA}$$

The base current is much smaller than the divider current, and the assumptions made in the design are valid.

In the actual design of a high-performance, high-volume product, many other considerations must be accounted for, which narrows the choice of the range and values of the components. Other considerations include economy, stability, power dissipation, and the ac characteristics of the circuit.

3.11 AN INTRODUCTION TO STABILITY FACTORS

A stability factor is a relationship expressing the rate of change of one variable with respect to another. With respect to bias circuits, stability factors express how much the dc operating or Q point will change for a given change in the transistor or circuit parameters. These factors quantitatively reflect the stability of the operating point, and they are extremely valuable in predicting the performance of the dc circuit over a wide temperature range.

The temperature of the environment of a circuit will vary. For commercial applications, the temperature variation will be small, typically about 10 to 20°C. For military and aerospace applications, the variation will be large, typically in excess of 100°C. When temperature changes, the values of most bipolar transistor parameters will change. Figure 3-13 illustrates the changes in V_{BE} and β graphically and in data form. The question is, "How much and to what degree do these changes impact the dc behavior of the circuit?" Stability factors provide the answers.

The *letter designation* for stability factor is S. The two subscripts associated with S ($S_{IC, VBE}$) indicate the two variables of concern. The stability factor $S_{IC, VBE}$ is defined as the rate of change of the collector current I_C with respect to V_{BE}. In other words, $S_{IC, VBE}$ relates how much I_C will change (ΔI_C) for a change in V_{BE} (ΔV_{BE}). Mathematically, it is defined as

$$S_{IC, VBE} = \frac{\Delta I_C}{\Delta V_{BE}}$$

The quantity $\Delta I_C/\Delta V_{BE}$ is the slope of the line relating I_C and V_{BE}. Other prominent stability

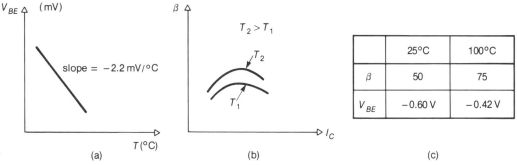

Figure 3-13. Transistor parameter changes: (a) V_{BE} and temperature; (b) β and temperature and I_C; (c) typical data.

factors are

$$S_{IC,\beta} = \frac{\Delta I_C}{\Delta \beta}$$

$$S_{VCE,VBE} = \frac{\Delta V_{CE}}{\Delta V_{BE}}$$

$$S_{VCE,\beta} = \frac{\Delta V_{CE}}{\Delta \beta}$$

$$S_{VCE,VCC} = \frac{\Delta V_{CE}}{\Delta V_{CC}}$$

To find a stability factor, the equation relating the two variables of concern must be known. Consider the base-driven model in Figure 3-14. To find $S_{IC,VBE}$, the expression for I_C and V_{BE} must be determined. For simplicity, let $I_C = I_E$. From a previous derivation,

$$I_C \cong \frac{V_{BB} - V_{BE}}{R_E + R_B/\beta} = \frac{V_{BB}}{R_E + R_B/\beta} - \frac{V_{BE}}{R_E + R_B/\beta}$$

All quantities except for V_{BE} are assumed to be constant. Those include V_{BB}, β, and V_{CC}. The above equation has the form of

$$y = b + mx$$

where y and x are the variables, b is a constant, and m is the slope of the linear relationship between y and x. The slope is the stability factor. Using this as a basis,

$$S_{IC,VBE} = -\frac{1}{R_E + R_B/\beta}$$

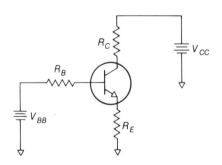

Figure 3-14. Base-driven bias circuit model.

The dimension or unit of measurement of this stability factor is typically expressed as milliamperes per millivolts. If $R_E = R_B = 10$ kΩ, and $\beta = 200$, then

$$S_{IC,VBE} = -\frac{1}{10.05 \text{ k}\Omega} \cong -0.1 \; \mu A/mV$$

The above relationship says that the collector current will decrease (minus sign) by about 0.1 μA for every 1 mV increase in V_{BE}. V_{BE} will decrease by 2.2 mV for every degree Celsius increase in temperature.

The other half of the Q point, V_{CE}, is also sensitive to changes in V_{BE}. The approximate expression for V_{CE} is

$$V_{CE} \cong V_{CC} - I_C(R_C + R_E)$$

I_C is equal to

$$I_C \cong \frac{V_{BB} - V_{BE}}{R_E + R_B/\beta}$$

Substituting for I_C in the V_{CE} expression produces

$$V_{CE} \cong V_{CC} - \left(\frac{V_{BB} - V_{BE}}{R_E + R_B/\beta}\right)(R_C + R_E)$$

$$\cong \left\{V_{CC} - \left(\frac{V_{BB}(R_C + R_E)}{R_E + R_B/\beta}\right)\right\}$$

$$+ \frac{V_{BE}(R_C + R_E)}{R_E + R_B/\beta}$$

The equation has the form of $y = b + mx$. The slope, and hence the stability factor, is

$$S_{VCE,VBE} = m = \frac{\Delta V_{CE}}{\Delta V_{BE}} = \frac{R_C + R_E}{R_E + R_B/\beta}$$

The dimension or unit of measurement of this stability factor is a dimensionless number or millivolt per millivolt. If $R_C = R_B = R_E = 10$ kΩ, $\beta = 200$, then

$$S_{VCE,VBE} = \frac{20 \text{ k}\Omega}{10.05 \text{ k}\Omega} \cong 2 \text{ mV/mV}$$

V_{CE} will increase (positive sign) by about 2 mV for every 1 mV increase in V_{BE}.

The operating point is not only sensitive to transistor parameters but also V_{CC}. The stability factor $S_{VCE,VCC}$ measures the sensitivity of V_{CE} to changes in the supply voltage. This stability factor is sometimes called regulation sensitivity and is important in voltage-regulator circuits. A similar technique is used in filter circuits to relate the stability of a filter characteristic to a change in a device or circuit parameter. In filter theory, they are called *sensitivity functions*.

3.12 JFET BIAS CIRCUITS

Self-Bias Circuit

The bipolar transistor is a current-controlled device, and the JFET is voltage controlled. These different operating modes translate to different approaches in dc biasing the devices. The circuit in Figure 3-15a is a popular, *single-supply JFET bias circuit* that uses an *n*-channel device. It is called a self-bias circuit. An explanation of what establishes the circuit values (design criteria) in a JFET bias circuit will be presented because it is more informative (and less complicated) than an analysis.

The drain family of curves (Figure 3-15b) plays a key role in establishing the Q point or I_D and V_{DS}. For a given or desired value of I_D, the value of V_{GS} that causes that amount of current must be known. It can be determined from the drain family of curves found in the device's data sheets or from the curve tracer's data. For an *n*-channel device, V_{GS} will vary from 0 V to $-V_{GS}$, which causes pinch off. For $V_{GS} = 0$ V, I_D is maximum, which is called I_{DSS}. For V_{GS} equal to the pinch-off voltage (minus), the drain current I_D is zero. The values of V_{GS} from 0 V to $-V_p$ (pinch off) correspond to values of I_D from I_{DSS} to 0 A, respectively.

The input resistance of the JFET is extremely high, and the drop across R_G will be zero volts because of the negligible gate current. Thus, the voltage at the gate with respect to ground is zero volts. The circuit must be designed such that the source voltage is the value of V_{GS} required to cause the device to conduct the desired value of I_D. The gate-to-source voltage will be negative because the gate is at zero volts and the source is at a positive voltage (with respect to ground).

$$V_{GS} = V_G - V_S = 0\text{ V} - I_D R_S = -I_D R_S$$

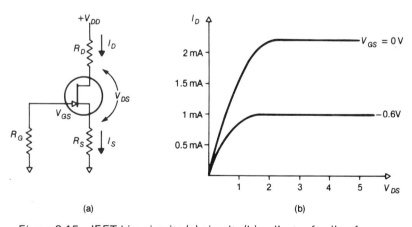

Figure 3-15. JFET bias circuit: (a) circuit; (b) collector family of curves.

The value of I_D is specified; the corresponding value of V_{GS} is found in the data sheet; and R_S must be chosen such that

$$R_S = \left|\frac{V_{GS}}{I_D}\right|$$

With R_S established, R_D is chosen to establish the required value of V_{DS}.

$$V_{DS} = V_{DD} - I_D(R_D + R_S)$$

Example

The circuit in Figure 3-15a will be designed for $I_D = 1$ mA and $V_{DS} = 5$ V. For this circuit, $V_{DD} = +12$ V, and from the drain family of curves, the gate-to-source voltage is -0.6 V for $I_D = 1$ mA. R_G does not impact the operating point and is usually chosen to establish the value of the circuit's input resistance. R_G is required to provide a voltage reference for the gate. Somewhat arbitrarily, it is specified at 100 kΩ. With the gate at zero volts (no gate current), the source must be at +0.6 V to provide the required V_{GS} (-0.6 V). If the source is at +0.6 V, the voltage drop across R_S must be the same. The current through R_S must be I_D ($= I_S$) or 1 mA, and thus the value of R_S is fixed to

$$R_S = \frac{0.60 \text{ V}}{1 \text{ mA}} = 600 \text{ Ω}$$

If V_{DS} is to be 5 V and the voltage at the source is +0.6 V, the drain voltage must be +5.6 V. With $V_D = +5.6$ V, the voltage drop across R_D will be 6.4 V. Since $I_D = 1$ mA,

$$R_D = \frac{6.4 \text{ V}}{1 \text{ mA}} = 6.4 \text{ kΩ}$$

Two-Supply Bias Circuit

A second *JFET biasing circuit* is shown in Figure 3-16. Normal variations in the device's relationship between V_{GS} and I_D minimally

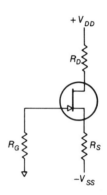

Figure 3-16. Two-supply JFET bias circuit.

affect the circuit's Q point; however, the circuit requires *two dc supplies*. The procedure for establishing the Q point in this circuit is similar to the self-bias case.

The circuit will be designed for a Q point at 1 mA and 5 V. The same device will be used as in the previous example, and $V_{DD} = V_{SS} = 12$ V. For $I_D = 1$ mA and $V_{GS} = -0.6$ V, then

$$V_S = +0.6 \text{ V}$$

The difference in potential across R_S is

$$V_{RS} = V_S - V_{SS} = 0.6 \text{ V} - (-12 \text{ V}) = 12.6 \text{ V}$$

Since the source current must be 1 mA, then R_S must be

$$R_S = \frac{V_{RS}}{I_D} = \frac{12.6 \text{ V}}{1 \text{ mA}} = 12.6 \text{ kΩ}$$

If $V_{DS} = 5$ V and $V_S = +0.6$ V, then $V_D = +5.6$ V. The difference in potential across R_D is

$$V_{RD} = V_{CC} - V_D = 12 \text{ V} - (+5.6 \text{ V}) = 6.4 \text{ V}$$

Since $I_D = 1$ mA,

$$R_D = \frac{V_{RD}}{I_D} = \frac{6.4 \text{ V}}{1 \text{ mA}} = 6.4 \text{ kΩ}$$

In actuality, the drain current is established by V_{GS} and the R_S, V_{SS} portion of the circuit. Changes in V_{GS} will result in very small changes in the voltage drop across R_S.

Problems

1. For the circuit in Figure 3-4a, $R_1 = 9.1$ kΩ, $R_2 = 4.3$ kΩ, $R_C = 2$ kΩ, $R_E = 2$ kΩ, $V_{CC} = 20$ V, $V_{BE} = 0.600$ V, and $\beta_{dc} = 200$. Find
    (a) I_C
    (b) V_{CE}
    (c) V_{RC} (the voltage drop across R_C)
    (d) V_C (the collector voltage with respect to ground)
    (e) R_{TH} (Figure 3-4b)
    (f) V_{TH} (Figure 3-4b)

2. Repeat problem 1, except change V_{CC} to +15 V.

3. The emitter resistor R_E in Figure 3-6 increases to 3 kΩ. All other component values and conditions remain the same. Determine
    (a) I_C
    (b) V_{CE}
    (c) V_C

4. The V_{CC} supply in Figure 3-6 is decreased to -10 V. All other component values and conditions remain the same. Determine
    (a) I_C
    (b) V_{CE}
    (c) V_E

5. For the circuit in Figure 3-7, $V_{CC} = 15$ V, $R_C = 9.1$ kΩ, $R_B = 680$ kΩ, $\beta = 250$, and $V_{BE} = 0.600$ V. Find
    (a) I_C
    (b) V_{CE}
    (c) Is the circuit midpoint biased?

6. In the circuit of problem 5, the transistor's β decreases to 25. All other conditions remain the same. Find
    (a) I_C
    (b) V_{CE}
    (c) Is the circuit midpoint biased?

7. Find I_C, V_{CE}, and V_C for the nV_{BE} bias circuit in Figure 3-8 if $V_{CC} = 12$ V, $V_{BE} = 0.720$ V, $\beta = 100$, $R_C = 2.7$ kΩ, $R_1 = 10$ kΩ, and $R_2 = 1$ kΩ.

8. The V_{BE} of the transistor in the circuit of problem 7 changes to 0.550 V (temperature increases). All other component values and conditions remain the same. Find I_C, V_{CE}, and V_C.

9. The resistor values in the emitter bias circuit of Figure 3-9 are all 18 kΩ. For the transistor, $V_{BE} = 0.600$ V and $\beta = 200$. Both dc supplies are 15 V. Calculate
    (a) I_C
    (b) V_{CE}
    (c) V_B
    (d) V_C
    (e) V_E

10. In Figure 3-9, $V_{EE} = -5$ V, $V_{CC} = +10$ V, $R_B = 510$ Ω, $R_E = 1$ kΩ, $R_C = 2$ kΩ, $V_{BE} = 0.500$ V, and $\beta_{dc} = 150$. Calculate the values of I_C, I_E, I_B, and V_{CE}.

11. Find the dc operating point for the transistor Q_1 in Figure 3-10. For the transistor, assume $V_{BE} = 0.600$ V and $\beta_{dc} = 200$.

12. In the second IF amplifier circuit (Q_4) in Figure 3-10, the β for the transistor is 200 and $V_{BE} = 0.650$ V. Calculate
    (a) I_C
    (b) V_{CE}
    (c) V_E

13. Design a voltage-divider bias circuit whose Q point is 2 mA (I_C) and 4 V (V_{CE}). A single 15 V dc supply is available. The transistor is an npn with a V_{BE} of 0.600 V and $\beta_{dc} = 150$.

14. Rearrange the circuit designed in problem 13 for a pnp transistor that has the same specifications as the npn.

15. Design a collector feedback bias circuit to meet the same specifications given in problem 13. Convert the circuit to a pnp version.

16. Find the stability factor $S_{IC,VBE}$ for the collector feedback bias circuit in Figure 3-7. Determine the value of this stability factor if $V_{CC} = 15$ V, $R_C = 9.1$ kΩ, $R_B = 680$ kΩ, $\beta = 75$, and $V_{BE} = 0.600$ V.

17 Repeat problem 16, except find the stability factor $S_{VCE,VBE}$.

18 For the JFET transistor in Figure 3-16, $V_{GS} = -0.6$ V at $I_D = 3$ mA, $V_{GS} = -0.4$ V at $I_D = 2$ mA, and $V_{GS} = -0.2$ V at $I_D = 1$ mA. In this circuit, $V_{SS} = -30$ V, $V_{DD} = +30$ V, $R_S = 18$ kΩ, $R_D = 9.1$ kΩ, and $R_G = 1$ MΩ. The gate leakage current I_{GSS} is 100 pA. Estimate the value of I_D and V_{DS}.

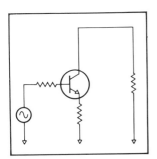

4

AMPLIFIER AC PRINCIPLES

The circuit whose behavior follows that of the amplification concept is called an amplifier. Superposition allows us to examine the dc and ac behavior of the circuit independently. This chapter discusses the ac principles of the amplifier.

The voltage, current, and power gain, and the input and output resistance of the amplifier are found by examining the amplifier ac equivalent circuit, including the ac model of the transistor. The initial gain stages or circuits in an amplifier system are described by its small-signal voltage gain. Voltage gain is the ratio of the input and output voltages. The ac emitter current in small-signal amplifiers is small compared to the dc emitter current. The key to finding the gain is to determine the emitter current and then the circuit and/or transistor voltages. An amplifier circuit has a voltage gain associated with the transistor and a voltage gain from source to load. There are a number of amplifier configurations, including the common-emitter, common-collector or emitter-follower, swamped-emitter, and common-base circuits.

The input and output resistances play an important role when connecting or interfacing amplifier circuits together or with other type circuits. The input resistance is the resistance seen between the amplifier's input terminals under ac conditions, and the output resistance is the resistance seen between the amplifier's output terminals under ac conditions.

The current and power gain of an amplifier is the ratio of the input and output currents and power values, respectively. Because of the large numbers associated with the different gains, logarithms are used, and the gain is called the Bel gain. The differential-amplifier circuit is a key building block in IC amplifiers.

4.1 INTRODUCTION

The primary characteristics of an amplifier are gain, input and output resistance, and lower and upper cutoff frequencies. To determine and evaluate these characteristics, the complete amplifier circuit must be analyzed. This analysis is quite detailed and involves the dc and ac responses of the circuit. Fortunately, through the superposition theorem, the dc and ac

53

responses of the amplifier can be separated and looked at individually, with the actual response being the sum of the two.

In this chapter the complete amplifier circuit will be reduced to an ac equivalent circuit and analyzed. From this analysis, the ac response of the amplifier will be determined and information extracted that is necessary to describe the amplifier's overall response. By using information previously determined from the dc response, the two will be combined, and the overall amplifier response can be described. Various circuit configurations will be examined.

4.2 AMPLIFIER AC EQUIVALENT CIRCUIT

Circuit Example

A *complete discrete component amplifier circuit* is shown in Figure 4-1a. From the analysis of this amplifier's dc equivalent circuit, all dc voltages and currents were determined, including the transistor's dc (Q) operating point. The *ac equivalent circuit* (Figure 4-1b) of this amplifier will be used to find the gain, input and output resistance, and the upper and lower cutoff frequencies.

Capacitances

Three types of capacitances are found in the amplifier circuit Figure 4-1a:

1. Discrete
2. Device
3. Distributed

Device and distributed capacitances are normally not shown on the schematic. They are identified from the components' equivalent circuits, and their values are found in the data sheets.

The *discrete capacitances* are the physical components in the circuit, that is, C_{C1}, C_{C2}, C_B, and C_S. These capacitors function as coupling and bypass capacitors, and their values are normally in the microfarad (μF) region. One of these capacitors (and an associated resistance) or their combination will establish the lower cutoff frequency. In the midband range of frequencies for the amplifier, these capacitors are equivalent to a short circuit. The coupling capacitors couple the ac signal in and out of the amplifier circuit. These capacitors are also dc blocking capacitors that isolate the dc bias voltages and currents of each circuit. The bypass capacitors bypass or ac short circuit part of the emitter resistance. Shorting part of R_E increases the circuit's voltage gain.

The *device capacitances* are those associated with the transistor. The two key transistor capacitances are the collector-to-base capacitance C'_c and the base-to-emitter capacitance C'_e. The device capacitances are very low in value, typically 5 to 200 pF. These low-value capacitances reflect the frequency limitation of the transistor and establish the maximum upper cutoff frequency of the circuit. In the midband range of frequencies for the amplifier, these low-value capacitances are equivalent to an open circuit. Frequently, low-value discrete capacitors are added to the circuit to establish the upper cutoff frequency.

The value and location of the *distributed capacitances* of a circuit depend on the construction and physical aspects of the actual circuit. Socket capacitance, wiring and coax capacitance, and interlead capacitance all can influence the value of the upper cutoff frequency. The value of these capacitances is low and in the same region as the device capacitances. For many cases, these capacitances directly add to the device capacitances. In the midband range of frequencies for the amplifier, these low-value capacitances are equivalent to an open circuit. Figure 4-2 shows the location and

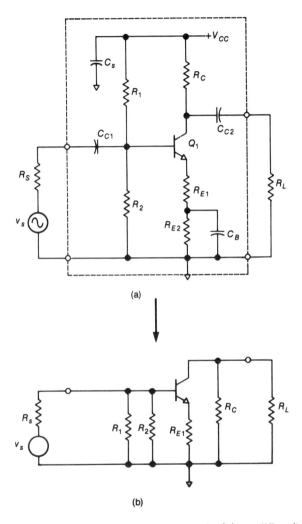

Figure 4-1. An amplifier and its ac equivalent circuit: (a) amplifier; (b) ac equivalent circuit.

type of common device and distributed and circuit capacitances found in the amplifier circuits.

In general, *capacitors in series with the signal path establish the lower cutoff frequency*. The input and output coupling capacitors are examples. In general, *capacitors in shunt with the signal path establish the upper cutoff frequency*. The input Miller capacitance version of C_c' is an example. Associated with each of the capacitances is a resistance, and the RC combinations actually establish the cutoff frequencies. These RC circuits, called lead and lag networks, are also known as simple high- and low-pass filters.

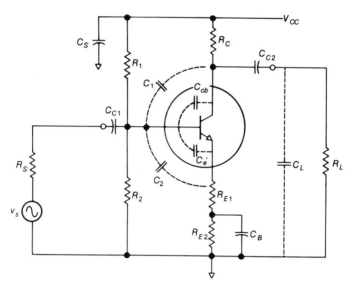

Figure 4-2. Device and distributed capacitances in an amplifier.

Example

What is the reactance of (a) a 10 μF coupling capacitor, (b) a C'_e of 250 pF, and (c) 1.5 ft of coax cable with a 30 pF/ft rating if the signal frequency is 10 kHz?

Solution

(a) $X_C = \dfrac{1}{2\pi f C_C}$

$= \dfrac{1}{(2\pi)(10 \cdot 10^{-6} \text{ F})(1 \cdot 10^4 \text{ Hz})}$

$= 1.59 \text{ }\Omega$

(b) $X_C = \dfrac{1}{2\pi f C'_e}$

$= \dfrac{1}{(2\pi)(250 \cdot 10^{-12} \text{ F})(1 \cdot 10^4 \text{ Hz})}$

$= 63.7 \text{ k}\Omega$

(c) $X_C = 354 \text{ k}\Omega$

Ac Ground

The ground in an ac equivalent circuit is called an ac ground. It may or may not be the same as the dc or system ground. Since large-valued capacitors are equivalent to a short circuit at the amplifier's midband range of frequencies, these capacitors can couple an ungrounded point to a grounded point. An example is the bypass capacitor in Figure 4-3a. Point A in the circuit is an ac ground because the bypass capacitor C_B is equivalent to a very low value of ac resistance and, in essence, short circuits point A to the circuit ground. Another example, Figure 4-3b, is the supply bypass capacitor. This large-valued capacitor makes the dc supply line look like an ac ground.

Example

What is the exact ac potential at point A in Figure 4-3a if $R_1 = 2$ kΩ, $R_2 = 2$ kΩ, $C_B = 10$ μF, and $V_s = 10$ V at a frequency of 10 kHz?

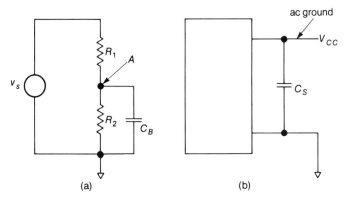

Figure 4-3. Bypass capacitors and ac ground.

Solution

The capacitive reactance of C_B at 10 kHz is

$$X_C = \frac{1}{2\pi f C} = \frac{1}{(2\pi)(1 \times 10^4 \text{ Hz})(1 \times 10^{-5} \text{ F})}$$
$$= 1.59 \text{ }\Omega$$

The parallel combination of C_B and R_2 is

$$X_C \| R_2 = 2000 \text{ }\Omega \| 1.59 \text{ }\Omega \cong 1.59 \text{ }\Omega$$

The voltage at point A is established by voltage-divider action:

$$V_A \cong \frac{Z_C}{Z_C + R_1} V_S = 0.008 \text{ V}$$

Finding the Ac Equivalent Circuit

To find the ac equivalent circuit of an amplifier,

1. Replace all high-value circuit capacitors with short circuits. Typically, coupling and bypass capacitors will fall in this category.

2. Replace all low-value capacitances with open circuits. Device and distributed capacitances (if schematically shown) fall into this category.

3. Replace all dc potential sources (ideal) with short circuits and ideal dc current sources with open circuits.

4. Replace high-value inductances with open circuits and low-value inductances with short circuits.

5. Identify all ac grounds.

Resistances in ac equivalent circuits are designated with lowercase letters. Generally, uppercase *subscripts* designate discrete resistors and lowercase subscripts designate device resistances. As an example, r_B is the discrete resistance in series with the base of a transistor in an ac equivalent circuit. It can represent a single component or the combination of several resistors. The resistance r_b' is the transistor's internal base resistance.

Base- and Emitter-Driven Models

Most amplifier circuits can be reduced or simplified, through analysis techniques, to one of *two circuit models or prototypes* shown in Figure 4-4. These prototype circuits are amplifier ac equivalent circuits, and they are called the base-driven and emitter-driven models. In the base-driven model, the signal source v_{bb} drives the base (through r_B) and is the more commonly found circuit. The circuits, along with the ac equivalent of the transistor, are

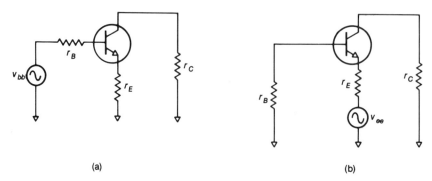

Figure 4-4. Amplifier circuit models: (a) base-driven model; (b) emitter-driven model.

used to determine voltage gain and input and output resistance. The capacitances are added to find the upper and lower cutoff frequencies.

4.3 TRANSISTOR AC MODEL OR EQUIVALENT CIRCUIT

First-Order or Ideal Model

The ideal or first-order and higher order ac models of the transistor are shown in Figure 4-5a. The circuit includes a dependent current source and the emitter diffusion resistance r'_e.

The current source is in series with the collector and it establishes the collector current i_c. The value of the current source depends on the value of the base current (i_b) and the ac current gain β_{ac}.

Emitter-Diode Ac Resistance, r'_e

The resistance r'_e is called the emitter diffusion resistance, the ac emitter resistance, or the incremental or dynamic resistance. It is associated with the forward-biased B-E junction.

A typical B-E diode curve relating I_E and

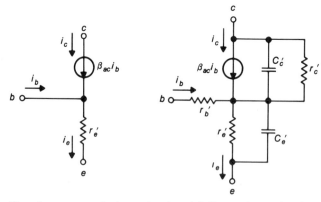

Figure 4-5 Transistor ac equivalent circuits: (a) first-order or ideal model; (b) higher-order model.

TRANSISTOR AC MODEL OR EQUIVALENT CIRCUIT

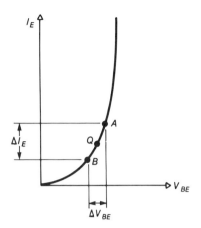

Figure 4-6. Graphical meaning of r'_e.

V_{BE} is shown in Figure 4-6. Point Q on this curve is the Q or quiescent point for the B-E pn junction. The coordinates of this point are the dc emitter current and the dc base-to-emitter voltage. The dc values of I_C and V_{CE} establish the transistor's Q point, and the dc values of I_E and V_{BE} establish the B-E junction's Q point.

The B-E junction is part of the transistor. The transistor is part of the circuit. If no ac signal is present in the circuit, the transistor will be dc biased at some value of I_C, V_{CE}, V_{BE}, and so on. These values establish the dc operating points. When an ac signal is introduced into the circuit, all the transistor's currents and voltages will vary slightly above and below the dc values. This movement, from point A to B, is illustrated for the B-E junction in Figure 4-6. The region along the curve from A to B can be approximated as a *straight line*, or we say the curve is piecewise linear. A straight line on an I-V graph models a resistor whose value is equal to the reciprocal of the line's slope. This B-E ac resistance is the emitter-diode ac resistance r'_e.

$$r'_e = \frac{1}{\text{slope}} = \frac{1}{\Delta I_E / \Delta V_{BE}} = \frac{\Delta V_{BE}}{\Delta I_E}$$

A change in V_{BE} (ΔV_{BE}) and a change in I_E (ΔI_E) can also be expressed as an ac voltage and ac current. In symbols,

$$\Delta I_E = i_e$$

$$\Delta V_{BE} = v_{be}$$

We can rewrite the equation for r'_e as

$$r'_e = \frac{v_{be}}{i_e}$$

Example

What is the value of r'_e in Figure 4-6 if the coordinates of point A are (0.500 V, 0.9 mA) and of point B are (0.505 V, 1.1 mA)?

Solution

$$r'_e = \frac{\Delta V_{BE}}{\Delta I_E} = \frac{V_{BE1} - V_{BE2}}{I_{E1} - I_{E2}}$$

$$= \frac{5 \text{ mV}}{0.2 \text{ mA}} = 25 \, \Omega$$

Determining the Value of r'_e

The value of r'_e will vary depending on the B-E dc operating point. *The value of r'_e is a function of the slope*, and the slope varies along the diode curve. For low values of I_E, the Q point is near the knee of the curve and the value of r'_e is typically in the hundreds of ohms region. At higher currents, the slope increases and r'_e decreases to the tens of ohms region or smaller.

The slope of the line, at a given point, can be determined using calculus. From calculus, the slope can be calculated using

$$r'_e = \frac{25 \text{ mV}}{I_E}$$

Thus, the emitter-diode ac resistance can be determined by dividing 0.025 V by the dc emitter current. A transistor whose emitter current is 1 mA will have an r'_e of 25 Ω.

r'_e and Temperature

The *emitter-diode ac resistance is temperature dependent*. As temperature increases, r'_e increases. The formula showing the temperature dependence is

$$r'_e = \left(\frac{T(°C) + 273°}{291°}\right)\frac{0.025\text{ V}}{I_E}$$

At 18° (room temperature),

$$r'_e = \frac{0.025\text{ V}}{I_E}$$

and at 100°C,

$$r'_e = \frac{0.032\text{ V}}{I_E}$$

Any amplifier characteristic whose value is a function of r'_e will be temperature dependent. Generally, this is not a desirable situation.

4.4 SMALL-SIGNAL VOLTAGE GAIN

Definition of Small Signal

Small signal means that the changes in a transistor's collector current, under ac conditions, are small compared to its quiescent or dc value. Small-signal amplifiers are used as the *initial stages* of receivers, amplifier systems, and measuring instruments. As an example, if a transistor's ac collector current was 20 µA and its dc collector current was 1 mA, then the circuit that the transistor was in would be called a small-signal amplifier.

The key amplifier characteristic is gain or amplification. This characteristic is determined from the amplifier and transistor's ac equivalent circuits. Figure 4-7b shows the ac equivalent circuit of the ac voltage amplifier. To ease the analysis, this ac circuit is further simplified by Theveninizing the input and output networks. The result is shown in Figure 4-7c. This circuit is called the base-driven model and serves as a prototype for many amplifier circuits. Before the voltage gain of this circuit can be determined, the ac emitter current must be found.

Ac Emitter Current

The *ac emitter current* plays a central or key role in the ac equivalent circuit. Once i_e is known, i_b can be calculated (using β) or i_c can be calculated (using α). With the transistor currents known, the ac voltage drops across resistors in series with the transistor leads can be found.

The ac emitter current is found by *applying KVL* to the base-emitter loop shown in Figure 4-8. The transistor schematic symbol is replaced by its ideal equivalent circuit. To assign signs to the algebraic terms in this expression, the circuit is analyzed during the positive half of the ac cycle. The results will be the same for either half-cycle. Applying KVL,

$$-v_{bb} + i_b r_B + i_e(r'_e + r_E) = 0$$

The currents i_b and i_e are related by

$$i_b = \frac{\alpha i_e}{\beta}$$

Substituting for i_b,

$$-v_{bb} + \frac{\alpha i_e}{\beta} + i_e(r'_e + r_E) = 0$$

Solving for i_e yields,

$$i_e = \frac{v_{bb}}{r_E + r'_e + \alpha r_B/\beta}$$

Circuit Ac Voltages

The ac voltage at the emitter, with respect to ground, is equal to the emitter ac current times the emitter resistance r_E.

$$v_e = i_e r_E$$

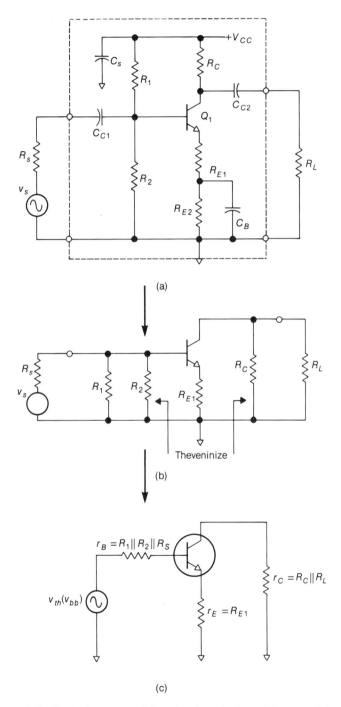

Figure 4-7. Reducing an amplifier circuit to its base-driven model.

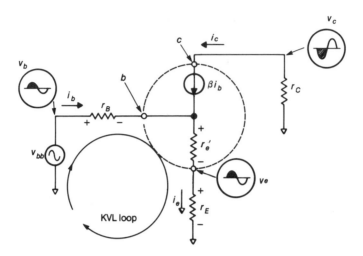

Figure 4-8. Finding i_e in the base-driven model.

This is the same as the ac voltage drop across r_E. The base voltage v_b is the sum of the voltage drops across r_E and r'_e.

$$v_b = i_e(r_E + r'_e)$$

If we let

$$i_c \cong i_e$$

then

$$v_c \cong i_e r_C$$

Phase Relationship of v_b, v_e, and v_c

The phase relationship among v_b, v_e, and v_c is shown in Figure 4-8. *The emitter and base ac voltages are in phase, and the collector ac voltage is 180° out of phase with respect to v_b and v_e.*

The emitter follows the base in voltage. As the base voltage becomes more positive, the emitter voltage becomes more positive. Since r_E is fixed value, as the emitter voltage increases, the emitter current increases. As i_e and hence i_c increase, the voltage drop across r_C increases. The collector side of r_C, however, gets larger in the negative direction owing to the polarity of the voltage drop caused by the direction of i_c. In the ac equivalent circuit, the collector voltage reaches its minimum value as the base voltage reaches its maximum value. Both waveforms are, of course, sinusoidal, but the collector voltage is 180° out of phase with respect to the base voltage. We say that v_c is inverted from v_b.

Transistor Voltage Gain, v_c/v_b

The voltage gain of a circuit is typically the ratio of the output voltage (at the load) to the input voltage (at the source). However, the ratio of the voltages at any two points can define a voltage gain.

Associated with the base and collector of a transistor in an amplifier circuit are ac voltages. *The ratio of these voltages, v_c to v_b, defines a voltage gain called the transistor voltage gain.*

$$\frac{v_c}{v_b} = A_v \text{ (transistor)}$$

SMALL-SIGNAL VOLTAGE GAIN

From previous derivations,

$$v_c \cong i_e r_C$$

and $v_b = i_e(r_E + r'_e)$

Thus, the ratio of v_c to v_b defines the transistor voltage gain.

$$\frac{v_c}{v_b} \cong \frac{r_C}{r_E + r'_e} = A_v \text{ (transistor)}$$

The resistances r_C, r_E, and r'_e are all from the ac equivalent circuit. They convert the current-amplifying transistor to a voltage amplifier. The transistor voltage gain (a dimensionless number) is equal to the ratio of two resistances. r_C and r_E are the circuit ac resistances associated with the collector and emitter and r'_e is the emitter diode ac resistance.

For many amplifier circuits, the voltage gain of the circuit will be equal to the transistor voltage gain v_c/v_b. This will be true when the source (or Thevenin) resistance equals zero or is extremely small compared to the circuit's input resistance. The circuit's maximum possible voltage gain is v_c/v_b. If the source resistance is high or the amplifier input resistance is low, an attenuation factor will be introduced that will reduce the circuit gain from its maximum value.

The voltage gain will be examined for several common amplifier circuits, including the common-emitter (CE), common-collector (CC), and swamped-emitter amplifiers.

The following example illustrates the procedure to find the dc quiescent point (I_C and V_{CE}) and the transistor voltage gain v_c/v_b.

Example

Find the voltage gain, v_L/v_S, for the circuit in Figure 4-9.

Solution

Since $R_S = 0$,

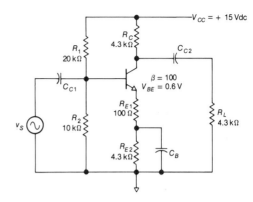

Figure 4-9. Voltage-amplifier circuit example, with $R_S = 0\,\Omega$.

$$v_b = v_S$$

and $\dfrac{v_L}{v_S} = \dfrac{v_c}{v_b}$

DC Analysis

The base voltage is found from the voltage divider action of R_1, R_2, and V_{CC}. (I_B is small compared to I_1.)

$$V_B = \left(\frac{R_2}{R_1 + R_2}\right) V_{CC} = +5.00 \text{ V}$$

The emitter voltage is less positive than the base by one V_{BE}.

$$V_E = V_B - V_{BE} = +4.40 \text{ V}$$

The dc emitter resistance is the sum of R_{E1} and R_{E2}.

$$R_E = R_{E1} + R_{E2} = 4.4 \text{ k}\Omega$$

The emitter current is determined from Ohm's law.

$$I_E = \frac{V_E}{R_E} = 1.0 \text{ mA}$$

Since β is reasonably high,

$$I_C \cong I_E = 1.0 \text{ mA}$$

The voltage drop across the collector resistor is

$V_{RC} = I_C R_C = 4.3$ V

The dc collector-to-emitter voltage can now be calculated using KVL.

$V_{CE} = V_{CC} - V_{RC} - V_{RE} = 6.3$ V

Ac Analysis

With the dc emitter current known,

$r'_e = \dfrac{0.025 \text{ V}}{I_E} = 25\ \Omega$

The collector ac resistance is the parallel combination of R_C and R_L.

$r_C = R_C \| R_L = 2.15$ kΩ

and $r_E = 100\ \Omega$

Thus, the voltage gain of the transistor and the circuit is

$\dfrac{v_L}{v_S} = \dfrac{r_C}{r_E + r'_e} = \dfrac{2150\ \Omega}{100\ \Omega + 25\ \Omega} = 17.2$

If v_s is a 10 mV peak ac signal, then the ac voltage at the load will be 172 mV peak.

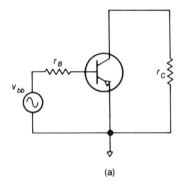

(a)

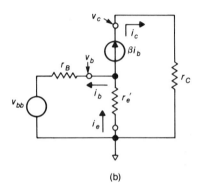

(b)

Figure 4-10. Common-emitter (CE) amplifier: (a) ac equivalent circuit; (b) with transistor ac equivalent circuit.

Transistor Voltage Gain of a Common-Emitter (CE) Amplifier

A special case of the base-driven model is when $r_E = 0\ \Omega$. When $r_E = 0\ \Omega$, the ac resistance between the emitter and ground is zero, and the emitter is at the circuit ac common or ground. This circuit is called a common-emitter (CE) amplifier.

For the CE amplifier (Figure 4-10) the collector ac voltage is

$v_c \cong i_e r_C$

and the base voltage is

$v_b = i_e r'_e$

The voltage gain from base to collector is

$\dfrac{v_c}{v_b} \cong \dfrac{r_C}{r'_e}$

This is the same as the relationship for v_c/v_b for the base-driven model, except $r_E = 0\ \Omega$.

In the example of Figure 4-9, if $r_E = 0\ \Omega$, the voltage gain goes from 17.2 to

$\dfrac{v_c}{v_b} = \dfrac{r_C}{r'_e} = \dfrac{2150}{25} = 86$

At first glance, this may appear to be a highly desirable circuit configuration due to the higher value of voltage gain. However, the resistance r'_e is temperature dependent, and since the gain

SMALL-SIGNAL VOLTAGE GAIN

is directly a function of r'_e, the gain is also temperature dependent. Ordinarily, this is *not* a desirable effect.

Example

Find the voltage gain v_L/v_S for the circuit in Figure 4-11.

Solution

Since $R_S = 0 \, \Omega$,

$$v_b = v_S$$

and $\dfrac{v_S}{v_L} = \dfrac{v_c}{v_b}$

Dc Analysis

The transistor base voltage will be less positive than V_{EE} by the voltage drop across R_2.

$$V_B = V_{EE} - \left(\frac{R_2}{R_1 + R_2}\right)(V_{EE} - V_{CC})$$

$$= +15 \text{ V} - 15.714 \text{ V} = -0.714 \text{ V}$$

The emitter voltage will be more positive (*pnp*) than the base by one V_{BE}.

$$V_E = V_B + V_{BE} = -0.714 \text{ V} + 0.71 \text{ V} \cong 0.0 \text{ V}$$

The emitter current is equal to the potential difference across R_E divided by R_E.

$$V_{RE} = V_{CC} - V_E = +15 \text{ V} - 0 \text{ V} = 15 \text{ V}$$

$$I_E = \frac{V_{RE}}{R_E} = 2 \text{ mA}$$

Since β is reasonably high,

$$I_C \cong I_E = 2 \text{ mA}$$

The voltage across R_C is

$$V_{RC} = I_C R_C = 7.2 \text{ V}$$

Using KVL,

$$V_{CE} = -V_{CC} - V_{EE} + V_{RC} + V_{RE}$$
$$= -15 \text{ V} - 15 \text{ V} + 7.2 \text{ V} + 15 \text{ V} = -7.8 \text{ V}$$

Ac Analysis

With I_E known,

$$r'_e = \frac{0.025 \text{ V}}{I_E} = 12.5 \, \Omega$$

The ac collector resistance is the parallel combination of R_C and R_L.

$$r_C = 3.6 \text{ k}\Omega \parallel 3.6 \text{ k}\Omega = 1.8 \text{ k}\Omega$$

Thus, $\dfrac{v_c}{v_b} = \dfrac{r_C}{r'_e} = \dfrac{1800 \, \Omega}{12.5 \, \Omega} = 144$

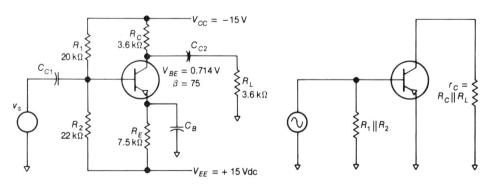

Figure 4-11. CE amplifier circuit example and ac equivalent circuit.

Transistor Voltage Gain of a Swamped-Emitter Amplifier

The base-driven model can also be called a swamped-emitter amplifier. It is called a *swamped-emitter amplifier* because the resistor r_E is usually added to *swamp out or minimize the temperature dependent effects of r'_e on the voltage gain*. The circuit usually represents a compromise or tradeoff between voltage gain and temperature stability (and also input resistance). Some circuit gain is sacrificed to allow for a reasonable degree of temperature stability.

The resistance r_E is made some multiple greater than r'_e. The greater r_E is made compared to r'_e the smaller the dependence of the voltage gain on temperature. For very large values of r_E,

$$r_E \gg r'_e,$$

the voltage gain is primarily established by r_C and r_E.

$$\frac{v_c}{v_b} = \frac{r_C}{r'_e + r_E} \cong \frac{r_C}{r_E} \quad \text{if } r_E \gg r'_e$$

In a CE amplifier, a 10% change in r'_e will cause a 10% change in the voltage gain. In the swamped-emitter amplifier, where $r_E = 10 r'_e$, the voltage gain would only change by about 1%.

Transistor Voltage Gain (v_e/v_b) for a Common-Collector (CC) Amplifier

For the circuits discussed thus far, the load resistor was connected through a coupling capacitor to the transistor's collector. We say that the output of the amplifier is at the collector and it drives a load, usually modeled as a resistor. For the base-driven model, the collector is not the only place for an amplifier output. A load resistor may also be driven from the emitter. For this case, the collector is typically connected to the dc supply and in the ac equivalent circuit is shown connected to the circuit common or ground. This circuit configuration (Figure 4-12) is called a *common-collector (CC) amplifier*.

The transistor voltage gain for the CC amplifier is the ratio v_e/v_b. From the ac equivalent circuit (Figure 4-12b) the base ac voltage is

$$v_b = i_e(r_E + r'_e)$$

and the emitter ac voltage is

$$v_e = i_e r_E$$

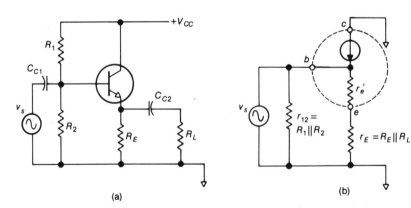

Figure 4-12. Common-collector (CC) amplifier: (a) circuit; (b) ac equivalent.

For the CC amplifier, the transistor voltage gain is

$$\frac{v_e}{v_b} = \frac{i_e r_E}{i_e(r'_e + r_E)}$$

Normally,

$$r_E \gg r'_e$$

and thus the voltage gain for the common-collector amplifier reduces to

$$\frac{v_e}{v_b} \cong \frac{r_E}{r_E} = 1$$

The common-collector amplifier is frequently referred to as an emitter follower. The ac voltage at the emitter (output) follows the ac voltage at the base (input) and both signals are in phase. If the maximum voltage gain of the circuit is 1, does it have any practical value? The answer is yes. As we shall find shortly, the circuit has an extremely high input resistance characteristic (and very low output resistance), making it an ideal circuit for interfacing.

4.5 INPUT RESISTANCE

Definition

The input resistance (r_{in}) of an amplifier is defined as the resistance between the amplifier's input terminals under ac conditions. For single-ended (one terminal is grounded) amplifiers, r_{in} is the ac resistance between the amplifier input and ground. The input resistance can be made up of several different components, depending on the configuration of a given circuit.

The circuit of Figure 4-13 illustrates how the input resistance of an amplifier may be determined. For this circuit,

$$r_{in} = \frac{v_{in}}{i_{in}} = \frac{10 \text{ mV}}{1 \text{ μA}} = 10 \text{ kΩ}$$

The voltage v_{in} and current i_{in} are ac quantities with both expressed as either peak, peak-to-peak, or rms values. The voltage v_{in} is the same as the driving source v_s, and the circuit's input current i_{in} is the same as i_S.

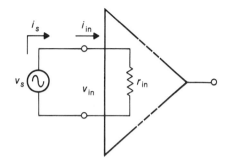

Figure 4-13. Determining the input resistance of an amplifier.

Resistance is a V/I ratio. If v_S is an ac voltage source and i_S is the source current, the ratio v_S/i_S defines the input resistance of the circuit that the source is driving. The input resistance represents the load for the source.

Input impedance would be a more appropriate term for input resistance. However, in the midband frequency range of the amplifier, the reactive components of the input circuit of the amplifier are small in value, and hence input resistance is generally acceptable. In general, high input resistance is a desirable characteristic.

The input resistance of an amplifier can also be found by examining the ac equivalent circuit and identifying the resistance(s) between the circuit's input terminals. If active devices are present, the devices will have to be represented by their ac equivalent circuits. Generally, the input resistance of an active device is found by finding the ratio of the input voltage and input current. This ratio will result in an expression containing circuit resistances and constants.

The designation r_{in} will be used to represent the input resistance of an amplifier *circuit*. The designation r_i will be used to represent the input resistance of the amplifier circuit's *active*

Input Resistance (r_{in}) of the Base-Driven Model

The input resistance of the base-driven model (Figure 4-14) is the sum of the base resistance r_B plus the input resistance of the transistor r_i.

$$r_{in} \text{ (circuit)} = r_B + r_i$$

The resistance r_i can be found from v_i/i_i using the ac equivalent circuit.

$$r_i \text{ (transistor)} = \frac{v_i}{i_i} = \frac{v_b}{i_b}$$

The input resistance of the transistor is equal to the ac base voltage divided by the transistor ac base current. As an example, if the base ac voltage is 2 mV and the base current is 1 μA, then

$$r_i = \frac{v_b}{i_b} = 2 \text{ k}\Omega$$

From the ac equivalent circuit, the base voltage is

$$v_b = i_e(r'_e + r_E)$$

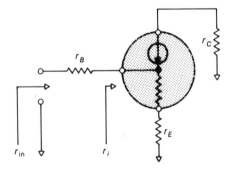

Figure 4-14. Input resistance of the base-driven model.

and the base current is

$$i_b = \frac{\alpha i_e}{\beta}$$

Thus, the ratio v_b/i_b will provide an alternate or equivalent expression for the input resistance.

$$r_i = \frac{v_b}{i_b} = \frac{i_e(r'_e + r_E)}{\alpha i_e/\beta} = \frac{\beta}{\alpha}(r'_e + r_E)$$

For transistors with reasonably high values of β,

$$r_i \cong \beta(r'_e + r_E)$$

This expression states that the resistance seen between the base and ground is β times the resistance in series with the emitter. The factor of β is introduced because of the differences in the magnitude of the base and emitter currents.

Summarizing, the input resistance of a transistor in an amplifier circuit is equal to the resistance in series with the base (later we will include r'_b) and β times the ac emitter resistances, both device and discrete.

Input Resistance (r_{in}) of the CE Amplifier

The CE amplifier is a special case of the base-driven model. In the model of the CE amplifier, the discrete emitter resistance is zero. For $r_E = 0 \ \Omega$, the input resistance becomes

$$r_{in} = r_B + r_i$$

where $r_i = \beta r'_e$

The input resistance of a CE stage is much smaller than that of the swamped-emitter circuit.

$$\beta r'_e \ll \beta(r'_e + r_E)$$

High voltage gain and high input resistance are desirable amplifier characteristics. The CE amplifier provides high voltage gain (compared to a swamped-emitter circuit), but it does this at the expense of temperature stability and

INPUT RESISTANCE

input resistance. Normally, a compromise is reached where some gain is sacrificed to make the circuit somewhat stable and with a higher input resistance.

Input Resistance of the CC Amplifier

The most frequent application of the CC amplifier or emitter follower is as an impedance-buffering circuit. Impedance-buffering circuits are used as buffers between circuits whose input and output impedances adversely affect each other. The emitter follower has the advantageous properties of high input resistance and low output resistance, although it possesses the disadvantage of low voltage gain. It is typically used as the first or initial stage in an amplifier system and buffers the signal transducer from the subsequent gain stages.

The input resistance of the CC amplifier is the same as the base-driven model. The ac equivalent circuits for both cases are the same except for the output. The output of a CE amplifier is at the collector, and the output of the CC stage is at the emitter.

$$r_{in} = r_B + r_i = r_B + \beta(r'_e + r_E)$$

When an emitter follower is used for its high input resistance characteristic, r_E is made very large. The resistance r_E has a different meaning for the swamped-emitter and common-collector circuits. For the swamped emitter,

$$r_E = R_{E1}$$

and in the common-collector circuit

$$r_E = R_E \| R_L$$

For practical discrete-component emitter followers, the input resistance reaches the 50-kΩ region. For many applications, this is still low, and an additional circuit technique (Darlington pair) has to be employed to further increase r_{in}.

Example

Find the input resistance r_{in} and the voltage gain v_e/v_b for the circuit in Figure 4-14? For this circuit, $r_B = 500~\Omega$, $r'_e = 50~\Omega$, $\beta = 80$, and $r_E = 750~\Omega$.

Solution

$$r_{in} = r_B + \beta(r'_e + r_E)$$
$$= 500~\Omega + 80(50~\Omega + 750~\Omega) = 64.5~\text{k}\Omega$$

$$\frac{v_e}{v_b} = \frac{r_E}{r'_e + r_E} = \frac{750}{800} = 0.938$$

Circuit Example

The voltage-divider biased amplifier circuit (Figure 4-15a) will again be used as a vehicle to illustrate an amplifier characteristic. The input resistance of this circuit will be found by first reducing the amplifier to its ac equivalent (Figure 4-15b) and then identifying the resistance(s) between the input terminal and ground. This amplifier is a member of the base-bias dc and base-driven ac models. It is also a swamped-emitter circuit.

The ac equivalent circuit shows two resistive components establishing the input resistance. The two resistances r_{12} and r_i are in parallel. The current source in the collector appears as a resistive open circuit.

$$r_{in} = r_{12} \| r_i$$

The resistance r_{12} is the parallel combination of R_1 and R_2, which are the base dc biasing resistors.

$$r_{12} = R_1 \| R_2$$

The resistance r_i is the transistor input resistance.

$$r_i = \beta(r'_e + r_E)$$

Thus, the circuit input resistance is

$$r_{in} = (R_1 \| R_2) \| \beta(r'_e + r_E)$$

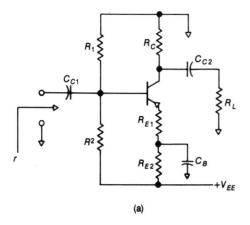

(a)

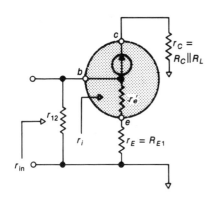

Figure 4-15. Input resistance circuit example: (a) circuit; (b) ac equivalent.

If $R_1 = 10$ kΩ, $R_2 = 20$ kΩ, and $r_i = 25$ kΩ, then

$r_{in} = 10$ kΩ || 20 kΩ || 25 kΩ = 5.26 kΩ

Since all the resistances are in parallel, the circuit's input resistance has to be less than the smallest one.

Darlington Pair

In an emitter follower, the transistor input resistance is

$r_i = \beta(r'_e + r_E)$

Generally, $r_E \gg r'_e$. Thus, $r_i \cong \beta r_E$.

To obtain a high input resistance requires making r_E large and/or β large. The maximum value of r_E is limited by dc current and voltage considerations. A transistor's β is a function of the device's physical characteristics and is also limited. *A circuit technique frequently used to obtain extremely high β, and hence extremely high input resistance, is the combination of two transistors in an arrangement called the Darlington pair.* This arrangement (Figure 4-16a) of the transistors behaves as a single transistor,

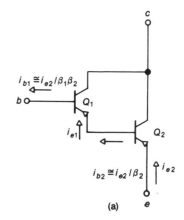

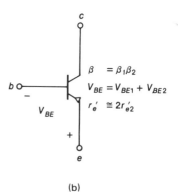

Figure 4-16. *pnp* Darlington pair: (a) circuit; (b) single-transistor equivalent.

INPUT RESISTANCE

with a composite β equal to the product of the individual β's.

$$\beta = \beta_1 \beta_2$$

In an amplifier circuit under forward-reverse bias conditions, the base current i_{b1} of the first transistor is multiplied by β_1 and becomes the emitter current i_{e1}. The emitter current drives the base of the second transistor and is again multiplied by the second transistor's current gain β_2. The resultant current i_{e2} is

$$i_{e2} \cong \beta_1 \beta_2 i_{b2}$$

For the Darlington pair, both B-E junctions are forward biased. Hence, the single-transistor equivalent (Figure 4-16b) has a V_{BE} of about 1.4 V or more precisely equals

$$V_{BE} = V_{BE1} + V_{BE2}$$

For a Darlington pair with similar transistors, the composite emitter ac resistance is twice the ac resistance of the second transistor.

$$r'_e \cong 2r'_{e2}$$

Darlington Pair Circuit Example

What is the magnitude of v_{o1} and v_{o2} in Figure 4-17 if $v_s = 10$ mV? What is the input resistance of the circuit?

Solution: Dc Analysis

$$V_B = V_{EE} + V_{R2} = -12 \text{ V}$$

$$V_E = V_B - V_{BE} = -13.3 \text{ V}$$

$$I_E = \frac{V_{RE}}{R_E} = 4.51 \text{ mA}$$

$$I_C \cong I_E = 4.51 \text{ mA}$$

$$V_{CE} = +V_{EE} - V_{RE} - V_{RC} = 8.8 \text{ V}$$

Ac Analysis

$$r'_e = 2r'_{e2} = 11.1 \text{ }\Omega$$

$$\frac{v_c}{v_b} = \frac{r_C}{r'_e + r_E} = \frac{1000}{54.1} = 18.5$$

$$\frac{v_e}{v_b} = \frac{r_E}{r'_e + r_e} = \frac{43}{54.1} = 0.795$$

$$v_s = v_b = 10 \text{ mV}$$

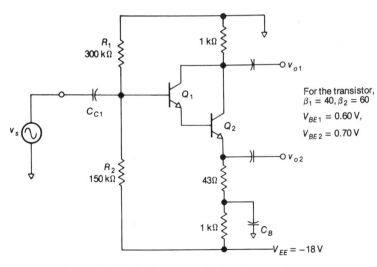

Figure 4-17. Darlington pair circuit example.

$$v_{o1} = v_c = \left(\frac{v_c}{v_b}\right) v_b = (18.5)(10 \text{ mV})$$

$$= 185 \text{ mV}$$

$$v_{o2} = v_e = \left(\frac{v_e}{v_b}\right) v_b = (0.795)(10 \text{ mV})$$

$$= 7.95 \text{ mV}$$

$$r_{12} = R_1 \| R_2 = 100 \text{ k}\Omega$$

$$r_i = \beta(r'_e + r_E) = 2400 \ (54.1 \ \Omega)$$

$$= 129.8 \text{ k}\Omega$$

$$r_{in} = r_{12} \| r_i = 56.5 \text{ k}\Omega$$

4.6 OUTPUT RESISTANCE

Definition

The output resistance (r_{out}) of an amplifier is defined as the resistance between the amplifier's output terminals under ac conditions. The output of most amplifiers is a simple terminal whose reference is the circuit's ground. The output resistance can be made up of several different components depending on the configuration of a given circuit. The output resistance does not include the load. Generally, the load is not considered part of the amplifier circuit and hence does not establish the amplifier's characteristics.

The circuit in Figure 4-18a illustrates how the output resistance of an amplifier may be determined. For this circuit

$$r_{out} = \frac{v_{out}}{i_{out}} = \frac{v_s}{i_s}$$

The source v_s must be an ac source (to determine an ac characteristic), and the input to the amplifier must be grounded. With the amplifier input grounded, the dependent voltage source (Av_{in}) in series with r_{out} is set to a value of 0 V. The ac output voltage v_{out} is then strictly a function of r_{out}. The circuit modeling the amplifier is a first-order equivalent ac circuit of a general-purpose amplifier. This test circuit is not practical because r_{out} is generally very low in value, causing an extremely high i_s. The circuit does however illustrate how forcing a voltage and measuring a current can determine a resistance.

A practical circuit used to measure r_{out} is shown in Figure 4-18b. The resistances R_{test} and r_{out} establish a voltage divider. With R_{test} and v_s known, and by measuring v_{out}, the output resistance is

$$r_{out} = \frac{v_{out}}{i_{out}} = \frac{v_{out}}{(v_s - v_{out})/R_{test}}$$

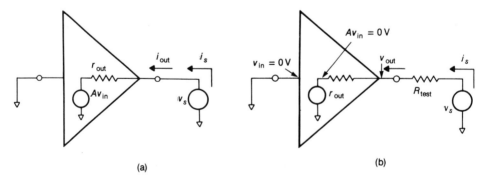

Figure 4-18. Determining the output resistance of an amplifier: (a) test circuit; (b) practical test circuit.

OUTPUT RESISTANCE

Output impedance would be a more appropriate term for output resistance. However, in the midband frequency range of the amplifier, the reactive components of the output circuit of the amplifier are small in value, and hence output resistance is generally acceptable. In general, low output resistance is a desirable characteristic.

The output resistance of an amplifier can also be found by examining the ac equivalent circuit and identifying the resistance(s) between the circuit's output terminals. If active devices are present, the devices will have to be represented by their ac equivalent circuits.

The *designation* r_{out} will be used to represent the output resistance of an amplifier *circuit*. The *designation* r_o will be used to represent the output resistance of the amplifier circuit's *active device*, that is, the transistor or operational amplifier.

Output Resistance (r_{out}) of the Base-Driven Model

To determine the *output resistance of the base-driven model* (Figure 4-19a), the ac collector resistance r_C must be split into the two parts that make it up.

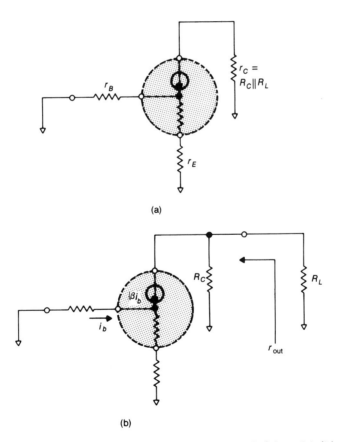

Figure 4-19. Output resistance of the base-driven model: (a) model; (b) separating the load resistor.

$r_C = R_C \| R_L$

The load resistor (Figure 4-19b) is separated from the amplifier's ac equivalent circuit. Looking back into the output terminal of the amplifier, there are two possible paths to ac ground. One is through the resistor R_C, and the other is through the current source. Since, ideally, the output resistance of a current source is $\infty \Omega$, the output resistance of the amplifier is

$r_{out} = R_C$

When second-order effects must be included, the transistor collector-to-emitter resistance r'_c will shunt the resistor R_C, and the parallel combination will establish r_{out}.

For discrete amplifiers with no feedback, R_C is in the hundreds of ohms or kilohm region and the output resistance is not low.

If the output is taken from the collector of the transistor, the ideal output resistance is independent of the resistances that are in the emitter or the base. The output resistance for the CE and swamped-emitter circuits is the same.

Output Resistance (r_{out}) of an Emitter Follower

The schematic of a *CC amplifier or emitter follower* is shown in Figure 4-20a. The output resistance for this amplifier circuit is determined from its ac equivalent circuit shown in Figure 4-20b. Looking into the output terminal of the amplifier shows three possible paths (all in parallel) to ac ground. One path is through R_E, another is through the emitter-base-r_B loop, and the third is through the emitter-collector loop. Since the output resistance of an ideal current source is $\infty \Omega$ then the R_E loop and the emitter-base-r_B loop establish the output resistance of the emitter follower.

The resistance in shunt with R_E is the sum of r'_e and the base resistance reflected to the

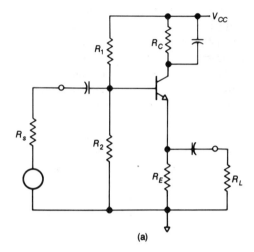

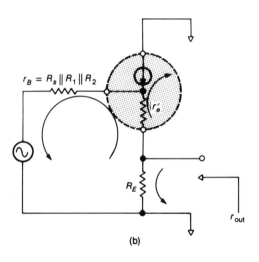

Figure 4-20. Output resistance of an emitter follower: (a) circuit; (b) ac equivalent.

emitter. Designating the resistance of this path as r_o, it is equal to

$r_o = r'_e + r_{BR}$

where r_{BR} is the *reflected base resistance*. The circuit output resistance is

$r_{out} = R_E \| r_o$

The reflected base resistance can be found using analytical techniques (KVL). The procedure would be similar to that employed to find the transistor input resistance. To find the transistor input resistance required reflecting the emitter resistances (r'_e and r_E) to the base. This was done by multiplying r'_e and r_E by β. To reflect the resistance in the base back to the emitter is the opposite procedure and requires dividing the base resistance by β. The resistance seen looking into the base of the transistor is beta times the resistance(s) in the emitter. The resistance seen looking into the emitter of a transistor is the base resistance (to ac ground) divided by beta (β). The reflected base resistance is

$$r_{BR} = \frac{r_B}{\beta}$$

and the output resistance of the emitter follower is

$$r_{out} = R_E \parallel \left(r'_e + \frac{r_B}{\beta} \right)$$

where $r_B = R_1 \parallel R_2 \parallel R_S$.

Circuit Example

What is the output resistance of the circuit shown in Figure 4-17?

Solution

For the emitter follower (v_{o2}), the output resistance is

$$r_{out} = R_E \parallel \left(r'_e + \frac{r_B}{\beta} \right)$$

Since $r_B = R_1 \parallel R_2 \parallel R_S$,
and $R_S = 0\ \Omega$
then $r_B = 0\ \Omega$

The ac emitter resistance of the Darlington pair is

$$r'_e = 2r'_{e1} = 2(5.5\ \Omega) = 11\ \Omega$$

The circuit output resistance equals

$$r_{out} = R_E \parallel r'_e = 43\ \Omega \parallel 11\ \Omega = 8.8\ \Omega$$

4.7 COMMON-BASE (CB) AMPLIFIER

The schematic of a *common-base or grounded-base (CB) amplifier* is shown in Figure 4-21a. The transistor is FR (forward–reverse) dc biased by the V_{EE} and V_{CC} potential supplies.

The fixed dc difference in potential ($V_{EE} - V_{BE}$) across R_E establishes the emitter (and I_C) current.

$$I_E = \frac{V_{EE} - V_{BE}}{R_E} = \frac{I_C}{\alpha}$$

Applying KVL to the V_{EE}-R_E-V_{CE}-R_C-V_{CC} loop provides the expression for the collector-to-emitter dc voltage.

$$V_{CE} = V_{CC} + V_{EE} - I_C(R_C + R_E)$$

The voltage gain and input and output resistance of the CB amplifier is determined using the ac equivalent circuit in Figure 4-21b. The transistor voltage gain is the ratio of the ac collector voltage v_c to the ac emitter voltage v_e. The ac voltage at the emitter is the product of i_e and r'_e.

$$v_e = i_e r'_e$$

Similarly, the ac collector voltage is

$$v_c \cong i_e R_C$$

Therefore, the transistor voltage gain is

$$\frac{v_c}{v_e} \cong \frac{R_C}{r'_e}$$

Looking into the CB circuit, the input resistance of the circuit is the parallel combination of R_E and the input resistance of the transistor.

$$r_{in} = R_E \parallel r_i$$

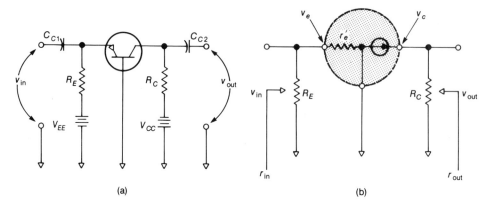

Figure 4-21. Common-base (CB) amplifier: (a) schematic; (b) ac equivalent circuit.

The input resistance of the transistor is the ac emitter resistance r'_e.

$$r_i = r'_e$$

Since the ac emitter resistance is generally small compared to R_E,

$$r_{in} = R_E \| r'_e \cong r'_e$$

The output resistance for the CB amplifier circuit is the same as the CE amplifier.

$$r_{out} = R_C$$

The CB amplifier has about the same voltage gain and output resistance as the CE amplifier. However, the input resistance of the circuit is very small. The low input resistance ($<50\ \Omega$) will overload most signal sources and greatly restricts the usage of this configuration. Occasionally, this circuit is used in high-frequency applications where low impedances are not uncommon.

4.8 INTERFACING

An amplifier circuit has an input and an output. The input and/or the output can be connected to a signal source, a transducer, or another circuit. The amplifier, by itself, has a certain level of performance, but this level can be modified or changed when the input and output circuits are connected. As an example, the transistor voltage gain of a CE stage is directly proportional to the ac collector resistance of which the load resistance is a part. Thus, the value of the voltage gain is determined, in part, by the load that the amplifier is driving.

The performance of two circuits joined together can be maximized by matching the input and output characteristics of the individual circuits. *The joining or uniting of two circuits to optimize certain performance criteria is called interfacing.* The effect of two circuits acting together will be demonstrated by considering the finite resistance associated with a voltage source driving the input of an amplifier. A voltage V_s and its output resistance R_s could also represent the Thevenin voltage and resistance of another circuit.

4.9 EFFECT OF SOURCE RESISTANCE ON VOLTAGE GAIN

The swamped-emitter amplifier circuit in Figure 4-22a is driven by a voltage V_s with a *finite output resistance R_s*. The voltage gain from

EFFECT OF SOURCE RESISTANCE ON VOLTAGE GAIN

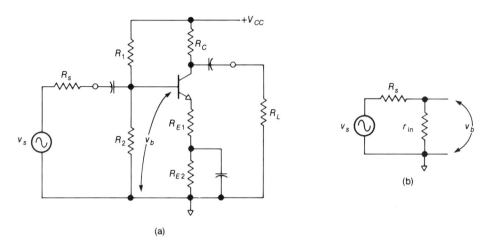

Figure 4-22. Swamped-emitter amplifier with finite R_s: (a) circuit; (b) attenuation circuit.

source (V_s) to collector (or the load since it is in parallel) is not the same as the voltage gain from base to collector. The resistance R_s is in series with the base, and it will reduce the amplitude of the signal at the base. The ac signal at the transistor base will be *less* than the signal source by the amount of the voltage drop across R_s. The amount of the voltage drop across R_s is a function of the value of R_s and the input resistance of the amplifier. The source resistance and input resistance form an ac voltage divider (Figure 4-22b) with the base voltage determined by

$$v_b = \frac{r_{in}}{R_s + r_{in}} v_s \quad \text{or} \quad \frac{v_b}{v_s} = \frac{r_{in}}{R_s + r_{in}}$$

The voltage gain from source to collector is the product of the voltage gain from source to base and the voltage gain from base to collector. Mathematically,

$$\frac{v_c}{v_s} = \frac{v_b}{v_s} \frac{v_c}{v_b}$$

The voltage gain v_c/v_b is the transistor gain, and v_b/v_s is an attenuation factor due to the finite source resistance. For a CE amplifier,

$$\frac{v_c}{v_s} = \frac{v_L}{v_s} = \frac{v_b}{v_s} \frac{v_c}{v_b} = \frac{r_{in}}{r_{in} + R_s} \frac{r_C}{r'_e}$$

For a swamped-emitter amplifier circuit, the voltage gain from source to load is

$$\frac{v_L}{v_s} = \frac{r_{in}}{r_{in} + R_s} \frac{r_C}{r_E + r'_e}$$

To *maximize the voltage gain*, the source resistance must be made small and/or the input resistance must be made large. The source resistance may also model the output resistance of another circuit. The voltage gain from source to load cannot be larger than the voltage gain from base to collector. Essentially, the transistor (an active device) provides the voltage gain in the circuit, and any passive circuit in series with the transistor circuit will reduce the overall gain. To maximize the voltage gain, the output resistance of one circuit must be low compared to the input resistance of the next stage. To *maximize the power gain*, the output resistance of one circuit must be equal to the input resistance of the next stage (circuit).

Circuit Example

Find the voltage gain from source to load for the circuit in Figure 4-22a if $R_1 = 20\ \text{k}\Omega$, $R_2 = 22\ \text{k}\Omega$, $R_C = 3.6\ \text{k}\Omega$, $R_L = 3.6\ \text{k}\Omega$, $R_{E1} = 91\ \Omega$, $R_{E2} = 7.5\ \text{k}\Omega$, $\beta = 75$, $V_{BE} = 0.714\ \text{V}$, and $R_s = 500\ \Omega$. How much less (in percent) is the overall gain compared to the transistor voltage gain? For the circuit, $V_{CC} = +30\ \text{V dc}$.

Solution

$$V_B = V_{CC} \frac{R_2}{R_1 + R_2} = +15.714\ \text{V}$$

$$V_E = V_B - V_{BE} = +15.00\ \text{V}$$

$$I_E = \frac{V_E}{R_E} \cong 2\ \text{mA}$$

$$r'_e = \frac{25\ \text{mV}}{I_E} = 12.5\ \Omega$$

$$r_C = R_C \| R_L = 1.8\ \text{k}\Omega$$

The transistor voltage gain is the ratio of the ac resistance in the collector to the ac resistance in the emitter.

$$\frac{v_c}{v_b} = \frac{r_C}{r_E + r'_e} = \frac{1800}{103.5} = 17.4$$

The input resistance of the circuit is the parallel combination of R_1, R_2, and the input resistance of the transistor.

$$r_{in} = R_1 \| R_2 \| \beta (r_E + r'_e)$$
$$= 20\ \text{k}\Omega \| 22\ \text{k}\Omega \| 7.8\ \text{k}\Omega = 4.46\ \text{k}\Omega$$

$$\frac{v_b}{v_s} = \frac{r_{in}}{r_{in} + R_s} = \frac{4460}{4960} = 0.90$$

$$\frac{v_L}{v_s} = (0.90)(17.4) = 15.66$$

The overall voltage gain is reduced by 10% owing to the finite source resistance.

4.10 CURRENT GAIN

Definition

The ratio of the load voltage to the source voltage defines the voltage gain of a circuit. For the same circuit, *the ratio of its load current to the source current defines the current gain.*

$$A_i = \frac{i_L}{i_s}$$

The performance of all the amplifier circuits previously discussed can be expressed not only in terms of voltage gain but also in terms of current and power gain.

The circuit in Figure 4-23 illustrates how to measure the current gain of an amplifier circuit. The source and load currents are measured with ammeters in series with the source and load, re-

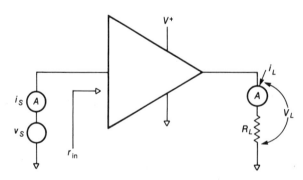

Figure 4-23. Determining the current gain of an amplifier.

spectively. The value of the source current will depend on v_S and the input resistance of the amplifier. The value of the load current is related to the load voltage and load resistance. For single-circuit amplifiers, the current gain can be expressed in an alternate way. Since

$$i_S = \frac{v_S}{r_{in}} \quad \text{and} \quad i_L = \frac{v_L}{R_L}$$

then

$$A_i = \frac{i_L}{i_S} = \frac{v_L/R_L}{v_S/R_S} = \frac{v_L}{v_S} \frac{r_{in}}{R_L}$$

or $A_i = \dfrac{A_v}{A_r}$

where A_v is the circuit voltage gain and A_r is a type of resistance gain.

The current gain of an amplifier is determined from its ac equivalent circuit. However, parallel resistances cannot be combined in most cases because of the current-division properties of a parallel circuit. For a single-transistor circuit, the *maximum* current gain will be equal to the transistor's current gain β. The overall circuit current gain will be less than β if current-dividing passive components are present in the circuit.

The laboratory measurement of an amplifier's current gain can present some *practical problems*. Current measurements must be made in series with the component whose current is being measured. For amplifiers on printed circuit boards, measuring a current requires the desoldering of one lead of a component to insert the ammeter. The output resistance of the ammeter must be very low to maximize the measurement accuracy and minimize loading problems. The value of input current of electronic ammeters must be very small compared to the circuit currents. For practical reasons, currents are often measured indirectly by measuring a potential difference across a resistor and then dividing by the value of resistance.

A_i of a Swamped-Emitter Amplifier

The ac equivalent circuit of a *voltage-divider biased swamped-emitter amplifier* will be used to determine the circuit's *current gain*. In the ac equivalent circuit (Figure 4-24) the parallel resistances establishing r_C are not combined. The load and collector resistances (R_L and R_C) are separated to identify the load and collector resistor currents. To find the expression or

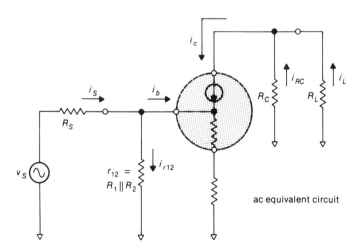

Figure 4-24. Current gain of a swamped-emitter amplifier.

equation for current gain requires finding the relationship between i_S and i_L. In the ac circuit, i_S is related to i_b by the current-divider action of r_{12} and r_i. The transistor collector current i_c is related to the transistor base current i_b by β, and the load current is related to i_c by the current-divider action of R_C and R_L. Thus, i_S is related to i_b, which is related to i_c, which is related to i_L. If these relationships, which eventually tie i_S to i_L, are expressed mathematically, the current gain expression will result. The base current is related to the source current by

$$i_b = \frac{g_i}{g_i + g_{12}} i_S$$

where $g_i = 1/r_i$ and $g_{12} = 1/r_{12}$. In a parallel circuit, *current division* is best expressed using conductances. The collector current is related to the base current by the transistor current gain.

$$i_c = \beta i_b = \beta \frac{g_i}{g_i + g_{12}} i_S$$

The ac collector current will divide into two components. One component will be i_{RC} and the other will be i_L. The load current, as a part of i_c, is found by current division.

$$i_L = \frac{G_L}{G_L + G_C} i_c = \frac{G_L}{G_L + G_C} \beta \frac{g_i}{g_i + g_{12}} i_S$$

In the above expression, G_L and G_C are the load and collector conductances ($1/R_L$ and $1/R_C$). The ratio of load to source current defines the circuit's current gain.

$$A_i = \frac{i_L}{i_S} = \frac{G_L}{G_L + G_C} \beta \frac{g_i}{g_i + g_{12}}$$

The current gain expression for a CE amplifier will be the same as for the swamped-emitter circuit, except the emitter resistance r_E will be zero. For the swamped-emitter circuit,

$$g_i = \frac{1}{r_i} = \frac{1}{\beta(r'_e + r_E)}$$

and for the CE circuit

$$g_i = \frac{1}{r_i} = \frac{1}{\beta r'_e}$$

A_i of an Emitter Follower

The ac equivalent circuit of a *voltage-divider biased emitter follower* is shown in Figure 4-25. The procedure to find the *current gain* of this circuit is similar to that used to find A_i for the

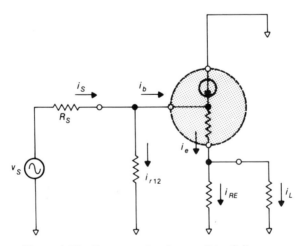

Figure 4-25. Current gain of an emitter follower.

CURRENT GAIN

swamped-emitter amplifier. The load and emitter resistances (R_L and R_E) are separated to identify the emitter resistor and load resistor currents. The relationship for i_S and i_L is found by relating i_S to i_b, then relating i_b to i_e, and finally relating i_e to i_L. The source current is related to i_b, which is related to i_e, which is related to i_L. If these relationships, which eventually tie i_L to i_S, are expressed mathematically, the current gain expression will result.

The base current is related to the source current by the current-divider action of r_{12} and r_i.

$$i_b = \frac{g_i}{g_i + g_{12}} i_S$$

The emitter current is related to the base current by β.

$$i_e = (\beta + 1) i_b = \frac{\beta}{\alpha} i_b$$

The emitter current will divide into two parts. One part will be i_{RE} and the other part will be i_L. The load current i_L, as a part of i_e, is found by current division.

$$i_L = \frac{G_L}{G_L + G_E} i_e = \frac{G_L}{G_L + G_E}(\beta + 1)\frac{g_i}{g_i + g_{12}}$$

In the above expression, G_L and G_E are the load and emitter conductances ($1/R_L$ and $1/R_{E1}$). The ratio of load to source current defines the circuit's current gain.

$$A_i = \frac{i_L}{i_S} = \frac{G_L}{G_L + G_E}(\beta + 1)\frac{g_i}{g_i + g_{12}}$$

The current gain expression for the emitter follower is similar to the gain expression for the CE amplifier. For both cases, the current gain can be significantly greater than 1. While the voltage gain of the emitter follower is slightly less than 1, the current gain and power gain can be high.

Circuit Example

Find the value of the current gain from source (i_S) to load (i_L) for the circuit in Figure 4-26.

Solution

The circuit uses a pair of dc supplies. It is an emitter-follower circuit with a Darlington pair

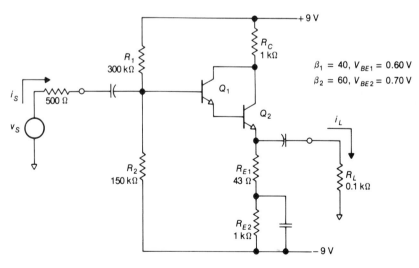

Figure 4-26. Current gain circuit example.

of transistors.

$$V_B = V_{EE} + (V_{CC} - V_{EE})\frac{R_2}{R_1 + R_2}$$
$$= -9\text{ V} + 6\text{ V} = -3\text{ V}$$

$$V_{BE} = V_{BE1} + V_{BE2} = 1.30\text{ V}$$

$$V_E = V_B - V_{BE} = -4.3\text{ V}$$

$$R_E = R_{E1} + R_{E2} = 1.043\text{ k}\Omega$$

$$I_E = \frac{V_{RE}}{R_E} = \frac{V_E - V_{EE}}{R_E} = 4.51\text{ mA}$$

$$r'_e = 2r'_{e2} = 2\left(\frac{25\text{ mV}}{I_E}\right) = 11.1\text{ }\Omega$$

The input resistance of the circuit is equal to the composite beta of the Darlington pair times the emitter ac resistances.

$$r_i = \beta(r'_e + r_E) = 2400(11.1\text{ }\Omega + 30.1\text{ }\Omega)$$
$$= 98.9\text{ k}\Omega$$

The emitter ac resistance r_E is the parallel combination of R_{E1} and R_L.

$$r_E = R_{E1} \| R_L = 30.1\text{ }\Omega$$

Conductance is the reciprocal of resistance.

$$g_i = \frac{1}{r_i} = \frac{1}{98.9\text{ k}\Omega}$$
$$= 10.1\text{ }\mu\text{S} \text{ (microsiemens)}$$

$$r_{12} = R_1 \| R_2 = 100\text{ k}\Omega$$

$$g_{12} = \frac{1}{r_{12}} = 10\text{ }\mu\text{S}$$

$$G_L = \frac{1}{R_L} = 10\text{ mS}$$

$$G_{E1} = \frac{1}{R_{E1}} = 23.3\text{ mS}$$

For this circuit, the current gain expression is

$$A_i = \frac{i_L}{i_S} = \frac{G_L}{G_L + G_{E1}}(\beta + 1)\frac{g_i}{g_i + g_{12}}$$

Substituting numerical values in the gain equation yields

$$A_i = \frac{10\text{ mS}}{23.3\text{ mS}}(2401)\frac{10.1\text{ }\mu\text{S}}{20.1\text{ }\mu\text{S}} = 518$$

The absolute maximum possible current gain for this circuit is the current gain of the transistors, or 2401. The current dividers in the base and emitter circuits reduce the gain to 518. The Darlington pair not only increases the input resistance but also the current and power gain.

4.11 POWER GAIN

Definition

The ratio of the load power to the source (input) power defines the power gain of a circuit.

$$A_p = \frac{p_L}{p_S}$$

The power dissipated by the load is equal to the product of the load current and load voltage.

$$p_L = i_L v_L$$

The amplifier's input power is equal to the product of the source current and source voltage.

$$p_S = i_S v_S = i_{in} v_{in}$$

Substituting for p_L and p_S in the power gain equation provides an expression for power gain in terms of voltage and current gain.

$$A_p = \frac{p_L}{p_S} = \frac{i_L v_L}{i_S v_S} = A_i A_v$$

If the voltage gain and the current gain of a circuit are known, the power gain can be determined from the above relationship.

In amplifier systems, several gain stages or circuits are usually cascaded (successive). The first stages and the last stages serve different

POWER GAIN

purposes. The input to the first stage is a very small potential, and the purpose of this stage is to increase the magnitude of the signal voltage. At the other end of the system, the purpose of the last stages is to increase the voltage *and* current gain, or the power gain of the system. The last or output stage must provide reasonably high levels of output voltage and output current to drive the load, which, typically, is a transducer. The latter amplifier stages are called power amplifiers and are designed to optimize the circuit's power gain.

Example

The input voltage and input current of an amplifier system are 10 mV and 1 µA, respectively. The load voltage and load current are 2 V and 20 mA. Find the voltage, current, and power gain of the system.

Solution

$$A_v = \frac{v_L}{v_{in}} = \frac{2 \text{ V}}{10 \text{ mV}} = 200$$

$$A_i = \frac{i_L}{i_{in}} = \frac{20 \text{ mA}}{1 \text{ µA}} = 20{,}000$$

$$A_p = A_v A_i = 4{,}000{,}000$$

$$p_{in} = i_{in} v_{in} = (10 \text{ mV})(1 \text{ µA})$$
$$= 10 \text{ nanowatts (nW)}$$

$$p_L = i_L v_L = (20 \text{ mA})(2 \text{ V}) = 40 \text{ mW}$$

$$A_p = \frac{p_L}{p_{in}} = \frac{40 \text{ mW}}{10 \text{ nW}} = 4{,}000{,}000$$

BEL GAIN

Logarithms

The numbers associated with voltage, current, and power gain can become very large. *These large numbers can be reduced to a smaller number of digits by using logarithms.* Logarithms represent a way of compressing numbers or a scale. By using logarithms, computations are also simplified. In logarithms, multiplication is reduced to addition, division is reduced to subtraction, raising a power is reduced to one multiplication, and extracting a root is reduced to one division. Simply stated, a logarithm is an exponent. An exponent is the power of another number called a base. The base is usually the number 10 (common logarithm) or the irrational number 2.71828..., called e (natural logarithm).

The data sheets for amplifiers will almost always express voltage gain using logarithms. Graphs on these data sheets will also use logarithmic gain. The gain-frequency response curve is a graph of voltage gain (using logarithms) versus frequency.

Logarithm Relationships

The logarithm of a quantity is the exponent of the power to which a base (a given number) must be raised to equal the quantity.

If $b^x = N$

then x is the logarithm of N to the base b, or

$$x = \log_b N$$

The following basic logarithmic relationships will help when performing gain calculations.

$$\log (MN) = \log M + \log N$$

$$\log \frac{M}{N} = \log M - \log N$$

$$\log x^p = p \log x$$

Remembering the logarithms (base 10) of the more common decimal numbers will help in most calculations.

$$\log 1 = 0$$
$$\log 2 = 0.30103\ldots \cong 0.3$$
$$\log 4 = 0.60206\ldots \cong 0.6$$
$$\log 8 = 0.90309\ldots \cong 0.9$$

log 10 = 1
log 100 = 2
log 1000 = 3

Bel Power Gain

Power gain is defined as the ratio of the load or output power to the source or input power.

$$A_p = \frac{p_L}{p_S}$$

This gain is called the *ordinary* power gain.

The bel power gain is defined as the logarithm (the base is 10) of the ordinary power gain. The designation for bel power gain is given a prime (') to differentiate it from the ordinary power gain.

$$A'_p = \log A_p$$

Bel power gain is a dimensionless number, but bel is attached to the answer to differentiate it from the ordinary power gain number. If the load power in a circuit is 10 W and the source power is 1 W, then the ordinary power gain is

$$A_p = \frac{10\ W}{1\ W} = 10$$

The bel power gain is the log of the ordinary power gain.

$$A'_p = \log A_p = \log 10 = 1\ \text{bel} = 1\ B$$

The bel as a unit of gain is relatively large, and the decibel (dB) is more commonly used. The prefix *deci* means one-tenth ($\frac{1}{10}$); thus

10 dB = 1 B = 10 decibels

Bel Voltage Gain

Since

$$p_L = v_L^2 r_L$$
and $p_S = v_S^2 r_i$

then the power gain can be expressed in terms of the voltage gain, the load resistance (r_L), and the amplifier input resistance (r_i).

$$A_p = \frac{p_L}{p_S} = \frac{(v_L)^2/r_L}{(v_S)^2/r_i} = (A_v)^2 \frac{r_i}{r_L}$$

If the load resistance equals the input resistance, the power and voltage gain are *directly* related.

$$A_p = (A_v)^2 \quad \text{for } r_L = r_i$$

An amplifier system whose input and load resistances are equal is called an *impedance-matched system*. Such systems are common in the communications area of electronics.

If we take the log of both sides of the above equation,

$$\log A_p = \log (A_v)^2$$

a bel gain relationship is established for impedance-matched systems.

$$A'_p = \log A_p = \log (A_v)^2 = 2 \log A_v \quad r_L = r_i$$

The bel voltage gain, by definition, is

$$A'_v = 2 \log A_v$$

In an impedance-matched system, the bel voltage gain and the bel power gain are equal. In *non-impedance-matched* systems, the bel voltage gain and the bel power gain must be computed independently.

The bel voltage gain, in decibels, equals

$$A'_v\ (\text{dB}) = 20 \log A_v$$

The bel power gain, in decibels, equals

$$A'_p\ (\text{dB}) = 10 \log A_p$$

4.12 POWER AND VOLTAGE GAIN OF CASCADED STAGES

Most amplifier systems are comprised of a number of cascaded amplifier circuits. The sys-

DIFFERENTIAL AMPLIFIER

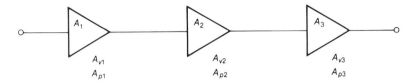

Figure 4-27. Cascaded amplifier circuits.

tem in Figure 4-27 contains three stages with ordinary voltage gains of $A_{v1}, A_{v2},$ and A_{v3} and ordinary power gain values of $A_{p1}, A_{p2},$ and A_{p3}. The overall ordinary voltage gain is the product of $A_{v1}, A_{v2},$ and A_{v3}.

$$A_v = A_{v1} A_{v2} A_{v3}$$

The overall ordinary power gain for the system is the product of $A_{p1}, A_{p2},$ and A_{p3}.

$$A_p = A_{p1} A_{p2} A_{p3}$$

If the gain of the individual circuits is expressed in terms of bels, then

$$A'_v = A'_{v1} + A'_{v2} + A'_{v3}$$
and $A'_p = A'_{p1} + A'_{p2} + A'_{p3}$

The use of logarithms (bel gain) changes the operation of finding the overall gain from multiplication to division.

Example

The operational amplifier IC consists of three amplification stages or circuits. The nominal values of the voltage gain of the first, second, and third stages are 100, 1000, and 1, respectively. Find the overall ordinary and bel voltage gain.

$A_v = A_{v1} A_{v2} A_{v3} = 100{,}000$
$A'_{v1} = 20 \log A_{v1} = 20 \log 100 = 40$ dB
$A'_{v2} = 20 \log A_{v2} = 20 \log 1000 = 60$ dB
$A'_{v3} = 20 \log A_{v3} = 20 \log 1 = 0$ dB
$A'_v = A'_{v1} + A'_{v2} + A'_{v3} = 100$ dB $= 100{,}000$

4.13 DIFFERENTIAL AMPLIFIER

The *differential*, or difference, *amplifier* is a small-signal, class A *circuit that amplifies the difference in potential between the circuit's two input terminals*. The *basic* differential amplifier circuit (Figure 4-28) uses two matched transistors whose emitters are tied together. The input signals are applied to the two bases, with two possible outputs taken at the two collectors (with respect to ground). The output can also be the difference in potential between the two collectors. Ideally, the circuit is symmetrical with each half identical to the other. This type of circuit is frequently used in linear integrated circuits and is the first stage of the operational amplifier IC. Integrated-circuit technology allows the two transistors and two

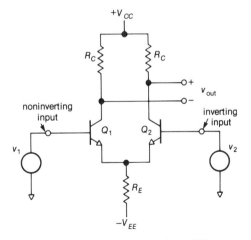

Figure 4-28. Differential amplifier.

collector resistors to be closely matched in value and performance. The output of the differential amplifier used in the operational amplifier is between the collectors of Q_1 and Q_2.

Dc Analysis

The dc equivalent circuit of the differential amplifier is shown in Figure 4-29a. The circuit is dc biased with two dc supplies. The emitters of the two transistors are near 0 V, and the current through R_E is basically determined by V_{EE} and R_E.

$$I_{RE} \cong \frac{V_{EE}}{R_E}$$

Since the two transistors' characteristics are matched or very similar, this current splits into the two equal emitter currents.

$$I_E \cong \frac{V_{EE}}{2R_E} \cong I_{C1,2}$$

The dc voltage at the collector of Q_1 is $+V_{CC}$ less the drop across R_C.

$$V_{C1} = V_{CC} - I_{C1}R_C$$

Similarly,

$$V_{C2} = V_{CC} - I_{C2}R_C$$

The output, generally, is equal to the difference in potential between the collectors.

$$V_{out} = V_{C2} - V_{C1} = -(I_{C2} - I_{C1})R_C$$

In practical differential amplifier circuits, the emitters are driven by a current source in place of the resistor R_E.

Ac Analysis

The ac equivalent circuit of the differential amplifier is shown in Figure 4-29b. There will be an output voltage when there is a difference between the input signal sources v_1 and v_2. If v_1 is more positive than v_2, the polarity of the differential output is as indicated in Figure 4-28. For $v_1 > v_2$, Q_1 conducts more current than Q_2 and causes a greater voltage drop across the collector resistor of Q_1 than the voltage drop across the collector resistor of Q_2. For this case, the collector of Q_2 is nearer V_{CC} and is more positive than the collector of Q_1.

Each half of the differential amplifier behaves like a common-emitter (CE) amplifier with a

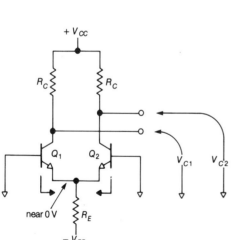
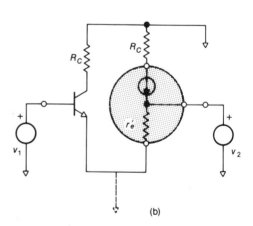

Figure 4-29. Equivalent circuits of the differential amplifier: (a) dc; (b) ac.

gain of R_C/r_e'. The emitter is near 0 Vdc and to the signal sources looks like the equivalent of a ground. Using superposition, it can be shown that

$$v_{out} \cong A(v_1 - v_2)$$

where $A \cong \dfrac{R_C}{r_e'}$

The potential difference between the collectors of the transistors Q_1 equals the potential difference between the bases multiplied by a constant established by the ratio of two resistances.

The inputs of the differential amplifier are known as double-ended or differential inputs. If one input is grounded, and the source drives the other input, the amplifier is said to have a single-ended input. The v_1 input is called the *noninverting input* because a positive v_1 acting alone will produce a positive output voltage ($V_{C2} > V_{C1}$). The v_2 input is called the *inverting input* because a positive v_2 acting alone will produce a negative output voltage ($V_{C2} < V_{C1}$).

The input resistance of the differential amplifier is similar to the CE amplifier. For the differential amplifier,

$$r_{in} = 2\beta r_e'$$

The 2 in the above relationship is due to the two emitter ac resistances seen from the base of any transistor with the other base grounded.

Example

For the circuit in Figure 4-28, $v_1 = +510$ mV and $v_2 = +507$ mV (p-p). The differential voltage gain ($\approx R_C/r_e'$) of the circuit is 60. Find the peak-to-peak value of v_{out}.

Solution

$v_{in} = v_1 - v_2 = 3$ mV

$v_{out} = A(v_1 - v_2) = 60(3 \text{ mV}) = 180$ mV

4.14 JFET COMMON SOURCE AMPLIFIER

Discrete component JFET amplifier circuits are frequently used as the initial stages in many amplifier systems. These circuits have an extremely high input impedance, which is needed to prevent the loading of the signal source or input transducer that has a high output impedance. The JFET is a device that is also frequently used as a solid-state switch.

A common-source amplifier is shown in Figure 4-30. The gate resistor R_G can be extremely high because of the low gate leakage current and the high transistor input resistance. The amplifier is self-biased (dc), and the gate to source voltage v_{gs} is the same as v_S. The so-called gain of the JFET is measured by its transconductance $g_m (v_{gs}/i_d)$. Transconductance transfers the input voltage v_{gs} to an output current i_d. This ac output current will develop an ac voltage that is the same as the load voltage v_L.

$$v_L = i_d r_D$$
$$i_d = g_m v_{gs}$$
$$v_{gs} = v_S.$$

Thus, $v_L = g_m r_D v_S$ and
$$A_v = v_L/v_S = g_m r_D$$

The resistance r_D is the parallel combination of R_D and R_L and is derived from ac equivalent circuit considerations.

$$r_D = R_D \| R_L$$

For the values listed on the schematic,

$r_D = 5$ kΩ, and

$A_v = (3000 \; \mu\text{S})(5 \text{ k}\Omega) = 15$

The circuit's input resistance r_{in} is approximately equal to R_G.

$$r_{in} \cong R_G = 10 \text{ M}\Omega$$

The output resistance of the circuit r_{out} is equal to the parallel combination of the drain resis-

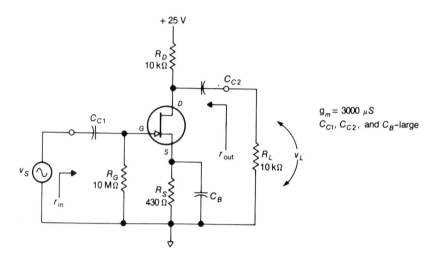

Figure 4-30. JFET common-source amplifier.

tance R_D and the output resistance (r_{ds}) of the JFET. The resistance r_{ds} (like r'_c) is usually high (100 kΩ) compared to R_D, and the circuit's output resistance is approximately equal to R_D.

$$r_{out} = R_D \| r_{ds} \cong R_D = 10 \text{ k}\Omega$$

Problems

1. How much will the base-to-emitter voltage of a transistor change if its emitter ac resistance is 100 Ω and the emitter current changes 100 μA?

2. What is the temperature increase above 18°C that will cause a 1% increase in r'_e?

3. Find I_B and V_C in the circuit of Figure 4-9.

4. Calculate the voltage gain v_c/v_b of the circuit in Figure 4-9 if R_{E1} is changed to 150 Ω.

5. What is the voltage gain v_c/v_b of the circuit in Figure 4-9 if C_B is reconnected to bypass both R_{E1} and R_{E2}?

6. Does the voltage gain of the circuit in Figure 4-9 increase or decrease if the load resistor R_L is made smaller?

7. What is the value of the dc load voltage in Figure 4-11?

8. Calculate the voltage gain v_c/v_b in the circuit of Figure 4-11 if temperature causes r'_e to increase from 12.5 to 15 Ω. What is the percent of change in gain?

9. Calculate the voltage gain of the circuit in Figure 4-11 if the bypass capacitor C_B becomes an open circuit.

10. What is the voltage gain v_e/v_b for the circuit in Figure 4-14 if r_E is shunted with a high-value bypass capacitor?

11. Calculate the input resistance r_{in} of the circuit in Figure 4-9.

12. Calculate the input resistance r_{in} of the circuit in Figure 4-11. In general, will swamped-emitter amplifiers or common-emitter amplifiers have a higher input resistance?

13. Theoretically, the composite current gain of a Darlington pair is $\beta_1 \beta_2$. In practice, the actual current gain is less. Why?

14. Describe, in words, what would happen if the Darlington pair in Figure 4-17 were

replaced with a single transistor with the specifications of Q_2.

15 In Figure 4-18b, v_s = 50 mV, R_{test} = 4 kΩ, and v_{out} = 10 mV. Find r_{out}.

16 Determine the output resistance r_{out} of the circuit in Figure 4-9. Assume r_c' is very high.

17 Determine the output resistance r_{out} of the circuit in Figure 4-11. Assume r_c' is very high.

18 What is the voltage gain v_L/v_S in the circuit of Figure 4-9 if the source has an output resistance of 100 Ω?

19 What is the voltage gain v_L/v_S in the circuit of Figure 4-11 if the source has an output resistance R_S of 100 Ω?

20 Calculate the current gain i_L/i_S in the circuit of Figure 4-9.

21 Calculate the current gain i_L/i_S in the circuit of Figure 4-11.

22 Find the ordinary and bel power gain of the circuit in Figure 4-26.

23 What is the input power of an amplifier's input signal if v_o = 2 V (rms), R_L = 8 Ω, and the circuit's power gain is 2.7 B?

24 What is the differential output voltage of the differential amplifier stage if R_C = 10 kΩ, v_S = 20 mV (rms), and the transistor's emitter current is 0.25 mA?

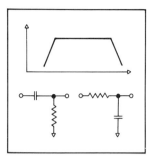

5
AMPLIFIER FREQUENCY RESPONSE

An amplifier is a circuit with gain. The behavior of the gain over the frequency spectrum is called the frequency response. *This chapter discusses the gain-frequency behavior of the amplifier.*

The frequency limits of the amplifier are called the upper *and* lower cutoff frequencies. *The range between these two frequencies is called the* passband, *and in this range the gain remains constant. For frequencies above the upper cutoff frequency and below the lower cutoff frequency, the gain will decrease. The upper cutoff frequency in an amplifier is established by the equivalent of one or more lag networks. A lag network is an RC network where the voltage gain over the frequency spectrum goes from 1 to 0 and the phase goes from 0° to -90°. The lower cutoff frequency is established by the equivalent of one or more lead networks. A lead network is an RC network where the voltage gain goes from 0 to 1 and the phase goes from +90° to 0°. The voltage gain for lead and lag networks is expressed in complex numbers. The complex numbers reflect the magnitude and phase information contained in the circuit's behavior. The Bode plot graphically illustrates the magnitude and phase response of both amplifiers and lead and lag networks.*

5.1 UPPER AND LOWER CUTOFF FREQUENCIES

An amplifier is a circuit whose primary characteristic is to increase the magnitude of a time-varying potential. The circuit is said to have gain. *The gain remains constant over a certain range of frequency. The upper and lower limits of this frequency range are called the upper and lower cutoff frequencies, and their difference defines the bandwidth of the amplifier.* If the gain becomes frequency dependent (near the cutoff frequencies), then, within the circuit, the characteristics of the amplifier's components must be frequency dependent. The frequency-dependent components are the distributed, discrete, and active device capacitance and inductance. Associated with each capacitance and inductance will be a resistance. Together, they form *RL* and *RC* circuits. The reactance of the capacitance and inductance varies with frequency, and when its value is near its associated resistance, the gain performance of the amplifier is significantly affected.

The *reactive components form ac voltage dividers* with the associated resistances, and they define the cutoff frequencies. RC networks dominate in low-frequency ($\lesssim 1$ MHz) amplifiers and there are two types: lag and lead. For *RC lag* networks, the capacitances are in shunt with the signal path, and the output signal diminishes in size as frequency increases. For *RC lead* networks, the capacitances are in series with the signal path, and the signal diminishes as the frequency decreases. In a single transistor amplifier circuit, there will be several lead and lag networks. Usually one of each will dominate in establishing the upper and lower cutoff frequencies.

The reactive components in an amplifier not only affect the *magnitude* of the output signal relative to the input, but also the *phase*. Near the cutoff frequencies, the phase of the output signal will begin to significantly vary relative to the input. This information is important to maintain stability in the system.

The gain-frequency performance of an amplifier, in terms of magnitude and phase, is graphically illustrated in a graph called a *Bode plot. This graph provides a picture of what the gain and phase of an amplifier do over the frequency spectrum.* It provides a visual picture of the amplifier's frequency response.

Before the cutoff frequencies of an amplifier can be determined, an analysis of the *RC* lead and lag networks will be made. These basic circuits model the frequency-response-determining networks in the amplifier.

5.2 LAG NETWORK

Basic Operation

The *RC* circuit in Figure 5-1 is called a *lag network*. The "lag" in lag network relates to the phase of the output signal relative to the input. The output falls behind, or lags, the input as the frequency of the input source increases. This circuit may also be viewed as a frequency selective network called a low-pass filter.

The capacitance C is in shunt with the signal path (input to output). As frequency increases, the capacitive reactance ($X_C = 1/2\pi fC$) decreases. Since the capacitor and resistor form an ac voltage divider, the potential difference across C (the output voltage) also decreases. As a limit, as frequency increases to infinite hertz, the capacitive reactance goes to zero ohms, and the output voltage goes to zero volts. At the other end of the frequency spectrum, as frequency decreases to zero hertz, the reactance goes to infinite ohms, and the output voltage equals the input.

The mathematical relationship between the input and output voltage will be made using complex numbers. *Complex numbers contain magnitude and phase information*, and they are used to represent the impedance of the components.

Voltage Gain

The voltage gain of the RC lag circuit can be determined using complex numbers. Since the lag network is an equivalent circuit found in

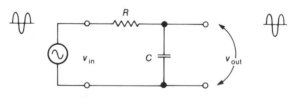

Figure 5-1. Lag network.

LAG NETWORK

the amplifier, the behavior of the amplifier will be similar to the behavior of the lag network. The impedance of the capacitor C is

$$Z_C = -jX_C$$

The impedance of the resistor is

$$Z_R = R$$

The output voltage is determined by the ac voltage-divider action of R and C.

$$V_{out} = V_{in} \frac{-jX_C}{R - jX_C} = V_{in} \frac{Z_C}{Z_R + Z_C}$$

or

$$\frac{V_{out}}{V_{in}} = \frac{-jX_C}{R - jX_C}$$

This gain equation is expressed with complex numbers in rectangular coordinate form. If the complex numbers in rectangular coordinate form are converted to their equivalent polar coordinate form, the magnitude and phase information is immediately discernable. For a complex number c, then

$$c = a + jb \quad \text{(rectangular coordinate form)}$$

and in polar coordinate form, the number c equals

$$c = (a^2 + b^2)^{1/2} \underline{/\tan^{-1}(b/a)}$$
$$= \text{magnitude} \underline{/\text{angle}}$$

The polar coordinate form of the gain equation is

$$\frac{V_{out}}{V_{in}} = \frac{X_C \underline{/-90°}}{\sqrt{R^2 + X_C^2} \underline{/-\tan^{-1}(X_C/R)}}$$

$$= \frac{\text{complex number } c_1}{\text{complex number } c_2}$$

The *magnitude of the gain function* is the ratio of the magnitudes of the two complex numbers.

$$\left|\frac{V_{out}}{V_{in}}\right| = \frac{X_C}{\sqrt{R^2 + X_C^2}}$$

The *phase of the output voltage relative to the input voltage* is designated by the Greek letter phi (ϕ).

$$\phi = -90° + \tan^{-1}\left(\frac{X_C}{R}\right)$$

As frequency varies, X_C varies, and the magnitude of the voltage gain and the phase between the input and output will vary. The variation in gain and phase, for a certain frequency range, is very small. The variation is most significant when the capacitive reactance equals the resistance.

$$X_C = R$$

Before we look at what happens to the magnitude and phase of the gain function at $X_C = R$, we will examine the *range of values* of the function over the frequency spectrum. The gain function's minimum and maximum values will occur at 0 Hz (dc) and at ∞ Hz. The magnitude function is

$$\left|\frac{V_{out}}{V_{in}}\right| = \frac{X_C}{\sqrt{R^2 + X_C^2}}$$

As $f \to 0$ Hz (as f approaches or goes to 0 Hz), $X_C \to \infty \, \Omega$ (X_C goes to infinite ohms), and for $X_C \gg R$,

$$\left|\frac{V_{out}}{V_{in}}\right| \longrightarrow \frac{X_C}{\sqrt{X_C^2}} = 1$$

At the other end of the frequency spectrum, as

$$f \longrightarrow \infty \text{ Hz}, \quad X_c \longrightarrow 0 \, \Omega,$$

and

$$\left|\frac{V_{out}}{V_{in}}\right| \longrightarrow 0$$

For *low frequencies*, the value of the *output* voltage equals the *input* voltage. For *high frequencies*, the magnitude of the *output* voltage decreases and goes to 0 V as a limit.

For the phase function, as

$$f \longrightarrow 0 \text{ Hz}, \quad X_C \longrightarrow \infty \, \Omega, \quad \frac{X_C}{R} \longrightarrow \infty$$

Then $\phi = -90° + \tan^{-1} \infty = -90° + 90° = 0°$

At the other end of the frequency spectrum, as

$$f \longrightarrow \infty \text{ Hz}, \quad X_C \longrightarrow 0 \, \Omega, \quad \frac{X_C}{R} \longrightarrow 0$$

Then $\phi = -90° + \tan^{-1} 0 = -90° + 0° = -90°$

The *phase* goes from 0° at low frequencies to -90° at high frequencies.

What happens at a frequency where the capacitive reactance equals the resistance? For the case of $X_C = R$,

$$\left|\frac{V_{out}}{V_{in}}\right| = \frac{X_C}{\sqrt{2X_C^2}} = \frac{1}{\sqrt{2}} = 0.707$$

and $|V_{out}| = 0.707 \, |V_{in}|$

For $X_C = R$, the phase function is

$$\phi = -90° + \tan^{-1} \frac{X_C}{X_C} = -90° + 45° = -45°$$

The frequency where the capacitive reactance equals the resistance is called the cutoff (f_c), critical, or corner frequency. At this frequency, the magnitude of the gain function is 0.707 of its low-frequency value (-3 dB) and the phase is -45°. At f_c, the output lags the input by -45°. When

$$X_C = R$$

then $\frac{1}{2\pi f C} = R$

and solving for f yields

$$f = \frac{1}{2\pi f C}$$

This value of frequency is the cutoff or critical frequency and is designated f_c. Its value is established by the circuit's resistance and capacitance.

Example

Find the cutoff frequency for the lag network in Figure 5-2.

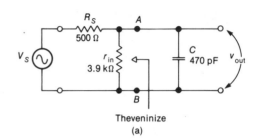

Theveninize
(a)

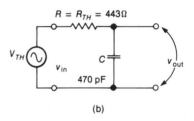

(b)

Figure 5-2. Lag network circuit example: (a) circuit; (b) Theveninized version.

Solution

The circuit to the left of terminals A and B must be Theveninized. The circuit, after Theveninizing, reduces to the standard lag network form. For the lag network,

$$R = R_{TH} = 500 \, \Omega \, \| \, 3.9 \, \text{k}\Omega = 443 \, \Omega$$

$$V_{TH} = \left(\frac{r_{in}}{R_S + r_{in}}\right) V_S = \frac{3900}{4400} \, (1 \text{ V})$$

$$= 0.886 \text{ V}$$

$$f_c = \frac{1}{2\pi RC} = 764 \text{ kHz}$$

The magnitude of the output voltage is 0.707 of its low-frequency value.

$$V_{out} = 0.707 \, (0.886 \text{ V}) = 0.626 \text{ V}$$

The output voltage, at 764 kHz, will lag the input voltage by 45°.

Voltage Gain Versus Frequency

The general *shape* of the gain–magnitude curve will be the same for all lag networks. The values of the circuit's resistance and capacitance shift the curve up or down the frequency scale. A general curve or graph can be made if the gain is expressed in terms of f and f_c. The behavior of the gain function is better understood if it is expressed in terms of frequency rather than reactance.

The cutoff or critical frequency is

$$f_c = \frac{1}{2\pi RC}$$

If both sides of the equation are divided by f, then

$$\frac{f_c}{f} = \frac{1}{2\pi RCf} = \left(\frac{1}{2\pi fC}\right)\left(\frac{1}{R}\right) = \frac{X_C}{R}$$

In a lag network, the ratio X_C/R is equivalent to the ratio f_c/f. The gain function can be expressed in terms of the ratio R/X_C and hence f/f_c.

$$|A_v| = \left|\frac{V_{out}}{V_{in}}\right| = \frac{X_C}{\sqrt{R^2 + X_C^2}} = \frac{1}{\frac{\sqrt{R^2 + X_C^2}}{X_C}}$$

$$= \frac{1}{\sqrt{1 + (R/X_C)^2}}$$

Hence

$$|A_v| = \frac{1}{\sqrt{1 + (f/f_c)^2}}$$

Since most graphs plot voltage gain in decibels, the bel voltage gain version of the above equation will be examined.

Bel Voltage Gain

The bel voltage gain, by definition, is two times the log of the ordinary voltage gain.

$$|A_v|_B = 2 \log A_v = 2 \log \left(1 + \frac{f}{f_c}\right)^{-1/2}$$

$$A_v' = -\log\left\{1 + \left(\frac{f}{f_c}\right)^2\right\}$$

The relationship for the bel voltage gain is in terms of the ratio f/f_c. The behavior of the gain-magnitude function (in bels) can be made by expressing the frequency f in terms of f_c, that is, $10 f_c$, $100 f_c$, or $0.1 f_c$. The general behavior of the gain expression as a function of frequency, relative to f_c, can be made. The limits for the gain-magnitude function are from 1 or 0 dB, at low frequencies, to 0.707 or -3 dB at f_c, to 0 at high frequencies. The reference for the relative terms low and high is the cutoff frequency f_c.

When $f = 0.1 f_c$ (a decade of frequency below f_c),

$$\frac{f}{f_c} = 0.1$$

and $A_v' = -\log\left\{1 + \left(\frac{f}{f_c}\right)^2\right\} = -\log 1.01$

$$= 0.00432 \text{ B}$$

The ordinary voltage gain is 1% less than its very low frequency value at a frequency one decade below f_c. For all practical purposes, the voltage gain is 0 B or 1 for all frequencies one decade below f_c or lower.

When $f = 10f_c$ (a decade of frequency above f_c),

$$\frac{f}{f_c} = 10$$

and $A_v' = -\log(1 + 100) = -2.00 \text{ B} = -20 \text{ dB}$

When $f = 100f_c$, the bel voltage gain is -4 B or -40 dB. The gain decreases by 20 dB for every decade increase above the cutoff frequency f_c.

For a lag network, all the action (so to speak) occurs near the cutoff frequency. A decade of frequency below f_c, the bel gain is nearly 0 dB or 1. At f_c, the gain is down 3 dB; and for every decade increase in frequency above f_c, the gain is reduced by 20 dB (a factor of 10).

If the input voltage of a lag network is 1 V and the cutoff frequency is 1 kHz, the magnitude of the output voltage at 1 kHz is 0.707 V. At 100 Hz, the voltage out is 0.99 V, and at 10 kHz the output is 0.099 V. The output decreases another factor of 10 at 100 kHz, or it falls to 0.0099 V.

Bode Plot of Bel Voltage Gain

A *straight-line approximation of a graph of a circuit's voltage gain (in decibels) versus the log of frequency is called a Bode plot.* The Bode plot of the magnitude of the voltage gain of the lag network is shown in Figure 5-3a. This graph is a straight-line *approximation* of what the lag network's gain does versus frequency. The Bode plot is made on semilog paper. The vertical axis is linear and is calibrated in decibels. The horizontal axis is logarithmic. The log scale compresses the frequency spectrum. The graph in Figure 5-3b is the *actual* gain versus frequency response. It is a picture of the lag network's gain equation

$$A'_v = -\log\left\{1 + \left(\frac{f}{f_c}\right)^2\right\}$$

The Bode plot is a very good approximation of the lag network's frequency response. The greatest error, -3 dB, is at f_c. At a decade of frequency above and below f_c, the error is about 1%. The cutoff frequency f_c is also called the corner or break frequency because the straight-line response breaks at this point.

Phase Angle Versus Frequency

The general shape of the phase-angle curve will be the same for all lag networks. The values of the circuit's resistance and capacitance will shift the curve up and down the scale. The phase function will be expressed in terms of the ratio f/f_c.

In place of

$$\phi = -90° + \tan^{-1}\left(\frac{X_C}{R}\right)$$

the phase-angle function can be expressed as

$$\phi = -90° + \tan^{-1}\left(\frac{f_c}{f}\right)$$

The phase angle for a lag network varies from 0° at low frequencies to $-90°$ at high frequencies. The reference for the relative terms high and low is f_c. When $f = 0.1 f_c$,

$$\frac{f_c}{f} = 10$$

and $\phi = -90° + \tan^{-1}(10) = -90° + 84.3°$
$= -5.7°$

When $f = f_c$,

$$\frac{f_c}{f} = 1$$

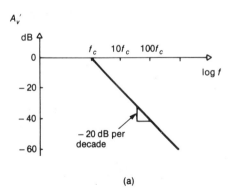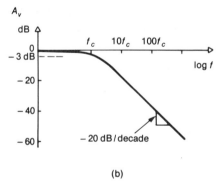

Figure 5-3. Frequency response of a lag network: (a) ideal (Bode plot); (b) actual.

and $\phi = -90° + \tan^{-1}(1) = -90° + 45° = -45°$

When $f = 10f_c$,

$$\frac{f_c}{f} = 0.1$$

and $\phi = -90° + \tan^{-1}(0.1) = -90° + 5.7°$
$= -84.3°$

When $f = 100f_c$,

$$\phi = -89.43°$$

For all practical purposes, for signal frequencies a decade or more *below* f_c, the phase angle between the output and input is 0°. For signal frequencies a decade or more *above* f_c, the phase between the output and input is -90°. Figure 5-4 graphically illustrates several phase relationships between an input and output signal.

If the corner frequency of a lag network is 1 kHz, then the phase lag at 10 kHz is -84.3°

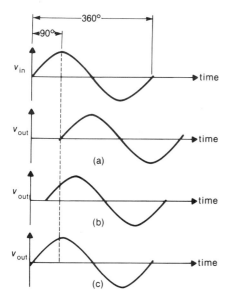

Figure 5-4. Output lagging the input: (a) by 90°; (b) by 45°; (c) by about 6°.

and at 100 Hz it is -5.7°. At 100 kHz, the phase angle is -89.43°.

Bode Plot of Phase Angle Versus Frequency

A straight-line approximation of a graph of a circuit's phase angle (in degrees) versus the logarithm of frequency is called a Bode plot.

The Bode plot of the phase relationship between the input and output voltage for the lag network is shown in Figure 5-5a. This graph is a straight-line approximation of what the phase angle does versus frequency. The Bode plot is made on semilog paper with the linear vertical axis calibrated in degrees and the logarithmic horizontal axis in frequency.

The graph in Figure 5-5b is the actual phase versus frequency response. The circuit's actual behavior is modeled by the following equation

$$\phi = -90° + \tan^{-1}\left(\frac{f_c}{f}\right)$$

The Bode plot is a good approximation of the lag network's phase response. The greatest error, about 6°, is at $0.1f_c$ and $10f_c$.

5.3 AMPLIFIER LAG NETWORKS

The amplifier circuit of Figure 3-1 contains *two lag networks*. They arise from the *two capacitances in shunt with the signal path*. As frequency increases, the reactance of these capacitances decreases. As the reactance decreases, the potential across the capacitances decreases, and the potential across the resistance in series with the signal path increases. The result is a smaller and smaller ac signal at the amplifier output as frequency continually increases.

With each capacitance C, there will be an associated resistance R, and each R and C forms a lag network. In well-designed amplifiers, one lag network will dominate, and it will establish

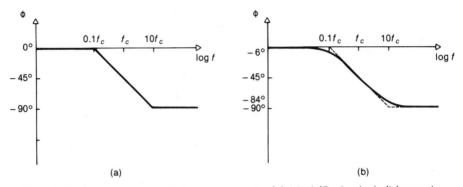

Figure 5-5. Phase response of the lag network: (a) ideal (Bode plot); (b) actual.

the upper cutoff frequency. The *two shunt capacitances* are

1. The base-to-emitter device capacitance C'_e in shunt with the input Miller base-to-collector capacitance. The base-to-collector capacitance is the device capacitance C_c, or it can be an externally connected capacitor. The lag network associated with this capacitance is called the *base lag network*.

2. The output *Millerized* base-to-collector *capacitance* in shunt with the input capacitance of the next stage (or the load capacitance). This network is called the *collector or output lag network*.

Base Lag Network

The ac equivalent circuit of the *voltage-divider-biased CE amplifier* is shown in Figure 5-6. The base-to-emitter capacitance C'_e is one part of the base lag network's capacitance. The other part is the input Miller capacitance of C_c.

C_c represents the *depletion-layer capacitance* of the reverse-biased collector–base junction. It has several names: C'_c, C_{cb}, C_{ob}, and C_{obo}. They are all slightly different, but any one is sufficiently accurate for our purposes. One of them is usually listed in the transistor's data sheets and will typically be less than 10 pF. The distributed capacitance associated with the physical circuit will mask out the individual differences.

The base-to-emitter capacitance C'_e represents the *diffusion capacitance* associated with the forward-biased base-to-emitter junction. A typical value of C'_e is from 200 to 400 pF. It is calculated from the following expression:

$$C'_e \cong \frac{1}{2\pi r'_e f_T}$$

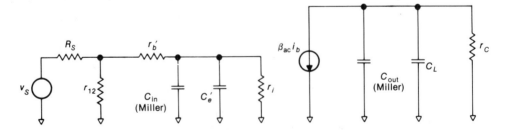

Figure 5-6. Ac equivalent circuit of a voltage-divider biased CE amplifier.

The frequency f_T is the current gain-bandwidth product of the transistor. It is the frequency where the current gain of the transistor is 1.

The *input Miller capacitance* equals the voltage gain of the transistor (plus 1) times the feedback capacitance. The voltage gain of the CE circuit is

$$\frac{v_c}{v_b} = \frac{r_C}{r'_e}$$

and the input Miller capacitance is

$$C_{\text{in}} \text{ (Miller)} = \left(\frac{r_C}{r'_e} + 1\right) C_c$$

The total capacitance is the parallel combination of C'_e and C_{in} (Miller).

$$C = C'_e + C_c \left(\frac{r_C}{r'_e} + 1\right)$$

The *internal base resistance* r'_b is a device parameter that cannot be neglected this time. Ordinarily its effect is small except in determining the base lag network's f_c. If the source resistance (R_S) is extremely small, r'_b will play a key role in establishing the value of f_c. The exact value of r'_b is very difficult to determine, and it is ordinarily between 50 and 200 Ω. If the value is not given, nor found in the data sheets, a value from 100 to 200 Ω can be assumed.

The resistance for the base lag network is comprised of four components: R_S, r'_b, r_{12}, and r_i. The lag network's resistance R (Figure 5-7) is arrived at when the resistive portion of the circuit is Theveninized. The source resistance R_S and the biasing resistance r_{12} are in parallel and are summed with the transistor's base spreading resistance r'_b. This combination is in parallel with the transistor's input resistance r_i. Normally R_S is small and r_{12} is large, and their parallel combination is very near R_S in value.

$$R = (R_S \| r_{12} + r'_b) \| r_i$$
$$= (R'_S + r'_b) \| \beta r'_e \quad \text{where } R'_S = R_S \| r_{12}$$

If the amplifier was a swamped-emitter circuit, the input resistance r_i would equal $\beta(r'_e + r_E)$. Because of the low values of r'_b and R_S relative to r_{12} and r_i, the resistance $(R_S + r'_b)$ usually dominates in establishing the base lag network's resistance R.

Collector Lag Network

The capacitance of the *collector lag network* is the parallel combination of the output Miller capacitance and the load capacitance. The value of the output Miller capacitance is about the same as the value of C'_e.

$$C_{\text{out}} \text{ (Miller)} = C_c \left(\frac{A_v + 1}{A_v}\right)$$
$$= C_c \left(\frac{r_C/r'_e + 1}{r_C/r'_e}\right)$$

The collector lag network's capacitance is

$$C = C_{\text{out}} \text{ (Miller)} + C_L$$

The transistor current source is in parallel with the collector resistance r_C $(R_C \| R_L)$. This parallel combination can be thought of as a

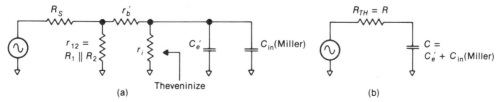

Figure 5-7. Base lag network: (a) ac equivalent circuit; (b) basic form.

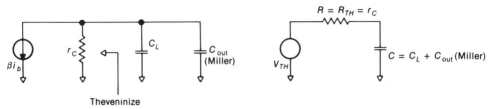

Figure 5-8. Collector lag network: (a) ac equivalent circuit; (b) basic form.

Norton circuit, which has a Thevenin equivalent. The Thevenin equivalent is shown in Figure 5-8, and the circuit is reduced to the basic RC lag circuit's form. The resistance associated with the collector lag network is the ac collector resistance r_C.

$$R = r_C = R_L \| R_C$$

The output resistance of the transistor (r_c') is assumed very large compared to the collector and load resistances. If this situation is not true, then the lag network resistance will be the parallel combination of r_c' and r_C.

The collector lag network's cutoff frequency is determined by the R and C values.

$$f_c = \frac{1}{2\pi RC}$$

There will be a cutoff, corner, or break frequency associated with each lag network. *The lower of the two f_c's will establish the upper cutoff frequency.* Normally, the second f_c will be at least a decade of frequency higher. If the two corner frequencies are near each other, the upper cutoff frequency will be determined by their combination. For this case, the composite f_c is best determined by graphically summing the responses of the individual lag networks. It is not good design practice to locate the corner frequencies near each other. If a feedback path is present (intentional or not) and the phase shift is excessive (contributed by the second lag network), frequency instability could result (the circuit could oscillate).

Lag Network Circuit Example

Find the corner frequencies of the base and collector lag networks in Figure 5-9. Identify the upper cutoff frequency of the amplifier.

Solution

The amplifier is a *voltage-divider-biased, swamped-emitter circuit.* The transistor in this circuit is a 2N2222A, and the appropriate transistor parameters are listed on the schematic.

The dc voltage at the emitter is

$$V_E = \left(\frac{R_1}{R_1 + R_2}\right) V_{CC} - V_{BE}$$

$$= +3.38 \text{ V} - 0.58 \text{ V} = +2.80 \text{ V}$$

$$I_E = \frac{V_E}{R_E} = 0.483 \text{ mA} \cong I_C$$

$$r_e' = \frac{25 \text{ mV}}{I_E} = 52 \text{ }\Omega$$

The ac collector resistance is the parallel combination of R_C and R_L.

$$r_C = R_C \| R_L = 6.2 \text{ k}\Omega \| 20 \text{ k}\Omega = 4.73 \text{ k}\Omega$$

The transitor voltage gain is

$$\frac{v_c}{v_b} = \frac{r_C}{r_E + r_e'} = 18.8$$

The voltage gain of the transistor is needed to find the Miller input and output capacitance. The parallel combination of R_1 and R_2 is r_{12}.

$$r_{12} = R_1 \| R_2 = 51 \text{ k}\Omega \| 20 \text{ k}\Omega = 14.4 \text{ k}\Omega$$

$$(r_{12} \gg R_S)$$

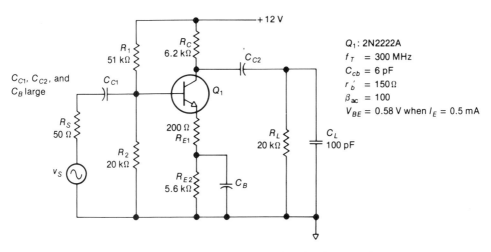

Figure 5-9. Lag and lead networks' circuit example.

The input resistance of the transistor is

$$r_i = \beta(r'_e + r_E) = 25.2 \text{ k}\Omega$$

The resistance R of the base lag network is the parallel combination of $(R_S + r'_b)$ and r_i. The resistance $(R_S + r'_b)$, for this case, dominates.

$$R \text{ (base lag)} = (R_S + r'_b) \| r_i = 200 \text{ }\Omega \| 25.2 \text{ k}\Omega$$
$$= 196 \text{ }\Omega$$

The capacitance of the base lag network is the sum of the Miller input capacitance and the base-to-emitter diffusion capacitance.

$$C'_e = \frac{1}{2\pi f_T r'_e} = 10.2 \text{ pF}$$

The Miller input capacitance is equal to the transistor voltage gain (plus 1) times the collector-to-base capacitance.

$$C_{in} \text{ (Miller)} = C_{cb} \left(\frac{v_c}{v_b} + 1\right) = 19.8 \text{ (6 pF)}$$
$$= 118.8 \text{ pF}$$

The capacitance of the base lag network is

$$C \text{ (base lag)} = C'_e + C_{in} \text{ (Miller)} = 129 \text{ pF}$$

An important observation! The distributed capacitance between the base and collector and the base and emitter *may be more* than the device capacitance (about 10 pF).

The corner or cutoff frequency of the base lag network is

$$f_c = \frac{1}{2\pi RC} = \frac{1}{6.28(196 \text{ }\Omega)(129 \text{ pF})}$$
$$= 6.3 \text{ MHz}$$

The resistance of the collector lag network is the parallel combination of R_C and R_L (r_C).

$$R \text{ (collector lag)} = 6.2 \text{ k}\Omega \| 20 \text{ k}\Omega = 4.73 \text{ k}\Omega$$

The capacitance of the collector lag network is the output Miller capacitance plus the load capacitance.

$$C = C_{out} \text{ (Miller)} + C_L =$$
$$6 \text{ pF} \left(\frac{19.8}{18.8}\right) + 100 \text{ pF} = 106.3 \text{ pF}$$

The cutoff frequency of the collector lag network is

$$f_c = \frac{1}{2\pi RC} = \frac{1}{(6.28)(4.73 \text{ k}\Omega)(106.3 \text{ pF})}$$
$$= 317 \text{ kHz}$$

which is the upper cutoff frequency.

Lower Cutoff Frequency

An amplifier's lower cutoff frequency is determined by one or more lag networks. In an *RC* lag network, the capacitance is in series with the signal path, and the output signal diminishes as the frequency of the input signal decreases. *The lower cutoff frequency is the lower frequency where the amplifier's passband gain is down 3 dB or 0.707 of the value of the ordinary gain.* Some amplifiers do not have capacitances in series with the signal path. These amplifiers are dc coupled, and their lower cutoff frequency is at 0 Hz or dc. Capacitances in series with the signal path are usually coupling capacitors. These capacitors couple the ac signal in and out of the amplifier circuit and provide dc isolation between circuits.

5.4 LEAD NETWORK

Basic Operation

The *RC* network in Figure 5-10 is called a *lead network*. The "lead" in its name refers to the phase of the output signal relative to the input. The *output voltage* of the lead network *leads* the input (more and more) *as* the *frequency* of the input source *decreases*. This circuit may also be viewed as a frequency selective network called a high-pass filter.

The capacitance *C* is in series with the signal path. As frequency decreases, the capacitive reactance and the potential difference across *C* also increase. This reduces the potential across *R* and in an amplifier will reduce the amplitude of the output signal.

The magnitude of the gain of the circuit varies from one to zero from one end of the frequency spectrum to the other. As $f \to \infty$ Hz (as frequency approaches infinite hertz), $X_C \to 0 \, \Omega$ (the capacitive reactance goes to zero ohms) and $|V_o| \to |V_{in}|$ (the magnitude of the output voltage approaches the magnitude of the input voltage).

At the other end of the frequency spectrum

as $\quad f \longrightarrow 0$ Hz

then $\quad X_C \longrightarrow \infty \, \Omega$

and $\quad |V_o| \longrightarrow 0$ V

Voltage Gain

The *voltage gain* of the *RC* lead network can be determined using *complex numbers*. Since the lead network is an equivalent circuit found in the amplifier, the behavior of the amplifier will be similar to the behavior of the lead network. The impedance of the capacitor *C* is

$$Z_C = -jX_C$$

and the impedance of the resistor *R* is

$$Z_R = R$$

The output voltage is determined by the ac voltage-divider action of *R* and *C*.

$$V_{out} = V_{in} \frac{Z_R}{Z_R + Z_C} = V_{in} \frac{R}{R - jX_C}$$

or $\quad \dfrac{V_{out}}{V_{in}} = \dfrac{R}{R - jX_C}$

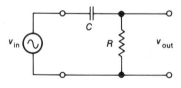

Figure 5-10. Lead network.

The above gain equation is expressed with complex numbers in rectangular coordinate form. If the complex numbers in rectangular form are converted to their equivalent polar coordinate form, the magnitude and phase information are immediately discernible. The polar coordinate form of the gain equation is

$$\frac{V_{out}}{V_{in}} = \frac{R \underline{/0°}}{\sqrt{R^2 + X_C^2} \underline{/\tan^{-1}(-X_C/R)}}$$

The *magnitude of the gain function* is the ratio of the magnitudes of the two complex numbers.

$$\left|\frac{V_{out}}{V_{in}}\right| = \frac{R}{\sqrt{R^2 + X_C^2}}$$

The *phase* ϕ is equal to the angle of the complex number in the numerator *minus* the angle of the complex number in the denominator.

$$\phi = \tan^{-1}\frac{X_C}{R}$$

As frequency varies, X_C varies, and the magnitude of the voltage gain and the phase between the input and output will vary. The variation in gain and phase, for a certain frequency range, is very small. The variation is most significant when the capacitive reactance equals the resistance.

The phase of the circuit varies from 0° at high frequencies to 90° at low frequencies. As

$$f \longrightarrow \infty \text{ Hz}$$
$$X_C \longrightarrow 0\, \Omega$$
and $\quad \phi \longrightarrow \tan^{-1} 0 \longrightarrow 0°$

At the other end of the frequency spectrum, as

$$f \longrightarrow 0 \text{ Hz}$$
$$X_C \longrightarrow \infty\, \Omega$$
and $\quad \phi \longrightarrow \tan^{-1} \infty \longrightarrow 90°$

$X_C = R$

What happens at a frequency where the capacitive reactance equals the resistance? For the case of $X_C = R$, the magnitude of the gain is

$$\left|\frac{V_{out}}{V_{in}}\right| = \frac{R}{\sqrt{X_C^2 + R^2}} = \frac{R}{\sqrt{2R^2}} = \frac{1}{\sqrt{2}}$$
$$= 0.707$$

and $\quad |V_{out}| = 0.707\, |V_{in}|$

For $X_C = R$, the phase function is

$$\phi = \tan^{-1}\left(\frac{X_C}{R}\right) = \tan^{-1} 1 = 45°$$

The frequency where the value of the capacitive reactance equals the value of resistance is called the cutoff (f_c), critical, or corner frequency. At this frequency, the magnitude of the gain function of the lead network is 0.707 of its high-frequency value, and the output voltage leads the input by 45°.

Bel Voltage Gain Versus Frequency

The general shape of the gain-magnitude curve will be similar for all lead networks. The values of the circuit's resistance and capacitance shift the curve up or down the frequency scale. In a lead or lag network, the ratio f_c/f is equivalent to X_C/R. The gain function can be expressed in terms of X_C/R and hence f_c/f.

$$|A_v| = \left|\frac{V_{out}}{V_{in}}\right| = \frac{R}{\sqrt{R^2 + X_C^2}}$$
$$= \frac{1}{\sqrt{1 + (X_C/R)^2}}$$

Hence,

$$|A_v| = \frac{1}{\sqrt{1 + (f_c/f)^2}}$$

The bel voltage gain is two times the log of the ordinary voltage gain.

$$|A_v|_B = 2 \log A_v = 2 \log \left\{1 + \left(\frac{f_c}{f}\right)^2\right\}^{-1/2}$$
$$= -\log\left\{1 + \left(\frac{f_c}{f}\right)^2\right\}$$

The gain-magnitude function varies from zero at low frequencies to one at high frequencies.
When $f = 10f_c$,

$$\frac{f_c}{f} = 0.1$$

and $A'_v = -\log(1 + 0.01) = -0.0043 \text{ B} \cong 0.0 \text{ dB}$

When $f = f_c$,

$$\frac{f_c}{f} = 1$$

and $A'_v = -\log(1 + 1) \cong -0.3 \text{ B} \cong 3.0 \text{ dB}$

When $f = 0.1f_c$,

$$\frac{f_c}{f} = 10$$

and $A'_v = -\log(1 + 100) \cong -20 \text{ dB}$

When $f = 0.01f_c$,

$$\frac{f_c}{f} = 100$$

and $A'_v = -\log(1 + 10,000) \cong -40 \text{ dB}$

The gain continually decreases by 20 dB (a factor of 10) for every decade decrease in frequency.

If the input voltage of a lead network is 1 V and the cutoff frequency is 1 kHz, the magnitude of the output voltage at 1 kHz is 0.707 V. At 10 kHz and higher, the output is 0.99 V or very near 1 V. At 100 Hz, the output is 0.099 V, and the output will decrease another factor of 10 at 10 Hz where its value is 0.0099 V.

The *gain-magnitude frequency response* of a lead network is shown in Figure 5-11. The Bode plot is the straight-line approximation, and the actual response follows the gain-magnitude function.

Phase Angle Versus Frequency

The *phase function* for the lead network is

$$\phi = \tan^{-1} \frac{X_C}{R}$$

Since $\dfrac{X_C}{R} = \dfrac{f_c}{f}$

then the phase function can be expressed in terms of frequency.

$$\phi = \tan^{-1} \frac{f_c}{f}$$

The phase angle for a lead network varies from $+90°$ at low frequencies to $0°$ at high frequencies. The reference for the relative terms high and low is f_c.

When $f = 10f_c$,

$$\frac{f_c}{f} = 0.1$$

and $\phi = \tan^{-1} 0.1 = 5.7°$

When $f = f_c$,

$$\frac{f_c}{f} = 1$$

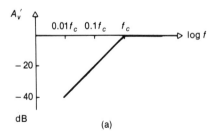 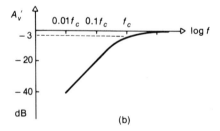

Figure 5-11. Gain-magnitude frequency response of a lead network: (a) ideal (Bode plot); (b) actual.

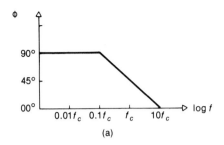
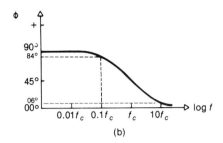

Figure 5-12. Phase response of the lead network: (a) Bode plot (ideal); (b) actual.

and $\phi = \tan^{-1} 1 = 45°$

When $f = 0.1 f_c$,

$$\frac{f_c}{f} = 10$$

and $\phi = \tan^{-1} 10 = 84.3°$

When $f = 0.01 f_c$,

$$\frac{f_c}{f} = 100$$

and $\phi = \tan^{-1} 100 = 89.4° \cong 90°$

The Bode plot of the phase response of the lead network is shown in Figure 5-12a. The graph in Figure 5-12b is the actual response of the phase angle versus frequency. The circuit's behavior is modeled by the following equation.

$$\phi = \tan^{-1} \frac{X_C}{R} = \tan^{-1} \frac{f_c}{f} \quad \text{(lead network)}$$

Modified Lead Network

The *modified lead network* in Figure 5-13 is a more accurate model of the lead networks found in the amplifier. The resistance R_{TH} can be the Thevenin resistance of the previous circuit or the resistance of the signal source (R_S). The resistance R_L can be the load resistance or the input resistance of the next circuit. The major difference between the two circuits is that the maximum gain of the modified lead network is less than 1. At high frequencies, the capacitive reactance is small, and the signal voltage is divided between R_{TH} and R_L.

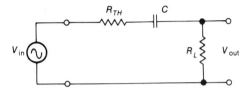

Figure 5-13. Modified lead network.

The complex voltage gain of the circuit is

$$\frac{V_{out}}{V_{in}} = \frac{R_L}{(R_L + R_{TH}) - jX_C}$$

In polar coordinate form, the gain and phase are

$$A_v = \left| \frac{V_{out}}{V_{in}} \right| = \frac{R_L}{\sqrt{(R_L + R_{TH})^2 + (X_C)^2}}$$

$$\phi = \tan^{-1} \left(\frac{X_C}{R_L + R_{TH}} \right)$$

The critical frequency is when X_C equals $(R_L + R_{TH})$ or

$$f_c = \frac{1}{2\pi (R_L + R_{TH}) C}$$

The *Bode plot of the gain and phase functions* is shown in Figure 5-14. The phase angle varies from $+90°$ at low frequencies to $0°$ at high frequencies. The gain function varies from 0 at low frequencies to $R_L/(R_L + R_{TH})$ at high fre-

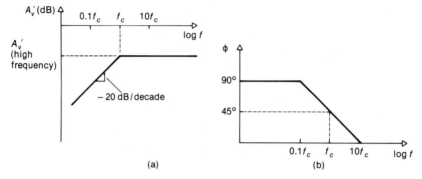

Figure 5-14. Bode plots of the modified lead network: (a) magnitude; (b) phase.

quencies. The bel gain at high frequencies is

A_v' (high frequency) = $2 \log A_v$

$$= 2 \log \frac{R_L}{R_L + R_{TH}}$$

Example

For the circuit in Figure 5-13, V_{in} = 1 V, R_{TH} = 500 Ω, R_L = 9.1 kΩ, and C = 0.2 μF. Find f_c, the high-frequency bel voltage gain, that is, A_v' at $100f_c$, and the output voltage at f_c.

Solution

The total lead network resistance is the sum of R_L and R_{TH}.

$R = R_L + R_{TH}$ = 9.60 kΩ

The cutoff or -3 dB frequency is

$$f_c = \frac{1}{2\pi(R_L + R_{TH})C}$$

$$= \frac{1}{2\pi(9.6 \text{ k}\Omega)(0.2 \text{ μF})} = 83 \text{ Hz}$$

At $100f_c$ or 8.3 kHz, the high-frequency ordinary voltage gain is

$$A_v \text{ (high frequency)} = \frac{R_L}{R_L + R_{TH}}$$

$$= \frac{9.1}{9.6} = 0.95$$

and A_v' = $2 \log A_v$ = $2 \log 0.95$

$= -0.044$ B $= -0.44$ dB

The output voltage at 8.3 kHz is the voltage gain of the circuit times the input voltage.

$V_{out} = A_v$ (high frequency) V_{in}

$= 0.95 (1 \text{ V}) = 0.95 \text{ V}$

The output voltage at 83 Hz is down 3 dB or 0.707 of the high-frequency value.

$V_{out} = A_v(f_c)V_{out}$ (high frequency)

$= (0.707)(0.95 \text{ V}) = 0.672 \text{ V}$

5.5 AMPLIFIER LEAD NETWORKS

The amplifier circuit of Figure 3-1 contains *two lead networks*. The lead networks *arise from the two capacitances*, C_{C1} and C_{C2}, in series with the signal path. As the frequency decreases, the reactance of the capacitance increases, and the potential across the capacitance increases. The result is a smaller and smaller ac signal at the amplifier output as frequency continually decreases.

With each series capacitance C, there will be an associated resistance R, and each R and C forms a lead network. One or both lead networks and/or the bypass capacitor network will establish the lower cutoff frequency. In well-

AMPLIFIER LEAD NETWORKS

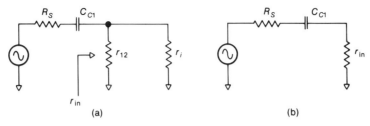

Figure 5-15. Input lead network: (a) ac equivalent circuit; (b) basic form.

designed amplifiers, one of the circuits will dominate. The *two series capacitances* are (1) the input coupling capacitor C_{C1}, and (2) the output coupling capacitor C_{C2}. The values of the coupling capacitors are generally high.

Input Lead Network

The circuit model of the *input lead network* is shown in Figure 5-15. The capacitance C is the input coupling capacitor C_{C1}.

$C = C_{C1}$ (input lead network)

The resistance in shunt with the signal path is the input resistance of the amplifier r_{in}. The resistance in series with the capacitance is the source resistance R_S. The input lead network is the modified RC lead circuit. The input lead network resistance is the sum of R_S and the amplifier's input resistance.

$R = R_S + r_{in} = R_S + (r_{12} \| r_i)$
$= R_S + \{r_{12} \| \beta(r'_e + R_{E1})\}$ (input lead network)

The corner or cutoff frequency associated with the input lead network is

$$f = \frac{1}{2\pi RC}$$

where R and C are as given above.

Output Lead Network

The portion of the ac equivalent amplifier circuit associated with the *output lead network* is shown in Figure 5-16a. The capacitance C is the output coupling capacitor C_{C2}.

$C = C_{C2}$ (output lead network)

The circuit can be reduced to the standard modifed lead network by Theveninizing the transistor current source and R_C portion of the circuit. The result is shown in Figure 5-16b. The total resistance of the output lead network is

$R = R_C + R_L$ (output lead network)

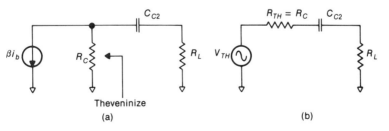

Figure 5-16. Output lead network: (a) ac equivalent circuit; (b) basic form.

Bypass Capacitor Network

The *bypass capacitor circuit and/or the coupling capacitor circuits* can establish the lower cutoff frequency. The bypass capacitor C_B (Figure 5-17) shunts the resistor R_{E2}. For high frequencies, the gain of the circuit is the ratio of the resistances in the collector and the emitter.

$$A_v \text{ (transistor)} = \frac{r_C}{r'_e + R_{E1}}$$

As the frequency of the signal source decreases, the gain decreases because R_{E2} is no longer shorted by the reactance of C_B. At very low values of frequency, the reactance of C_B does not equal a short circuit. For very, very low frequencies, the gain decreases because of the effect of R_{E2}.

$$A_v \text{ (transistor)} = \frac{r_C}{r'_e + R_{E1} + R_{E2}}$$

The bypass capacitor circuit produces the same effect as a lead network. The capacitance establishing the corner frequency will be the bypass capacitor C_B.

$$C = C_B \quad \text{(bypass capacitor network)}$$

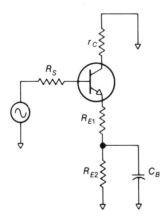

Figure 5-17. Bypass capacitor network.

This capacitance "sees" a resistance. The resistance R it sees is R_{E2} in parallel with the resistance seen looking into the emitter circuit. In the emitter circuit, the resistance is R_{E1} and r'_e plus the resistance in the base circuit divided by β. Thus R equals

$$R = R_{E2} \left\| \left(R_{E1} + r'_e + \frac{R_S}{\beta} \right) \right. \quad \text{(bypass capacitor network)}$$

Of course, the cutoff frequency is given by

$$f_c = \frac{1}{2\pi RC}$$

where R and C are as listed above.

Lead Network Circuit Example

Find the corner frequencies of the input and output lead networks and the bypass capacitor network for the amplifier circuit in Figure 5-9. For this circuit, $C_{C1} = 0.2$ μF, $C_{C2} = 2$ μF, and $C_B = 100$ μF. Identify the lower cutoff frequency of the amplifier.

Solution: Input Lead Network

The resistance of the input lead network is

$$R = R_S + r_{in}$$

From the lag network example, the input resistance of the transistor is 25.2 kΩ, and r_{12} is 14.4 kΩ. The input resistance of the circuit is the parallel combination of r_{12} and r_i.

$$r_{in} = r_{12} \| r_i = 14.4 \text{ k}\Omega \| 25.2 \text{ k}\Omega = 9.16 \text{ k}\Omega$$

Thus,

$$R = R_S + r_{in} = 50 \, \Omega + 9.16 \text{ k}\Omega$$
$$= 9.210 \text{ k}\Omega$$

and $C = 0.2$ μF

The corner frequency of the input lead network is

$$f_c = \frac{1}{2\pi RC} = \frac{1}{2\pi (9.21 \text{ k}\Omega)(0.2 \text{ μF})} = 86.4 \text{ Hz}$$

Output Lead Network

The resistance R of the output lead network is the sum of R_C and R_L.

$$R = R_C + R_L = 6.2 \text{ k}\Omega + 20 \text{ k}\Omega = 26.2 \text{ k}\Omega$$

The capacitance C of the output lead network is the coupling capacitor C_{C2} whose value for this circuit is 2 μF. The load capacitance C_L affects the upper cutoff frequency, and its value is very small compared to C_{C2}.

The corner frequency of the output lead network is

$$f_c = \frac{1}{2\pi(26.2 \text{ k}\Omega)(2 \text{ μF})} = 3.04 \text{ Hz}$$

Bypass Capacitor Network

The resistance associated with the 100 μF bypass capacitor is R_{E2} (5.6 kΩ) in parallel with the sum of R_{E1}, r'_e, and the reflected resistance in the base ($r_{12} \gg R_S$).

$$R \cong R_{E2} \| \left(R_{E1} + r'_e + \frac{R_S + r'_b}{\beta} \right); r_{12} \gg R_S$$

$$\cong 5.6 \text{ k}\Omega \| \left(200 \text{ }\Omega + 52 \text{ }\Omega + \frac{250 \text{ }\Omega}{100} \right)$$

$$\cong 244 \text{ }\Omega$$

The corner frequency of the bypass capacitor network is

$$f_c = \frac{1}{2\pi(244 \text{ }\Omega)(100 \text{ μF})} = 6.5 \text{ Hz}$$

Lower Cutoff Frequency

The corner frequencies of the amplifier are at 87, 3, and 6.5 Hz. The lower cutoff frequency of the amplifier is at 87 Hz. The nondominant cutoff frequencies are at least a decade of frequency lower than the dominant f_c.

5.6 BANDWIDTH

An amplifier's bandwidth is defined as the difference between the upper and lower cutoff frequencies.

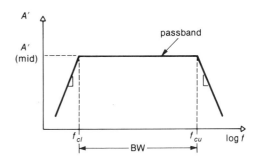

Figure 5-18. Bandwidth.

$$BW = f_{cu} - f_{cl}$$

A Bode plot of the bandwidth of an amplifier is illustrated in Figure 5-18. The bandwidth, ideally, defines the frequency range where the gain remains constant. It is referred to as the passband of the amplifier.

An amplifier contains several RC lag networks. Typically, one will dominate and establish the upper cutoff frequency. An ac amplifier contains several RC lead networks, and typically one will dominate and establish the lower cutoff frequency. A dc amplifier does not have any coupling capacitors or lead networks, and its lower cutoff frequency is at 0 Hz.

The typical amplifier found in a home stereo system has an upper cutoff frequency of 20 kHz and a lower cutoff frequency from 50 to 150 Hz. Because of the low f_{cl}, the bandwidth is approximately equal to the value of the upper cutoff frequency. The bandwidth of the stereo amplifier closely matches the bandwidth of a person's aural response.

5.7 SINUSOIDAL AND PULSED RESPONSES

If the input voltage of an amplifier is a sinusoidal signal, the frequency limits of the amplifier's response are established by the upper and lower cutoff frequencies.

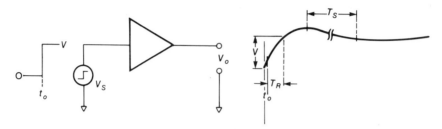

Figure 5-19. Pulsed response of an amplifier.

If the input voltage of an amplifier is a voltage step Figure (5-19), the time it takes the amplifier output to *rise* (10% to 90%) and *settle* (10% to 90%) is determined by the amplifier's rise time (T_R) and sag time (T_S).

If an amplifier has one dominant lead and lag network, the *sinusoidal and pulsed responses are related*. The rise time (T_R) is related to the upper cutoff frequency (f_{cu}), and the sag time (T_S) is related to the lower cutoff frequency (f_{cl}).

It can be shown that the time it takes for the output waveform to rise from 10% to 90% of the input's step of voltage is

$$T_R = 2.2\, RC$$

From the sinusoidal case, the upper cutoff frequency is also related to the factor RC.

$$f_{cu} = \frac{1}{2\pi RC}$$

Substituting for RC will relate the upper cutoff frequency to the rise time.

$$f_{cu} = \frac{1}{(2\pi T_R)/2.2} = \frac{0.35}{T_R}$$

In a similar manner, the lower cutoff frequency is related to the sag time.

$$f_{cl} = \frac{0.35}{T_S}$$

For practical reasons, the sag time-lower cutoff frequency relationship is seldom used.

If an amplifier's rise time was measured to be 17.5 μs due to a voltage step input, its upper cutoff frequency under sinusoidal conditions would be

$$f_{cu} = \frac{0.35}{17.5\ \mu s} = 20\ \text{kHz}$$

Problems

1. Express the impedance of a series RLC (resistor–inductor–capacitor) circuit in terms of complex numbers in rectangular and polar coordinate form.

2. Express the admittance of a parallel RLC circuit in terms of complex numbers in rectangular and polar coordinate form.

3. What is the exact magnitude and phase angle of the output voltage v_{out} of the circuit in Figure 5-2 if
   (a) $f = 7.64$ MHz
   (b) $f = 1$ MHz

4. The resistor and capacitor values for the lag network in Figure 5-1 are 4.7 kΩ and .001 μF respectively. Calculate
   (a) A_V at 20 kHz
   (b) A'_V at 40 kHz
   (c) ϕ at 40 kHz

5. The source resistance R_S of the circuit in Figure 5-2 is increased from 500 Ω to 1 kΩ. If $v_s = 1$ V (peak), calculate
   (a) f_c
   (b) $A_v\ (v_L/v_s)$ at f_c (exact to four figures)
   (c) v_{out} at $10 f_c$ (exact)
   (d) v_{out} at $0.1 f_c$ (exact)

6 Find the corner or cutoff frequencies of the base and collector lag networks in Figure 5-9 if R_s is increased to 500 Ω and the load resistor is increased to 1 MΩ. Identify the upper cutoff frequency of the amplifier.

7 Calculate the corner frequencies of the base and collector lag networks for the circuit in Figure 5-9 if an external 100 pF capacitor is added from the base to the collector of the transistor. Identify the upper cutoff frequency of the amplifier.

8 The resistor and capacitor values for the lead network in Figure 5-10 are 4.7 kΩ and 1 μF respectively. Calculate
 (a) f_c
 (b) A_v at 20 Hz
 (c) A'_v at 40 Hz
 (d) ϕ at 40 Hz

9 In Figure 5-13, R_{TH} = 2 kΩ, R_L = 2.7 kΩ, and C = 1 μF. Calculate
 (a) f_c
 (b) A_v at 20 Hz
 (c) A'_v at 40 Hz
 (d) ϕ at 40 Hz

10 Find the cutoff frequencies of the input and output lead networks, and the bypass capacitor network for the amplifier circuit in Figure 5-9 if C_{C1} = 0.2 μF, C_{C2} = 2 μF, and C_B = 100 μF, and the load capacitor is increased from 100 pF to 200 pF. Identify the lower cutoff frequency of the amplifier.

11 Find the cutoff frequencies of the input and output lead networks, and the bypass capacitor network for the amplifier circuit in Figure 5-9. For this circuit, C_{C1} = 0.15 μF, C_{C2} = 1 μF, and C_B = 68 μF. Identify the lower cutoff frequency of the amplifier.

12 Find the cutoff frequencies of the input and output lead networks, and the bypass capacitor network for the amplifier circuit in Figure 5-9. For this circuit C_{C1} = 2.2 μF, C_{C2} = .022 μF, and C_B = 200 μF. Identify the lower cutoff frequency of the amplifier.

13 What effect does a low value of r'_c have on the corner frequency associated with the output lead network?

14 What effect does a low value of r'_c have on the corner frequency associated with the output lag network?

15 An audio amplifier has an upper cutoff frequency of 18 kHz and a lower cutoff frequency of 120 Hz. Estimate the amplifier's bandwidth.

16 The f_{cu} of the amplifier in problem 15 is determined by a single or dominant RC lag network. What is the capacitance of the lag network if the circuit's resistance is 1 kΩ?

17 The resistor and capacitor values for the lag network in Figure 5-1 are 4.7 kΩ and .001 μF respectively. Calculate
 (a) f_c
 (b) T_R

18 Find T_R for the circuit in Figure 5-2.

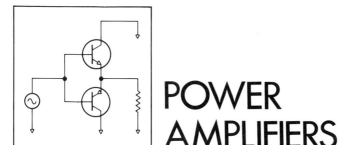

6
POWER AMPLIFIERS

Power amplifiers are amplifier circuits in which the power gain has been maximized. These circuits are used as the latter stages in amplifier systems.

The transistor's behavior in an amplifier is related to I_C and V_{CE}, which define the Q or quiescent point. The values of the Q point under ac conditions are graphically illustrated by the ac load line. A small movement of the Q point on the load line is a small-signal requirement, but a movement over the entire load line indicates large-signal behavior. The optimum location for the Q point is at the center of the load line. A transistor in a class A power amplifier conducts current for 360° of the input signal. The transistor's power dissipation is a limiting factor in most class A amplifiers. A transistor in a class B amplifier conducts current for 180° of the input signal; however the transistor dissipates much less power in class B circuits compared to class A. The class AB circuit is a modified class B amplifier that overcomes the crossover distortion associated with the class B circuit. Transistors dissipate the least power in class C amplifiers, which are frequently used in high-frequency applications.

6.1 DEFINITION

Power amplifiers are circuits in which the power gain has been maximized and are typically used as the latter stages in amplifier systems. Power amplifiers are capable of delivering a relatively large amount of power to the load. The transistors in power-amplifier circuits must be capable of dissipating much more power than the transistors in small-signal amplifiers. In power amplifiers, power transistors are used, and their power rating exceeds 600 to 800 mW.

Power amplifiers are also called *large-signal amplifiers*. The amplitude of the signal in an amplifier's latter stages is reasonably large; however, the meaning of large signal is better explained by examining the Q point and the amplifier's load lines.

6.2 DC LOAD LINE

The values of I_C and V_{CE} in an amplifier define the quiescent (at rest) or Q point. With no signal present, the values of I_{CQ} and V_{CEQ}

identify a point on a graph of I_C versus V_{CE}. When the transistor's collector family of curves is drawn on this graph, the Q point also identifies a point of operation of the transistor. The Q point will be one point on a straight line called the dc load line.

The dc load line represents all possible dc operating points for the circuit it is drawn for. The two limits or end points of the load line are called the *saturation point* and the *cutoff point*. Each limit is identified by a value of I_C and V_{CE}. At the saturation point, the transistor's current is maximum and is designated $I_C(\text{sat})$. The value of V_{CE} is zero volts (approximately).

Saturation point = (V_{CE}, I_C) = $\{0 \text{ V}, I_C(\text{sat})\}$

At the cutoff point, the transistor voltage is maximum and is designated $V_{CE}(\text{cutoff})$. The value of I_C is zero amperes (approximately).

Cutoff point = (V_{CE}, I_C) = $\{V_{CE}(\text{cutoff}), 0 \text{ A}\}$

At the cutoff point, the transistor's current is cut off. At the saturation point, the transistor's current is at the maximum value the circuit will allow.

The Q point and the upper limit, or saturation point, and the lower limit, or cutoff point, are shown on the dc load line in Figure 6-1. The value of V_{CE} when the transistor is operated at the saturation point is approximately zero volts. When $V_{CE} \cong 0$ V, the voltage drops across the emitter and collector resistors must equal the supply voltage (KVL). The collector current for this condition is called the *saturation collector current*.

$$V_{CC} = \{I_C(\text{sat})\}(R_E + R_C)$$

or $I_C(\text{sat}) = \dfrac{V_{CC}}{R_C + R_E}$

When the transistor is operated at the cutoff point, the value of I_C is near zero amperes, and the collector-to-emitter voltage will equal the dc supply voltage V_{CC}. With $I_C = 0$ A, the voltage drops across R_C and R_E will be zero, and

$$V_{CE}(\text{cutoff}) = V_{CC}$$

V_{CE} cannot go to zero volts in an actual circuit. Its minimum value is $V_{CE}(\text{sat})$ and will vary slightly with collector current. It is approximately 1 V. At the cutoff point, the transistor current will be the device's leakage current, which is ordinarily very small compared to the circuit current levels. With the two end points known, the dc load line can be drawn, on which will fall all possible Q points.

6.3 AC LOAD LINE

What happens to the Q point when an ac signal is introduced or coupled into the amplifier? The answer is that the Q point moves. The ac signal is algebraically added to the dc values,

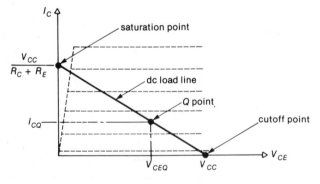

Figure 6-1. Dc load line.

AC LOAD LINE

causing the transistor to operate at a different value of I_C and a different value of V_{CE}. The two values are related because of the circuit that surrounds the transistor. Graphically, the operating point of the transistor is defined by a new set of values of I_C and V_{CE}, and we say that the Q point moved. As the magnitude of the ac input signal changes in real time, the Q point moves in real time.

How far can the Q point move and in what direction? What the Q point can do under ac conditions is graphically shown by the ac load line. *The ac load line represents all possible ac operating points.* At any instant of time, the collector current and collector-to-emitter voltage (together) will identify a point on the ac load line.

What determines the ac load line? The base-driven model in Figure 6-2a will be used to find the ac load line and its characteristics. Using KVL, an equation is written for the collector to emitter, r_C, and r_E loop.

$$v_{ce} + i_e r_E + i_c r_C = 0$$

Since β is high, $i_e \cong i_c$, and i_c can be substituted for i_e in the above equation. Solving for i_c produces the relationship

$$i_c \cong \frac{-v_{ce}}{r_C + r_E}$$

The ac quantities i_c and v_{ce} can be expressed as changes in dc values.

$$i_c = I_C - I_{CQ}$$

$$v_{ce} = V_{CE} - V_{CEQ}$$

Substituting for i_c and v_{ce} gives

$$I_C - I_{CQ} = -\frac{V_{CE} - V_{CEQ}}{r_C + r_E}$$

If this equation is solved for I_C and arranged in the form of a linear equation, the result is

$$I_C = -\frac{V_{CE}}{r_C + r_E} + I_{CQ} + \frac{V_{CEQ}}{r_C + r_E}$$

This equation relates the variable I_C and the variable V_{CE} and is a linear equation. The graph of a linear equation is a straight line that can be described by a slope (m) and a vertical intercept (b). The general mathematical form of the linear equation is

$$y = mx + b$$

The collector current I_C corresponds to y, V_{CE} corresponds to x, $-1/(r_C + r_E)$ corresponds to m, and $I_{CQ} + V_{CEQ}/(r_C + r_E)$ corresponds to the vertical intercept b. The equation is graphically illustrated in Figure 6-2b. This straight line is the ac load line. It represents all possible ac operating points. The Q point will be a point on

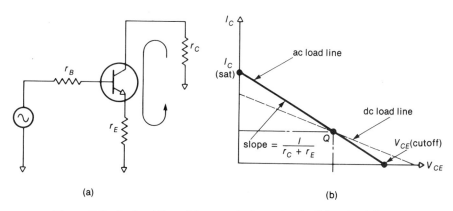

Figure 6-2. Ac load line: (a) ac equivalent circuit; (b) ac load line.

this line also, and it is the point of intersection of the dc and ac load lines.

The maximum collector current that the circuit will allow the transistor to conduct is $I_C(\text{sat})$. Its value can be found by setting $V_{CE} = 0$ V in the equation

$$I_C = -\frac{V_{CE}}{r_C + r_E} + \left(I_{CQ} + \frac{V_{CEQ}}{r_C + r_E}\right)$$

The result is

$$I_C = I_{CQ} + \frac{V_{CEQ}}{r_C + r_E} = I_C(\text{sat})$$

The largest value of V_{CE} occurs when I_C equals zero. For this case,

$$0 = -\frac{V_{CE}}{r_C + r_E} + \left(I_{CQ} + \frac{V_{CEQ}}{r_C + r_E}\right)$$

and

$$V_{CE}(\text{cutoff}) = V_{CEQ} + I_{CQ}(r_C + r_E)$$

Circuit Example

Draw the ac and dc load lines for the circuit in Figure 6-3. Identify the values of the ac and dc load line limits.

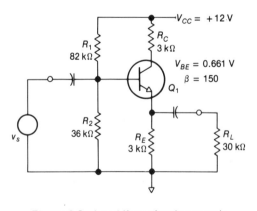

Figure 6-3. Load lines circuit example.

Solution

From the dc analysis,

$$I_{CQ} = 1 \text{ mA}$$
$$\text{and} \quad V_{CEQ} = 6 \text{ V}$$

From the amplifier's ac equivalent circuit,

$$r_C = 3 \text{ k}\Omega$$
$$\text{and} \quad r_E = R_E \| R_L = 2.73 \text{ k}\Omega$$

For the dc load line, the maximum value of V_{CE} is equal to V_{CC}.

$$V_{CE}(\text{cutoff}) = V_{CC} = 12 \text{ V}$$

The maximum possible dc value of I_C is a function of R_E and R_C.

$$I_C(\text{sat}) = \frac{V_{CC}}{R_C + R_E} = 2 \text{ mA}$$

The dc load line and its end points are shown in Figure 6-4a.

The slope of the ac load line is a function of the emitter and collector ac resistances.

$$\text{slope} = -\frac{1}{r_C + r_E} = -174.5 \text{ } \mu\text{s}$$

The saturation current is

$$I_C(\text{sat}) = I_{CQ} + \frac{V_{CEQ}}{r_C + r_E} = 2.05 \text{ mA}$$

The cutoff voltage is

$$V_{CE}(\text{cutoff}) = V_{CEQ} + I_{CQ}(r_C + r_E)$$
$$= 6 \text{ V} + 5.73 \text{ V} = 11.73 \text{ V}$$

The ac load line and its end points are shown in Figure 6-4b.

6.4 CLASSES OF OPERATION

There are *four different classes of operation* for power amplifiers: A, B, AB, and C. These classes basically describe the degree a transistor

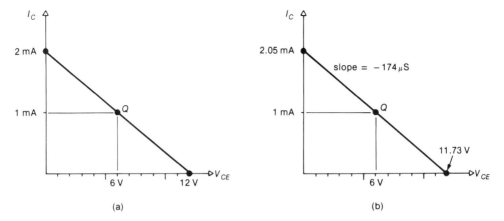

Figure 6-4. Load lines: (a) dc; (b) ac.

conducts relative to the input signal's cycle. In a class A amplifier, the transistor conducts current (and is in its active region) for 360° of the input signal waveform. In a class B amplifier, the transistor conducts for 180° of the input cycle. For class C operation, the transistor conducts for less than 180°, and for class AB the transistor conducts between 180° and 360°. The classes of operation are graphically illustrated in Figure 6-5. The amplifiers discussed thus far have all been class A amplifiers. The power dissipated by a transistor in a class

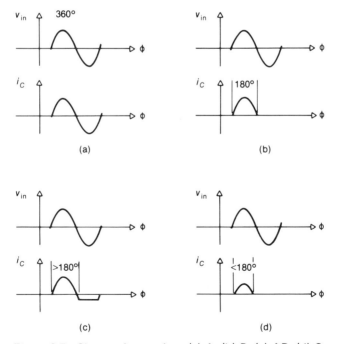

Figure 6-5. Classes of operation: (a) A; (b) B; (c) AB; (d) C.

A amplifier is high relative to the power delivered to the load. The other classes of operation have a higher load power to transistor power dissipation ratio. All small-signal linear amplifiers, and some power amplifiers, are operated class A.

6.5 CLASS A POWER AMPLIFIERS

Class A Operation

The transistor in a class A amplifier conducts for 360° of the input signal cycle. In terms of the ac load line, class A operation means that the Q point is not driven to (or beyond) the load line limits. If the amplifier is overdriven, the output signal will be clipped, distortion results, and the transistor *will not* be operating in its active region. Normal class A operation and saturation and cutoff clipping are shown in Figure 6-6.

When an input signal is introduced into the amplifier, the Q point will move. If the input signal is sinusoidal, the transistor's current and voltage will change sinusoidally. The relationship of the sinusoidal voltage and current in a transistor and the movement of the Q point are illustrated in Figure 6-6a.

Definition of Large and Small Signal

Small signal and large signal are defined according to the degree of movement of the Q point along the ac load line in an amplifier. If the Q point moves 10% (approximately) from its dc position when an ac signal is present, the amplifier is a small-signal amplifier. If the Q point moves over 90% (approximately) of the load line, the amplifier is a large-signal amplifier. The definition of small- and large-signal amplifiers is *graphically illustrated* in Figure 6-7.

Centered Q Point

The optimum location of the Q point in a class A amplifier is in the center of the load line. The reason is obvious. When the Q point is in the middle, the circuit can amplify the largest possible signal without clipping. If the Q point is offset from the center, the maximum peak-to-peak undistorted output will be less than the midpoint (centered) biased case. For a centered Q point, the saturation current and cutoff voltage are twice their quiescent values.

$$I_C(\text{sat}) = 2I_{CQ}$$
$$V_{CE}(\text{cutoff}) = 2V_{CEQ}$$

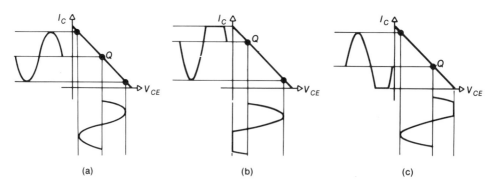

Figure 6-6. Class A operation: (a) normal; (b) saturation clipping; (c) cutoff clipping.

CLASS A POWER AMPLIFIERS

Power Considerations

A number of power-related factors must be considered in power amplifiers. Besides *power gain*, they include *maximum output power, transistor power dissipation, output efficiency*, and *derating factors*.

In a base-driven amplifier with a centered Q point, the *maximum possible output power* is the product of the maximum output rms voltage and maximum output rms current. If the ac resistances r_E and r_C (Figure 6-8) represent the load (seldom true), the maximum peak-to-peak voltage and current associated with these resistances are $2I_{CQ}$ and $2V_{CEQ}$. The rms value of a peak-to-peak signal is $1/\sqrt{2}$ times the peak value. Hence, the maximum output power is

$$p_o(\max) = V_{\text{rms}} I_{\text{rms}} = \frac{V_{CEQ} I_{CQ}}{\sqrt{2} \sqrt{2}}$$

$$= \frac{V_{CEQ} I_{CQ}}{2}$$

The power dissipated by a transistor under no-signal conditions is the product of the quiescent I_C and quiescent V_{CE}. With an input signal present, the collector current and voltage will change, but a fixed amount of heat is produced. The average power dissipated by the transistor will be less than under no-signal conditions. The *maximum* power dissipated by the transistor is

$$P_D(\max) = V_{CEQ} I_{CQ}$$

Efficiency is a measure of how well an amplifier converts dc power (from the dc supply) into ac output power. High efficiency is especially important in battery-operated electronic equipment. The *output efficiency is defined as the ratio of ac output power to dc input power*. It is symbolized by the Greek letter eta (η).

$$\eta = \frac{p_o}{P_{\text{dc}}}$$

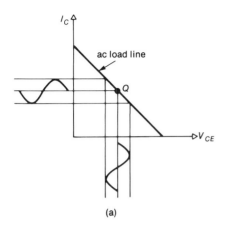

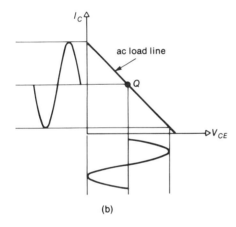

Figure 6-7. Small- and large-signal operation: (a) small signal; (b) large signal.

In an amplifier with a centered Q point, the sum of the ac resistances in the emitter (r_E) and collector (r_C) must equal the ratio of V_{CEQ} to I_{CQ}.

$$r_C + r_E = \frac{V_{CEQ}}{I_{CQ}} \quad \text{(centered } Q \text{ point)}$$

The above relationship can be used as a test to determine whether an amplifier has a midpoint biased Q point or not.

POWER AMPLIFIERS

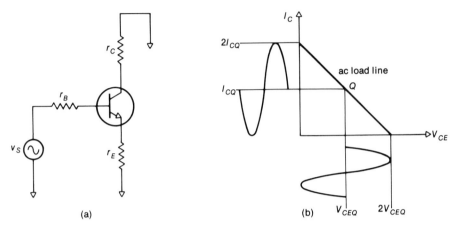

Figure 6-8. Maximum power out, class A: (a) base-driven model; (b) large-signal, centered Q point.

Power Amplifier Circuit Example

For the circuit in Figure 6-9, find (a) I_{CQ} and V_{CEQ}, (b) V_{CE}(cutoff) and I_C(sat), (c) A_p, (d) P_D, and (e) determine if the Q point is centered.

Solution

From the dc equivalent circuit,

$$I_E = \frac{V_{EE} - V_{BE}}{R_E + R_1/\beta} = \frac{9.3 \text{ V}}{220 \text{ }\Omega} = 42.3 \text{ mA}$$

$$= I_E \cong I_C$$

$$V_E = -(I_B R_1' + V_{BE}) = -(0.85 \text{ V} + 0.7 \text{ V})$$

$$= -1.55 \text{ V}$$

$$V_C = V_{CC} - I_C R_C = 10 \text{ V} - 4.23 \text{ V}$$

$$= +5.77 \text{ V}$$

$$V_{CEQ} = V_C - V_E = 7.32 \text{ V}$$

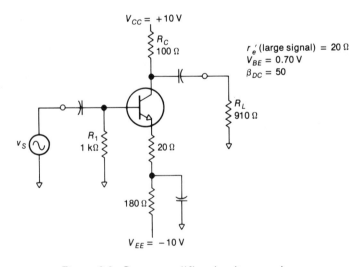

Figure 6-9. Power amplifier circuit example.

The Q point is at a collector current of 42.3 mA and a V_{CE} of 7.32 V. The quiescent current level in power amplifiers is higher than in small-signal amplifiers.

$$r_C = R_C \| R_L = 90\ \Omega$$
$$r_E = 20\ \Omega$$
$$V_{CE}(\text{cutoff}) = V_{CEQ} + I_{CQ}(r_C + r_E)$$
$$= 7.32\ \text{V} + 4.65\ \text{V} = 11.97\ \text{V}$$
$$I_C(\text{sat}) = I_{CQ} + \frac{V_{CEQ}}{r_C + r_E} =$$
$$42.3\ \text{mA} + \frac{7.32\ \text{V}}{110} = 108.8\ \text{mA}$$

$V_{CE}(\text{cutoff})$ is a differential voltage and represents the *ideal* maximum peak-to-peak signal the amplifier can process *if* the Q point is mid-biased.

$$A_V = \frac{v_c}{v_b} = \frac{v_L}{v_S} = \frac{r_C}{r_E + r'_e}\ \text{(large signal)}$$
$$= \frac{90}{22} = 4.1$$
$$r_{in} = \beta(r_E + r'_e) = 50(22\ \Omega) = 1100\ \Omega$$
$$A_i = \frac{R_1}{R_1 + r_{in}} \beta \frac{R_C}{R_C + R_L}$$
$$= \frac{1000}{2100} 50 \frac{100}{1010} = 2.36$$
$$A_p = A_i A_v = 9.68$$

For the Q point to be centered, the following must be true.

$$\frac{V_{CEQ}}{I_{CQ}} = r_C + r_E$$
$$\frac{V_{CEQ}}{I_{CQ}} = \frac{7.32\ \text{V}}{42.3\ \text{mA}} = 173\ \Omega$$
$$r_C + r_E = 110\ \Omega$$

The Q point is *not* centered on the ac load line.
The maximum power dissipated by the transistor will be under no-signal conditions.

$$P_D = I_{CQ} V_{CEQ} = (42.3\ \text{mA})(7.32\ \text{V})$$
$$= 0.310\ \text{W}$$

The *emitter ac resistance for large-signal* amplifiers is estimated from the transconductance $(I-V)$ curve of the base–emitter junction. This curve is usually given in the specifications sheets for the transistor. Under large-signal conditions, the Q point associated with this junction moves over an appreciable amount of the curve. The ac resistance is averaged over the entire range of operation and is defined as the ratio of the change in base-to-emitter voltage to the change in emitter (collector) current. Mathematically,

$$r'_e\ \text{(large signal)} = \frac{\Delta V_{BE}}{\Delta I_E}$$

The large-signal emitter resistance is graphically shown in Figure 6-10. The base–emitter junction's operating point moves from the quiescent state to points A and B when an ac signal is introduced into the circuit. At point A (defined by I_{C1} and V_{BE1}), the transistor is conducting the maximum current [near $I_C(\text{sat})$]. At point B (defined by I_{C2} and V_{BE2}), the transistor is conducting a minimum current (near zero). For these two points of operation, the large-signal ac resistance is

$$r'_e\ \text{(large signal)} = \frac{V_{BE1} - V_{BE2}}{I_{C1} - I_{C2}}$$

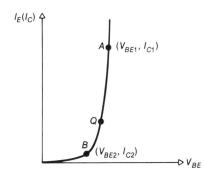

Figure 6-10. Large-signal emitter ac resistance.

Transistor Power Dissipation

The active and passive components in power amplifiers are required to dissipate more power than those in small-signal circuits. *The maximum power a transistor can dissipate is established by the transistor's maximum junction temperature.* A transistor is destroyed when it is forced to dissipate a power level such that its junction (collector) raises above some maximum value. The collector junction is the point where the most power is dissipated. A typical maximum junction temperature is from 150° to 200°C. *The transistor's junction temperature is established by*

1. The power being dissipated by the transistor.
2. The thermal conductivity of the transistor case.
3. The heat sink that is being used.

Device manufacturers usually specify the power dissipation rating of a transistor using graphs of power versus temperature. These graphs provide power dissipation data versus ambient temperature and case temperature for several packages or transistor housings.

Thermal Resistance

A transistor can dissipate a high power level if the heat generated at the junction can be quickly transferred to the environment where it can be dispersed. *Thermal resistance, symbolized by θ, is the resistance to heat flow from the junction to the surrounding air.* It is defined as

$$\theta = \frac{\Delta T}{P} = \frac{\text{temperature difference across thermal conductor}}{\text{power flowing through thermal conductor}}$$

The unit of measurement for θ is *degrees Celsius per watt (°C/W)*, and a typical value is 300°C/W for a transistor.

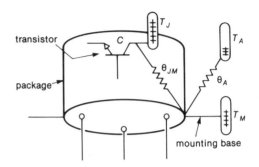

Figure 6-11. Transistor thermal circuit.

A thermal circuit for a transistor surrounded by free air is shown in Figure 6-11. The electrical resistance schematic symbol is also used for thermal resistance. The resistance to heat flow from the junction to the mounting base is the thermal resistance θ_{JM}; θ_A is the thermal resistance of that portion of air in contact with the mounting base. The temperature of the junction is T_J, of the mounting base, T_M, and of the environment, T_A.

$$T_J > T_M > T_A$$

The junction temperature is the highest and the ambient is usually the lowest. Heat will flow from the high temperature to the low.

Derating Factor and Heat Sinks

The power dissipation graph in Figure 6-12 shows that, as the ambient temperature increases, the maximum allowable power dissipation decreases. In many applications (especially military and aerospace), the temperature of the device's surroundings can be relatively high. For these cases, the maximum rating of the transistor must be derated or decreased. The maximum power dissipation at a particular ambient temperature can be determined from the graph or from the following equation.

$$P_D(\max) = P_D(\max \text{ at } 25°C) - \frac{\Delta T}{\theta_{JA}}$$

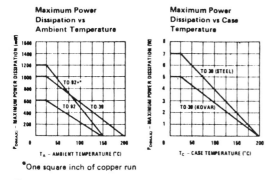

Figure 6-12. Transistor power dissipation data.

where $\Delta T = T_A - 25°C$

θ_{JA} = thermal resistance from junction to ambient

The reciprocal of θ_{JA} is called the derating factor.

derating factor (from junction to ambient)

$$= \frac{1}{\theta_{JA}}$$

A derating factor from the junction to the mounting base is also usually provided in the transistor's data sheets.

In Figure 6-11, the total thermal resistance from the junction to the environment is the sum of

$$\theta = \theta_{JM} + \theta_A$$

The thermal resistance from the mounting base to the environment (θ_A) can be larger than θ_{JM}. Ideally, θ_A can be made zero if the transistor is mounted to an infinite heat sink. A heat sink is a mass of metal mounted to the transistor to aid in the transfer of heat from the mounting base (or case) to ambient. An infinite heat sink is a mass of metal of such a size that θ_A is zero. In practice, of course, an infinite heat sink is not possible. Common *transistor heat sinks* (Figure 6-13) are the flat, vertical, finned aluminum heat sink and the cylindrical heat sink that slides over the transistor. The thermal resistance ratings for these heat sinks are 2°C/W and 8°C/W, respectively.

Example

The maximum operating junction temperature of a 2N3019 is 200°C. What is the maximum power dissipation if the junction-to-ambient thermal resistance is 217°C/W and the environment is at 25°C?

$$P_D(\text{max}) = P_D \text{ at } T_J = 200°C$$
$$= \frac{T_J - T_A}{\theta_{JA}} = \frac{200°C - 25°C}{217°C/W}$$
$$= 0.806 \text{ W}$$

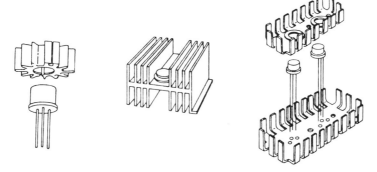

Figure 6-13. Transistor heat sinks.

What is the junction temperature if the transistor is dissipating 400 mW and the ambient temperature is 25°C?

$$T_J = T_A + P_D \theta_{JA}$$
$$= 25°C + \frac{217°C}{W}(0.40 \text{ W}) = 112°C.$$

How much is the maximum power dissipation reduced if the ambient temperature is increased to 50°C?

$$P_D(\text{max}) = \frac{T_J - T_A}{\theta_{JA}} = \frac{200°C - 50°C}{217°C/W}$$
$$= 0.691 \text{ W}$$

The maximum power dissipation is reduced by 14% when the ambient temperature rises to 50°C.

Other Class A Power Amplifier Circuits

Many types of power output stages (circuits) are used. The circuits and the knowledge surrounding power amplifiers are almost a topic by themselves. A few of the more popular class A stages are shown in Figure 6-14. Single-ended stages drive only one end of the load; the other end is at ground potential. Push-pull stages drive both ends of the load, but each is driven in the opposite direction.

6.6 CLASS B AND CLASS AB POWER AMPLIFIERS

Definition

A transistor in a class B amplifier conducts current for exactly one half-cycle of a sinusoidal input signal. Another transistor conducts current for the other half-cycle, with the two transistors and their associated circuitry forming the complete class B stage.

A transistor in a class AB amplifier conducts current (Figure 6-5c) *for slightly more than one half-cycle of the input signal.* Class AB amplifiers typically use two transistors, with the second transistor also conducting for slightly more than one-half of the input signal. For a portion of the input signal's cycle, both transistors will be on. This is done to reduce crossover distortion.

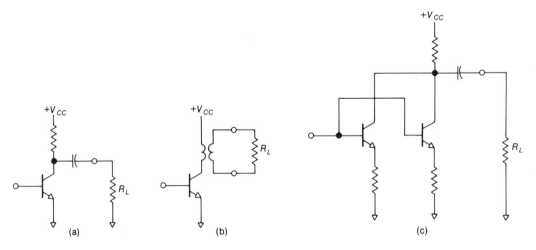

Figure 6-14. Class A power amplifiers: (a) resistive load; (b) transformer output; (c) parallel output.

The *basic reason* for using class B, AB, and C amplifiers is to *reduce the amount of power dissipated by the transistors* in the amplifier and the power dissipated by the amplifier itself. In class A amplifiers, the transistor dissipates twice the power delivered to the load. In class B push-pull amplifiers, the transistor dissipates one-fifth of the load power. Another important advantage of the class B amplifier is its low no-signal current drain or quiescent current. The current drain of a class B or class AB amplifier is typically 1% of a class A circuit. The low no-signal current drain is important in low-power and battery-powered applications. The class AB amplifier is basically a modified version of the class B amplifier. The output or last stage of the integrated-circuit operational amplifier is a class AB circuit.

Class B Push–Pull Action

The ac equivalent circuit of a *class B push-pull amplifier* is shown in Figure 6-15. The dc biasing is not shown, but each transistor is biased at cutoff. The transistors operate as emitter followers with the upper (*npn*) transistor following the positive half-cycle, and the lower (*pnp*) transistor following the negative half-cycle. The two transistors push and pull current to duplicate the input waveform at the output. During the positive half-cycle of the input signal, transistor Q_1 will conduct (Q_2

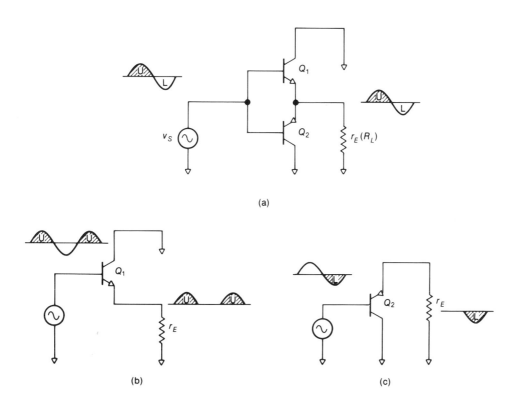

Figure 6-15. Class B push-pull amplifier: (a) ac equivalent circuit; (b) upper half of the input signal; (c) lower half of the input signal.

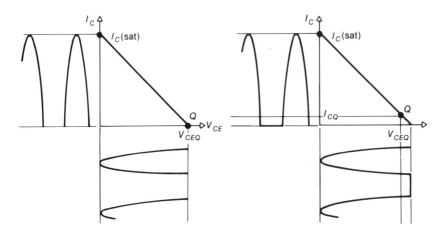

Figure 6-16. Ac load lines, classes B and AB: (a) class B; (b) class AB.

is off), and since the emitter follows the base, this half of the input waveform will be duplicated at the load (r_E or R_L). During the negative half-cycle of the input, Q_1 will be off and Q_2 will be on. The lower half of the input waveform will also be duplicated at the load.

Since the transistors are dc biased at cutoff, the Q point for a class B stage is at $I_{CQ} = 0$ A (Figure 6-16a). When the transistor does conduct, the quiescent current will go from zero to $I_C(\text{sat})$, where

$$I_C(\text{sat}) = \frac{V_{CEQ}}{r_C + r_E}$$

For the ac equivalent circuit shown, $r_C = 0 \,\Omega$. Remember, the class B amplifier is a large-signal amplifier and will operate over the entire load line (approximately). The Q point for the class AB circuit (Figure 6-16b) is at some value of I_C, typically 1% (to 5%) of $I_C(\text{sat})$.

A more accurately drawn output waveform of the transistorized push-pull amplifier is shown in Figure 6-17. The output waveform is distorted. The distortion, called *crossover distortion*, is at the zero crossover point or the region where the output waveform makes the transition through zero volts. The cause of the distortion is the barrier potential of the base-emitter junction of each of the transistors. As the input waveform goes from 0 to +0.6 V or from 0 to -0.6 V, both transistors are off, and the output is at zero volts. The output does not begin to follow the input until the input signal exceeds 0.6 V or the transistor's barrier potential. The distortion can be eliminated by having both transistors turned on when the input is at zero volts. This means the Q point (Figure 6-16b) will be slightly above cutoff. This small quiescent current is called a *trickle bias* and is 1 to 5% of $I_C(\text{sat})$. Each transistor will conduct for slightly greater than half of the input waveform, and the circuit operates, with this type biasing, class AB.

Figure 6-17. Distorted push-pull amplifier output.

Class AB Dc Biasing

A complete *class AB, push-pull, emitter-follower amplifier* is illustrated in Figure 6-18. The resistor-diode (R_1-D_1) voltage divider sets

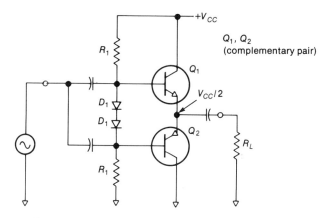

Figure 6-18. Class AB push-pull emitter follower.

the Q point slightly above cutoff to prevent crossover distortion. The *npn* and *pnp* transistors must be fully complementary, meaning the performance of the two devices must be matched or have the same performance characteristics.

Since the transistors are complementary and they are in series,

$$V_{CEQ} = \frac{V_{CC}}{2}$$

The quiescent current I_{CQ} is determined using a circuit technique called a *current mirror*. The basic current mirror circuit is drawn in Figure 6-19a. The diode D_1 is forward biased with a current of approximately

$$I_{D1} \cong \frac{V_{CC} - V_F}{R_1}$$

The transistor base current will be very small compared to I_1.

$$I_B \ll I_1$$

Hence, $I_{D1} \cong I_1$

The supply voltage $V_{CC}/2$ is *very large* compared to V_F, and small changes in V_F do not significantly affect the value of I_D. The forward voltage of D_1 is impressed across the base to the emitter of the transistor. The diode and tran-

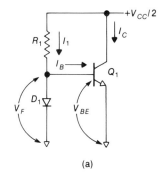
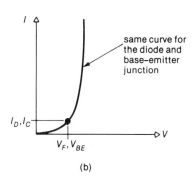

Figure 6-19. Current mirror: (a) circuit (*npn*); (b) *I-V* curves for D_1 and Q_1's B-E junction.

sistor are selected such that the current–voltage relationships of the diode *and* the base-to-emitter junction are the same (Figure 6-19b). If V_{BE} is the same as V_F, then I_C will be the same as I_D. The collector current is said to be the *mirror* of the current I_{D1}. The matching of the characteristics of the diode and the base-emitter junction is somewhat difficult for discrete components, but in integrated circuits it is not difficult at all. In ICs, the diode's geometry or physical characteristics are identical to the *B-E* geometry, and they are placed near each other on the same chip. The result is two junctions whose behavior is nearly identical. The current mirror biasing technique is a popular circuit scheme in linear integrated circuits, and it forms the basis for the current-differencing input stage of the Norton amplifier. An important advantage of the current mirror is its minimal dependence on temperature. The supply voltage is so large compared to V_F that changes in the barrier potential due to temperature (-2.2 mV/°C) minimally affect the collector current I_C.

For the class AB amplifier in Figure 6-18, the quiescent current is given as follows:

$$I_{CQ} = \frac{V_{CC} - 2V_{BE}}{2R_1}$$

Class AB Circuit Characteristics

Since the class AB stage is an emitter follower, its characteristics are similar to the class B emitter follower previously discussed. Each transistor behaves like an emitter follower for one-half (plus) of the input signal's cycle. The *input resistance* of the transistor that is on is

$$r_i \text{ (transistor)} = \beta_{dc} \{R_L + r'_e \text{ (large signal)}\}$$

For the *circuit*, the transistor's input resistance is shunted by the two R_1 resistors. The diodes are always on, and in the base-bias circuit they are ideally equivalent to a short circuit.

$$r_{in} \text{ (circuit)} = \frac{R}{2} \| r_i$$

The large-signal voltage gain is approximately one, since R_L will be much greater than r'_e (large signal).

$$A_v = \frac{v_e}{v_b} = \frac{R_L}{R_L + r'_e} \text{ (large signal)} \cong 1$$

The *peak value of the output voltage* for the push–pull circuit is V_{CEQ}. The *ac peak value of the output current* is $I_C(\text{sat})$. The maximum output power is the product of the rms values of the current and voltage.

$$p_O(\max) = (0.707 I_{CQ})(0.707 V_{CEQ})$$
$$= \frac{I_C(\text{sat}) V_{CEQ}}{2}$$

The no-signal power dissipation of the transistors is small because of the low value of I_{CQ}. The *worst-case power dissipation* under large-signal conditions can be found using calculus. It is given as follows:

$$P_D(\max) = \frac{V_{CEQ} I_C(\text{sat})}{10}$$

The power dissipated by a transistor in a push–pull class AB (or B) stage is *five times less* than the power delivered to the load. The power dissipated by a transistor in a class A stage is *two times more* than the power delivered to the load.

The efficiency ($\eta = p_O/P_{dc}$) of class B and AB stages is very high and is near 75%. This high efficiency is important in battery-operated electronics equipment. The transistor radio is a common example.

Circuit Example of a Class AB Amplifier

The circuit in Figure 6-20 is a *class AB emitter-follower, push–pull power amplifier*. The amplifier is biased with two symmetric dc

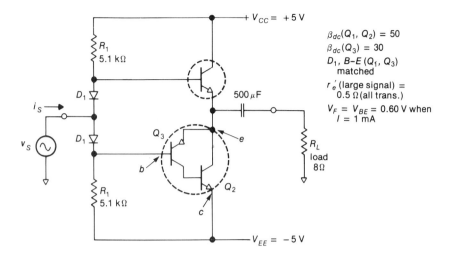

Figure 6-20. Class AB power amplifier circuit example.

supplies. For this circuit,

$$V_{CEQ} = \frac{V_{CC} - (-V_{EE})}{2} = 5\text{ V}$$

and $I_{CQ} = \dfrac{V_{CC} - V_{BE}}{R_1} = 0.863$ mA

Transistors Q_2 (npn) and Q_3 (pnp) form a complementary pair, which together simulate a pnp transistor. This arrangement is common in linear ICs because the *pnp*'s performance is not nearly as good as the *npn*'s. In integrated circuits, the *npn* is cheaper to manufacture and is the best performing transistor. The *pnp* requires one more assembly step, which accounts for the loss in performance and the extra cost.

The maximum collector current is determined by V_{CEQ} and the load resistor.

$$I_C(\text{sat}) = \frac{V_{CEQ}}{R_L} = 625\text{ mA}$$

The maximum power dissipated by the transistors is related to the dc values of V_{CE} and $I_C(\text{sat})$

$$P_D(\text{max}) = \frac{I_C(\text{sat})V_{CEQ}}{10} = \frac{(0.625\text{ A})(5\text{ V})}{10}$$
$$= 0.313\text{ W}$$

The power-dissipating capability of the transistor must exceed 313 mW, which normally is not a problem. The stage can ideally deliver five times this amount of power to the load.

$$P_O(\text{max}) = \frac{I_C(\text{sat})V_{CEQ}}{2} = 1.6\text{ W}$$

The voltage gain for the circuit is approximately 1.

$$A_v = \frac{v_b}{v_e} = \frac{R_L}{R_L + r_e'} \text{ (large signal)}$$
$$= \frac{8}{8.5} = 0.94$$

The input resistance of the transistor is

$$r_i = \beta\{R_L + r_e' \text{ (large signal)}\} = 50(8.5\text{ }\Omega)$$
$$= 425\text{ }\Omega$$

For the circuit,

$$r_{in} = \frac{R_1}{2}\|r_i = 2550\text{ }\Omega\|425\text{ }\Omega = 364\text{ }\Omega$$

The maximum current gain for this type of circuit is β; however, the base-bias circuit reduces this number. The supply current i_S splits into two parts, the base and the base-bias currents. The base current is found using current division.

$$i_b = i_S \frac{R_1/2}{r_i + R_1/2}$$

The base current is amplified by β_{dc} and delivered to the load.

$$i_c \cong i_L = \beta i_b = i_S \beta \frac{R_1/2}{r_i + R_1/2}$$

The current gain is the ratio of i_L to i_S.

$$A_i = \frac{i_L}{i_S} = \beta \frac{R_1/2}{r_i + R_1/2}$$

For the values listed in the circuit,

$$A_i = (50) \frac{2550}{425 + 2550} = 42.8$$

Other Class AB (B) Amplifier Circuits

The class AB emitter-follower, push–pull circuit was highlighted and emphasized to (1) illustrate class B and AB operation and because of its (2) importance in linear integrated circuits. It is by no means the only circuit used. Just as with the class A power-amplifier circuits, class B (and AB) amplifiers are almost a topic by themselves. The two circuits in Figure 6-21 are other frequently used class AB (B) amplifiers.

6.7 POWER AMPLIFIER ICs AND APPLICATIONS

A circuit of a commercial, monolithic integrated circuit, $\tfrac{1}{2}$ W *power amplifier* is shown in Figure 6-22. The IC contains a preamplifier stage and a class AB output stage. Large-value capacitors cannot be integrated, and all stages within ICs are dc coupled.

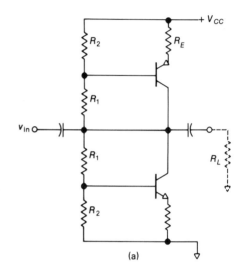

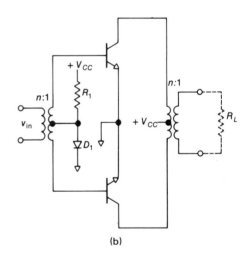

Figure 6-21. Other class AB (B) amplifier circuits: (a) swamped emitter, push-pull; (b) transformer coupled, push-pull.

An *application of this power amplifier* IC is shown in Figure 6-23. The circuit is used as a power amplifier to drive an 8 Ω speaker. The external (to the IC) components are used for

1. AC coupling.
2. Frequency response control.

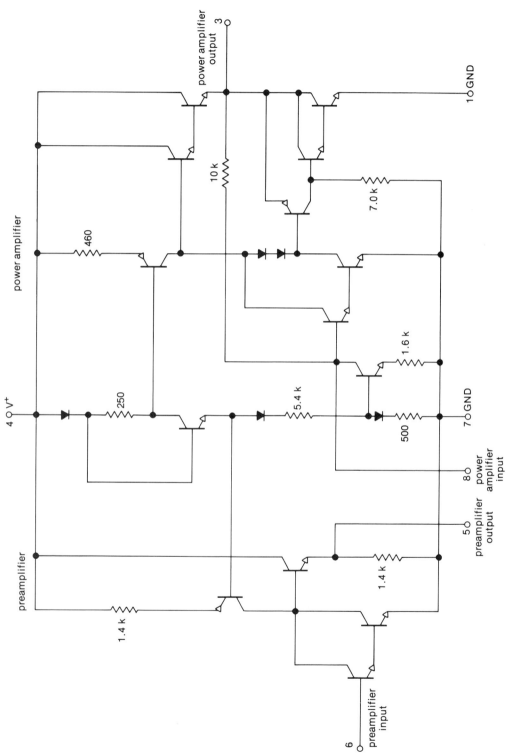

Figure 6-22. Schematic of the MC1306P audio power amplifier.

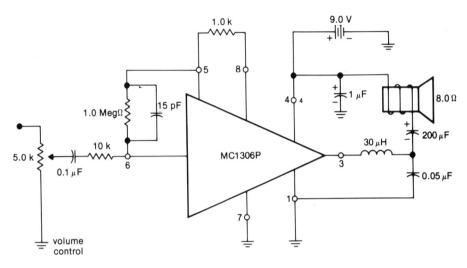

Figure 6-23. Audio power amplifier circuit.

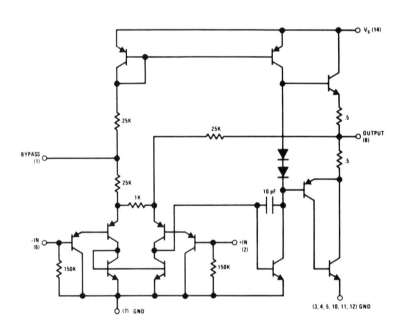

Figure 6-24. Simplified schematic of the LM 380 power amplifier.

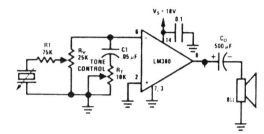

Figure 6-25. Power amplifier application—ceramic phonograph amplifier.

3. DC biasing.
4. Tone control.

The simplified schematic of *another commercial power amplifier IC* is shown in Fig. 6-24. It contains a differential input stage, a common-emitter gain stage, and a class AB output circuit. This amplifier has a fixed gain of 50, which is established internally.

The output is a quasi-complementary emitter follower. The output is class AB diode biased by D_1 and D_2, which are biased on by Q_{11} acting as a current source. This current source and the diodes are the collector load for the common-emitter gain transistor Q_{12}. The Q_{12} transistor is driven by the *pnp* differential stage of the input (Q_3 and Q_4). The amplifier voltage gain is twice the ratio of R_2/R_3. The resistors R_2 and R_1 bias the output to one-half of the value of the dc supply.

A *simple application* of this IC as a phonograph amplifier is illustrated in Figure 6-25. Volume control is provided by the potentiometer, which functions as a voltage divider, and the tone control is accomplished by varying the amplifier's high-frequency rolloff.

6.8 CLASS C OPERATION

Definition

A transistor in a class C amplifier conducts current for less than 180° of the sinusoidal input signal. In a practical class C circuit, the transistor conducts for only a very small portion of the input signal's waveform. The class C power amplifier can deliver more load power than a class B amplifier, but it typically must operate at a single frequency. This type of amplifier is commonly used in the RF (radio frequency) applications area.

Class C Circuit

A *tuned class C amplifier* is shown in Figure 6-26. The parallel capacitance and inductance

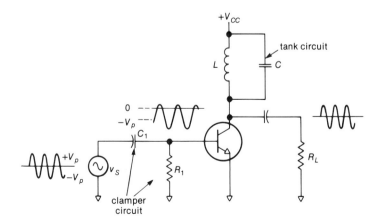

Figure 6-26. Class C amplifier circuit.

in the collector circuit is a tuned tank whose output is a sinusoidal voltage of a frequency approximately equal to

$$f_o \cong \frac{1}{2\pi \sqrt{LC}}$$

The resistance associated with the physical inductor and capacitor and the load resistance will cause the oscillations to decay. However, this is prevented by having the transistor conduct for a short period and reenergize the tank circuit to make up for the losses.

The transistor is biased *off except* for a small period by the clamping action of the R_1-C_1 base circuit. The time constant ($T = R_1 C_1$) of this network is very long compared to the input signal's period. After several cycles of the input signal, the coupling capacitor C will charge to the peak value of the input signal (it cannot discharge quickly enough). The average voltage at the base is $-V_p$. The transistor will only conduct for the top portion of the positive half-cycles. During this time the base voltage goes positive enough to forward bias the base-emitter junction. When the transistor conducts, it will energize the collector tank circuit, and the output will be a constant-amplitude, fixed-frequency sinusoidal voltage.

Problems

1. The V_{CC} supply in Figure 6-3 is changed to 15 V dc. For the ac load line, find
   (a) I_{CQ} and $I_C(\text{sat})$
   (b) V_{CEQ} and $V_{CE}(\text{cutoff})$
   For the dc load line, find
   (c) I_{CQ} and $I_C(\text{sat})$
   (d) V_{CEQ} and $V_{CE}(\text{cutoff})$

2. The load resistor R_L in Figure 6-3 is changed to 10 kΩ. For the ac load line, find
   (a) I_{CQ} and $I_C(\text{sat})$
   (b) V_{CEQ} and $V_{CE}(\text{cutoff})$

3. Does the amplifier circuit in Figure 6-3 have a midpoint biased Q point? Is it *near* midpoint biased (within 10%)?

4. Does the amplifier circuit in Figure 6-9 have a midpoint biased Q point? Is it near midpoint biased?

5. Does the amplifier circuit in Figure 4-17 have a midpoint biased Q point? Is it near midpoint biased?

6. What is the maximum power dissipated P_D by the transistor in Figure 6-3? Is the device operating within safe limits if the maximum power it can dissipate is 300 mW? What is the maximum power that can be delivered to the load in this circuit?

7. Calculate $P_D(\text{max})$ and $p_o(\text{max})$ for the circuit in Figure 6-20 if V_{CC} and V_{EE} are changed to 8 V supplies.

8. A single 15 V dc supply biases an amplifier, and a current of 75 mA is drawn from it. The amplifier can deliver 500 mW to a load. What is the circuit's efficiency?

9. Find the power gain A_p for the circuit in Figure 6-9 if the load R_L is 100 Ω.

10. What is the large signal emitter ac resistance if the coordinates of its large signal operation are (0.720 V, 75 mA) and (0.550 V, 2 mA)?

11. The maximum operating junction temperature of a 2N3019 is 200°C. What is the maximum power dissipation of the transistor if the junction to ambient thermal resistance is 217°C/W and the environment of the device is at 40°C?

12. If the transistor in problem 10 is dissipating 600 mW, what is its junction temperature? Will the device burn out?

13. Specify the change in the circuit of Figure 6-20 for a I_{CQ} of 2 mA.

14. The load resistor R_L in Figure 6-20 is increased to 16 Ω. Find $I_C(\text{sat}), A_v,$ and A_i.

15 Find the lower cutoff frequency of the load (C_o and R_L) in Figure 6-25.

16 Identify the position of the volume potentiometer (R_v) for maximum volume in the circuit of Figure 6-25.

17 What is the frequency of oscillation of the circuit in Figure 6-26 if $L = 10$ μH and $C = .001$ μF.

18 The frequency of the signal source v_s in Figure 6-26 is 500 kHz. If R_1 is 10 kΩ, specify the minimum value of C_1 such that the clamper's time constant is ten times the time period of the input signal.

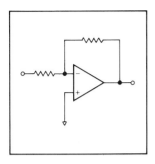

7
AMPLIFIER INTEGRATED CIRCUITS AND APPLICATIONS

Amplifier ICs are monolithic circuits that amplify. This chapter looks at several popular amplifier ICs and illustrates their applications.

The operational amplifier is a voltage amplifier IC with the characteristics of high voltage gain, high input resistance, low output resistance, and moderate bandwidth. It is probably the most important linear IC, and it serves as a fundamental building block in most analog systems. It is used in feedback control circuits. Because of its high voltage gain, the characteristics of these circuits are stable, and they are established by the discrete components surrounding the amplifier. Practical amplifier circuits must contend with the errors introduced by the finite values of the amplifier's characteristics and the data sheet parameters. Frequency plays a key role in limiting the use of the amplifier in circuit applications.

The op amp is a diversely applied device. It is used in circuits that amplify, compute, convert, filter, generate, and regulate voltage signals. A large number of basic one-amplifier circuits serve as analog building blocks for circuits that perform higher level functions. The two most fundamental op amp circuit configurations are the inverting and noninverting circuits.

A large applications area is op amp circuits that perform a mathematical operation or function. The common operations simulated by these circuits include summing, differencing, integration, log, and antilog functions. Multi-amplifier, advanced applications circuits, which utilize the basic configurations, include the simulated inductor and current source circuits.

The conventional operational amplifier is a voltage-in, voltage-out device. Another class of monolithic amplifiers exists called current-mode or Norton operational amplifiers. The Norton amplifier is a current-in, voltage-out device and is characterized by its current-mirror input circuit. The applications of this device are similar to the conventional op amp; however, it was designed to operate from a single voltage supply.

The comparator is a specially designed op amp that detects or compares two input voltages and provides an output that has two discrete states. These states indicate the relative magnitude of the two inputs. The device is important in interface applications and forms the basis for analog-to-digital (A/D) and digital-to-analog (D/A) convertors.

7.1 OPERATIONAL AMPLIFIER FUNDAMENTALS

What Is an Operational Amplifier?

In general, *an operational amplifier implies a voltage amplifier with the characteristics of high voltage gain, high input resistance, low output resistance, and moderate bandwidth.*

History

The most popular and widely used linear IC in the world is a low-frequency amplifier, the monolithic operational amplifier or op amp. It is sold by the tens of millions of devices per year. The op amp can be found in diverse applications and derives its name from those amplifier circuits that perform mathematical *operations*.

The circuit, as a functional block, was first used in the 1940s and was implemented then with vacuum tubes. Analog computers, with vacuum tube amplifier circuits, were used to simulate nonelectrical systems whose behavior was mathematically modeled using differential equations. Amplifier circuits were configured to perform the addition, subtraction, scaling, and integration *operations* for the equations.

In the early 1960s, small, transistorized, plastic encapsulated op amp modules were introduced. These plug-in discrete amplifier units were quickly followed by integrated versions, and the amplifiers' rise to dominance in the design of analog systems followed.

Presently, the small, lightweight, and well-performing IC amplifier circuit can be purchased for less than a dollar and can be found in such diverse applications as filters, oscillators, regulators, convertors, and signal-processing and control circuits. The device comes in many sizes, packages, and configurations and is used as a key building block in most analog subsystems.

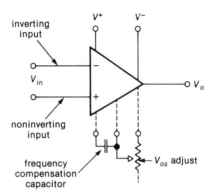

Figure 7-1. IC operational amplifier schematic symbol.

Symbology

The *symbol* for the operational amplifier is the *amplifier triangle*. It is a *two-port device*; that is, it has two input lines that sense a difference of potential and one output whose reference is the supply/circuit reference or ground. The amplifier (Figure 7-1) requires a minimum of five terminals or pins. Two are for the inputs, and they are designated inverting (−) and noninverting (+). Two pins are used for supplies, and the fifth pin is the output. The amplifier is typically powered by a positive and a negative voltage source, which are connected to the pins designated V^+ and V^-. The reference for the output is the supplies' common or ground. Additional pins, usually three, are provided to allow the user to compensate for inherent amplifier deficiencies (V_{os}) or to control its frequency response and bandwidth.

Characteristics

The op amp is a high-gain voltage amplifier with high input resistance, low output resistance, and moderate bandwidth. Table 7-1 lists the numbers associated with these characteristics for five design generations of monolithic

Table 7-1. Specifications of five generations of IC amplifiers.

Characteristic	Amplifier					
	709C	741C	301A	308	156	
Voltage gain	45,000	160,000	160,000	300,000	200,000	Typical specifications
Input resistance	250 kΩ	1 MΩ	2 MΩ	40 MΩ	10^6 MΩ	
Output resistance	150 Ω	75 Ω	75 Ω	75 Ω	75 Ω	
Bandwith (GBW)	5 MHz	1 MHz	1 MHz	1 MHz	5 MHz	

amplifiers. A circuit model or equivalent circuit for the device is shown in Figure 7-2. In this model, the source AV_{in} is an ideal, dependent voltage source, where A is the amplifier's voltage gain. The source's value depends on the value of V_{in}, where V_{in} is the differential input voltage. The resistance between the two input terminals is r_i, and the resistance in series with the output is the output resistance r_o.

If the input voltage to the amplifier is such that the inverting input (−) is positive with respect to the noninverting input (+), the output will be a negative voltage with respect to ground. If the noninverting (+) input is positive with respect to the inverting input (−), the output voltage will be positive with respect to ground. The magnitude of the output voltage will be A times the difference of the inputs.

The operational amplifier, generally speaking, is a low-frequency amplifier. It has an extremely high gain near dc, but the gain decreases as frequency increases beyond the amplifier's low cutoff frequency. The gain-frequency performance of the amplifier is *not* measured by its cutoff frequency but by a figure of merit called the gain-bandwidth (GBW) product. The value of the gain-bandwidth product is the frequency, on a Bode plot, where the amplifier's gain is 1 or 0 dB.

Input-Output Relationship

The *input-output relationship* for the amplifier can be expressed two ways. It can be expressed graphically by a two-dimensional graph called a *transfer characteristic* and by a mathe-

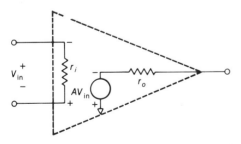

Figure 7-2. IC operational amplifier equivalent circuit.

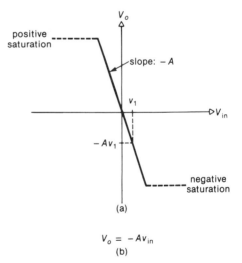

Figure 7-3. Input-output relationships: (a) graphic representation; (b) mathematical representation.

matical equation. Both forms are shown in Figure 7-3.

The *transfer characteristic* (Figure 7-3a) is a plot of V_o versus V_{in}. The polarity of the output is a function of the polarity of the applied input voltage relative to the amplifier's input terminals. For the transfer characteristic, the more positive input voltage is applied to the inverting (-) input. The slope of the diagonal line is $-A$ and illustrates the proportional or linear relationship of the input and output. This linear relationship is maintained until the amplifier output is voltage limited. The limit in the positive direction is called *positive saturation*, and in the negative direction it is called *negative saturation*. Saturation refers to the operating state of the transistors of the amplifier's output stage at these limits.

The input and output is *mathematically related* by

$$V_o = -A_v V_{in}$$

where A_v is the voltage gain of the amplifier and V_{in} is the potential difference between the inputs. The output, a single terminal, is referenced to the power supplies' ground or common measurement point and is capable of swinging to a positive or negative voltage.

every decade increase in frequency until the open-loop amplifier gain is 1 at about 1 MHz. The amplifier is not used for signals whose frequencies are above this point.

The gain-versus-frequency amplifier characteristic is illustrated in the Bode plot of Figure 7-4. The *Bode plot is a straight-line approximation of the amplifier's gain and frequency behavior*. The horizontal axis of the graph is the *log of frequency*, and the vertical axis is on a linear scale calibrated in *decibels* of *voltage gain*. The gain curve is flat until the cutoff frequency and then decreases at a rate of 20 dB per decade of frequency. The point where the curve intersects the 0 dB axis identifies the device's gain–bandwidth product (GBW) and is typically 1 MHz. This figure of merit is a fair measurement of the amplifier's frequency performance. Most of the present, large-selling monolithic amplifiers exhibit this type of characteristic.

For small signals, the amplifier's gain-versus-frequency performance follows that of a lag network. Internally, the amplifier's frequency-sensitive circuits can be reduced to a single *RC* lag network. The frequency-compensation capacitor (connected either internally or externally) is used to establish the capacitance of the

Gain Versus Frequency

Small Signal

In Table 7-1 the voltage gain is given as A_v, and it is very high, typically 200,000 or 106 dB. However, this number varies with frequency, and the input–output relationship is more accurately expressed by

$$V_o = -A_v(f)V_{in}$$

which shows the *frequency dependence of the voltage gain*. The gain is high (106 dB) near dc, but decreases as frequency increases beyond the amplifier's cutoff or -3 dB frequency, which is at 5 to 50 Hz. The gain decreases by 20 dB for

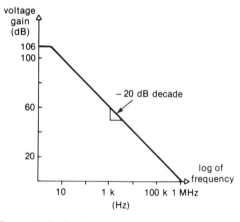

Figure 7-4. Small-signal frequency response (ideal).

RC network. The phase lag is 90° beyond 50 Hz and increases again at approximately 500 kHz, where a second nondominant lag network takes effect. Extreme care must be used in amplifier circuits to avoid positive feedback brought about by the additive effects of secondary amplifier and circuit lag networks. *The frequency response of an amplifier is limited by the small-signal response and by the amplifier's slewing-rate capabilities in the large-signal case.*

Large Signal

Slew-rate limit is the maximum rate of change (slope) of the amplifier's output voltage. Physically, this specification represents the ability of an amplifier to charge and discharge its compensation capacitor because of the finite amount of available current. A current source charging a capacitor models the frequency behavior of an amplifier under large-signal conditions. Slewing rate is abbreviated S_R and is specified in volts per microsecond (μs). The typical range of values for S_R is

$$0.5 \text{ V}/\mu\text{s} \leqslant S_R \leqslant 75 \text{ V}/\mu\text{s}$$

A sinusoidal signal will cease to be a small signal when its maximum rate of change exceeds the slew rate (S_R) limits of the amplifier. The highest distortion-free (<5%) signal that the amplifier can follow is

$$f_{\max} = \frac{S_R}{2\pi V_{pk}}$$

and is a function of the S_R specification of the amplifier and the signal's peak voltage (V_{pk}).

The large-signal frequency response curve is shown in Figure 7-5. The GBW and S_R ratings of IC amplifiers are key ac parameters.

7.2 INTEGRATED OPERATIONAL AMPLIFIER

Block Diagram

The *block diagram for a typical monolithic op amp* is shown in Figure 7-6. The first stage consists of a differential amplifier whose function is to amplify the difference between two input voltages. This stage is characterized by two *pnp* transistors with a common emitter. The second stage consists of voltage-amplifying, frequency-compensation, and dc level-shifting circuits. It is characterized by a common-emitter transistor with an unbypassed emitter resistor. The third or output stage is the workhorse of the amplifier. Its output must swing positive and negative with respect to ground and deliver the relatively heavy output current. Its output resistance must be low, and the stage must be protected against adverse loading conditions. It is characterized by the class AB output stage implemented with an *npn* and *pnp* transistor.

Four frequently used amplifiers in the industry are the 709, 741, 301A, and the 308. All major IC manufacturers produce them.

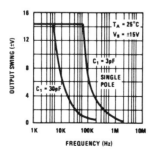

Figure 7-5. Large-signal frequency response.

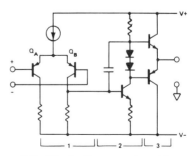

Figure 7-6. Block diagram of an IC operational amplifier.

Schematic Details

The *schematic for the monolithic 301A* is shown in Figure 7-7. The 301A contains 22 transistors and 16 resistors.

The Q_1-Q_3 and Q_2-Q_4 combinations are the equivalent of the *pnp* differential amplifier (Q_A-Q_B) shown in the block diagram. This stage must amplify voltage differences for all values of input voltages (common mode), have a high input resistance, a low offset voltage error (ΔV_{BE}), and have a small input current (base current I_B). The collector load resistors for the differential amplifier are simulated with active devices (transistors). Q_5 and Q_6 and associated circuitry form these active loads.

Transistors Q_9 and Q_{10} form a Darlington common-emitter amplifier stage with an unbypassed emitter resistor (R_8). The active load for this stage is the Q_{17} circuit. Transistor Q_{17} is one of three multiple-collector *pnp*s in the 301A. These devices are made by splitting the collector of a *pnp* transistor into two or more segments. By connecting one segment back to the base, the current gain of the transistor is then determined by the relative size of the segments. This circuit/device technique is extensively used in linear IC designs to provide the less sensitive current-source circuit bias. Frequency-response compensation is accomplished by connecting a capacitor (internally or externally) from the collector to base of the Darlington (Q_9-Q_{10}) pair. Externally, this is done through device pins 1 and 8 of a standard eight-pin T05 package. Pin 1 is used in conjunction with device pin 5 to correct for the offset voltage error in the input differential amplifier. This error is primarily caused by the mismatch in the base-to-emitter voltages of the two amplifier input transistors. In terms of amplifier performance, offset voltage means that, at zero volts input, the output will not be zero. Typically, the offset voltage is less than 5 mV. The offset correction is made by unbalancing the differential amplifier active loads, which causes the B-E voltages to shift.

Transistors Q_{16} and Q_{12} and associated circuitry form a class AB output stage. This type of output stage uses separate transistors to provide the two polarities of output current; however, both output transistors are on near zero volts output. For most of the output voltage range, one or the other transistor is off. Q_{15} and R_{11} form a current-limiting circuit. The output current flows through R_{11}. When this current begins to exceed 20 mA, the voltage drop across R_{11} approaches 500 mV, which begins to turn the base-to-emitter junction of Q_{15} on. As Q_{15} turns on, it shunts the base current of the output driver Q_{16}. As Q_{16} turns off, current limiting occurs. The current available at the base of Q_{16}, which is shunted to the output load, is fixed.

7.3 FEEDBACK AND THE AMPLIFIER

High Gain Implies Small Input Voltage

The op amp has high gain, where "high" translates to approximately 200,000. The maximum output voltage from the amplifier is typically ±15 V. If you divide the maximum output voltage by the gain, you get the maximum input voltage required, which is ±75 μV. Signals at this level are extremely difficult to process because of noise and interference. In addition, gains of 200,000 are seldom required. *Why, then, is the op amp so popular and how is it really used?*

Stability Through Feedback

Op amps are used primarily in circuits that contain feedback, usually negative feedback. Feedback, generally speaking, implies a situation in which output quantities in a system are

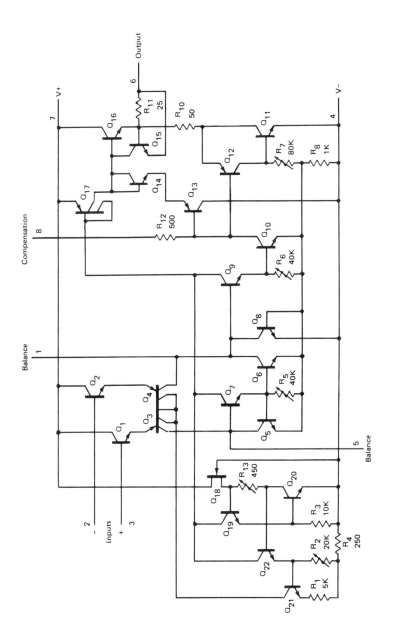

Figure 7-7. Schematic of the 301A IC operational amplifier.

allowed to affect the input quantities that caused them in the first place. Negative feedback is applicable when the output signal is purposely allowed to reduce the effect of the input signal. This, of course, results in the reduction of the net gain of the circuit. Feedback is introduced in an actual circuit by connecting a component, usually a resistor, from the output back to the inverting (−) input.

What do we receive and what do we lose in amplifier feedback circuits? We lose the maximum obtainable gain, which would be the gain of the amplifier itself. *We receive stability*, primarily gain stability. Most applications do not require high gain, but they do require the gain to be precise and stable. On an op amp data sheet, the amplifier gain is guaranteed to be a certain minimum amount, but the maximum is not specified, and for a given amplifier type, it will vary 10% to 20%. In addition to this relatively large variation, the gain of any specific amplifier will vary with temperature, age, and operating conditions. In fact, temperature is a crucial factor in all integrated circuits. Bipolar ICs use bipolar transistors whose base-to-emitter voltage is temperature dependent (−2 mV/°C) and whose current gain β is also temperature dependent. These two factors directly contribute to amplifier gain variations. A feedback configuration reduces these effects by several orders of magnitude. Besides gain stability, negative feedback tends to improve linearity, to reduce output resistance, and, in some configurations, to increase input resistance. If the amplifier has a gain of 200,000 and we need a circuit gain of only 20, the price paid for the advantages received is indeed modest. For the excess gain that we do not require, we receive gain stability. Most voltage-amplification requirements vary from 0.1 to 50. The number 200,000 is referred to as the *open-loop gain* of the amplifier. If we configure the amplifier in a circuit employing negative feedback so that the circuit's gain is 20, then 20 is called the *closed-loop gain*. Similarly, we can talk about open- and closed-loop input resistance, output resistance, and bandwidth.

The application of negative feedback to an operational amplifier yields an amplifier circuit with the closed-loop characteristics determined primarily by the values of the discrete feedback components whose accuracy and stability are easier to control.

The Feedback Concept

A *control system* is a system that regulates an output variable with the objective of producing a given relationship between it and the input variable or of maintaining the output at a fixed value. In a *feedback control system*, at least part of the information used to change the output variable is derived from measurements performed on the output variable itself.

A feedback control system, in *block diagram form*, is illustrated in Figure 7-8. The input variable is V_s, and the output variable is V_o. The system contains an amplifier with a voltage gain of A and a feedback network whose transfer function or input and output relationship is defined as F. The feedback network is usually implemented with discrete, passive components. The circles are summation points. To illustrate the stability characteristic of the feedback system, a voltage disturbance V_d is introduced in series with the amplifier's output. V_d can represent an externally induced voltage or a change in the output voltage due to changing circuit characteristics. We seek to find an expression relating the input and output variables, V_d, and circuit constants.

V_e, an error voltage, is a function of V_s and the feedback voltage V_f. It is the difference between the two; that is,

$$V_e = V_s - V_f \quad \text{(negative feedback)}$$

The output voltage, V_o, is a function of V_e, the gain of the amplifier A, and the disturbance voltage V_d.

$$V_o = AV_e + V_d$$

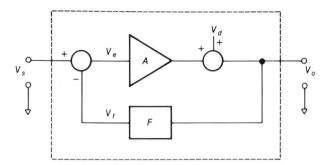

Figure 7-8. Negative feedback control system model.

The signal path for the feedback network is from the amplifier's output back to the input's summation point. The feedback network's input signal is V_o, and its output is V_f. They are related by

$$V_f = FV_o$$

If we solve the above equations for V_o, V_i, and V_d, we arrive at

$$V_o = \left(\frac{A}{1+AF}\right)V_i + \left(\frac{1}{1+AF}\right)V_d$$

Before we look at what this relationship means, let us look at, define, and examine its terms. *A is the voltage gain* (open loop) of the amplifier. *F is called the feedback factor*, and as we shall find out shortly, it is ideally equal to the reciprocal of the circuit's ideal voltage gain (V_o/V_i). The *product AF is called the loop gain*, or loop transmission, and is typically a very high number. A is typically in the hundreds of thousands region. F varies from 0.01 to 10. Thus, AF varies from about 1000 to 1,000,000. The quantity

$$1 + AF \cong AF \quad \text{for } AF \gg 1$$

Now, V_o is equal to some constant $A/(1+AF)$ times V_i, and summed with this term is another constant $1/(1+AF)$ times V_d. If the variable V_d has a magnitude of V_d volts, its contribution to the output voltage will be *attenuated* or reduced by $1/(1+AF)$. Hence, the overall circuit characteristics only change by an extremely small percentage ($\ll 1\%$) of the changes within the system itself. The amount will vary depending where the change occurs.

Neglecting V_d, the circuit's input and output voltage is related by

$$V_o = \left(\frac{A}{1+AF}\right)V_i$$

Using the argument,

$$1 + AF \cong AF \quad \text{for } AF \gg 1$$

then

$$V_o \cong \left(\frac{A}{AF}\right)V_i \cong \left(\frac{1}{F}\right)V_i$$

F is a dimensionless number and is a function, usually, of discrete components whose size, weight, performance, stability, and cost are easy to control. The ratio

$$\frac{V_o}{V_i} \cong \frac{1}{F}$$

defines the gain of the circuit (not the amplifier) and is called the closed-loop gain.

$$A_{CL} = \frac{V_o}{V_i} \cong \frac{1}{F}$$

The gain of the amplifier itself is called the open-loop gain and will now be identified as A_{OL}. *The key to having the circuit gain be a function of external (to the amplifier) components is that the voltage gain of the amplifier must be relatively high.*

In summary, using amplifiers in feedback configurations provides (1) a circuit gain relatively independent of the amplifier's gain, and (2) a desensitivity to changes within the circuit.

Amplifier feedback circuits also possess other advantages, but they will be identified as they arise. Operational amplifier applications abound and *most* of them use amplifiers in feedback configurations. Regulators, oscillators, and filters employ amplifiers using this circuit technique. The feedback circuit theory, relative to these applications, is discussed under the individual topics.

Feedback Factor

The *feedback factor F* is a special name for the *voltage gain of the discrete-component feedback circuit* associated with an amplifier in a feedback control system. The input signal to the feedback network is the output signal or voltage of the amplifier. The feedback network's output signal is summed with the circuit's input signal to produce a difference (negative feedback) or error voltage, which serves as the amplifier's input voltage.

The feedback factor derives its name from the fact that it feeds back a portion or factor of the output voltage. It is sometimes called the feedback network's transfer function, because this quantity transfers the input signal (when multiplied by F) to the output. Normally, it is nothing more than a resistor or impedance voltage divider. The components in this network are typically resistors and capacitors, but occasionally inductors and active devices are used.

7.4 OPERATIONAL AMPLIFIER APPLICATIONS

The Ideal Amplifier

The easiest way to analyze op amp circuits is to treat the amplifier as "ideal." *The ideal amplifier has infinite gain, infinite input resistance, infinite bandwidth, and zero output resistance.* Here the term *resistance* is used for simplicity. *Impedance* is the more proper term to use, since there are distributed reactive components associated with the input and output, but their role is small in the typical amplifier frequency range. These idealized properties allow us to say:

1. *The amplifier differential-input voltage is zero*. This is not an improper assumption, since the real amplifier is required to have a maximum of ±75 μV, and circuit voltages are in the high millivolts and volts regions. This statement means that both amplifier terminals must track each other. If a voltage is applied to the noninverting input, the amplifier will drive the inverting input, via the feedback components, to the same value. If one input is at a fixed voltage or ground, the other input will be very near that value.

2. *The current entering or leaving the amplifier inputs is zero*. This is not an improper assumption either, since the input currents are less than 100 nA, and circuit currents are in the high microamperes region or greater. In a real amplifier these input currents are called bias currents, and they are the base currents of high-gain transistors in the input differential amplifier stage.

Noninverting Amplifier Circuit

The *noninverting, unity gain amplifier circuit* of Figure 7-9 is the most basic of amplifier applications. A voltage source, V_s, is connected to the noninverting (+) amplifier input, and the output is connected back to the inverting (−) input. It is a negative feedback circuit configuration. The relationship between V_s and V_o will be derived three ways:

1. *Analysis using KVL:* Applying KVL to the loop indicated in Figure 7-9 yields

$$-V_s + V_{in} + V_o = 0$$

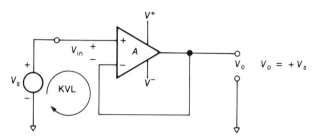

Figure 7-9. Noninverting amplifier circuit.

where V_{in} is the amplifier input voltage. The amplifier is not yet idealized. From the input-output voltage relationship of the amplifier itself.

$$AV_{in} = V_o$$

Eliminating V_{in} from the above equations and solving for V_o yields

$$V_o = \left(\frac{A}{1+A}\right) V_s$$

If $A \gg 1$

then $V_o \cong V_s$

One could also have arrived at this answer if V_{in} was approximated as zero because of the high gain.

2. *Analysis using the feedback model:* The feedback model fits the noninverting circuit (only). The model's input summation point is analogous to the amplifier's differential inputs. Figure 7-10 shows the analogy. From the feedback model,

$$V_o = \left(\frac{A}{1+AF}\right) V_s$$

The feedback factor F is 1. If A is high, then

$$AF \gg 1,$$

and $V_o \cong \left(\dfrac{A}{AF}\right) V_s \cong V_s$

3. *Analysis by inspection:* The noninverting terminal of the amplifier is connected to a voltage source and is at a potential of V_s volts. Since the difference in potential between the amplifier's inputs is *ideally* zero, the inverting input or the amplifier's output will be at a potential V_s. Hence,

$$V_o = V_s \text{ (ideally)}$$

Expressed another way, the amplifier output will drive the inverting input to a voltage of the same value as that at the noninverting input.

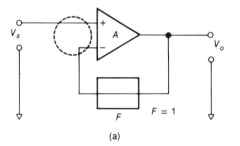
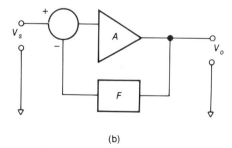

(a) (b)

Figure 7-10. Comparison of the model and the circuit: (a) noninverting amplifier circuit; (b) feedback model.

Many fundamental amplifier circuits can be analyzed by inspection. The key to finding the relationship between the input and output voltage is to apply the idealized amplifier's input properties. These properties, (1) the amplifier differential input voltage is zero, and (2) the current entering or leaving the amplifier inputs is zero, are very useful in developing an initial understanding of basic amplifier circuit principles.

Voltage Gain

The ratio of the circuit's output and input voltages defines the circuit's voltage gain. It is referred to as the *closed-loop gain* (A_{CL}). The voltage gain of the amplifier itself is called the open-loop gain (A_{OL}).

The closed-loop voltage gain for the noninverting, unity gain amplifier is

$$A_{CL} = \frac{V_O}{V_S} = 1$$

This relationship is for the ideal case. The closed-loop gain is exactly 1 if the open-loop gain of the amplifier is infinite.

The *exact* value of the closed-loop gain depends on the open-loop gain. For noninverting amplifier circuits, including the unity gain buffer amplifier, the two gains are related by

$$A_{CL} = \frac{A_{OL}}{1 + A_{OL}F}$$

The *feedback factor F* is a function of the components surrounding the amplifier and is different for each type of noninverting circuit configuration. For the unity gain buffer circuit, F equals 1.

As long as the open-loop gain is high, the closed-loop gain for the buffer circuit will remain near 1. For cases where the open-loop gain is *not* high, the closed-loop gain must be calculated using the above relationship. As an example, the amplifier's gain may not be considered high at high frequencies.

The + and − signs in the gain equation represent polarity or phase information. In noninverting amplifier circuits, the sign is positive (+) and the input and output ac signals are in phase. In inverting amplifier circuits, the minus (−) sign indicates a 180° phase difference between the input and output. For inverting circuits, the input and output dc voltages will be of the opposite polarity.

Input and Output Resistance

The *amplifier*, as a component, has a voltage gain, an input resistance, and an output resistance. The *circuit*, which contains an amplifier, also has a voltage gain and an input and output resistance.

The *amplifier's characteristics are designated* A_{OL}, r_i, and r_o. The *circuit's characteristics are designated* A_{CL}, r_{in}, and r_{out}. It can be shown that the input resistance of the circuit is

$$r_{in} \cong (AF)r_i$$

where r_i is the input resistance of the amplifier and AF is the loop gain.

The output resistance of the circuit is

$$r_{out} \cong \frac{r_o}{AF}$$

For typical values (see Figure 7-1) of $A = 100{,}000$, $r_i = 2$ MΩ, $r_o = 75$ Ω, and $F = 1$,

$$r_{in} \cong 2 \cdot 10^{12} \ \Omega$$

and $r_{out} = 7.5 \ \mu\Omega$

The feedback configuration makes the circuit input resistance significantly larger than that of the amplifier and the circuit output resistance significantly smaller. In practice, these high and low extremes cannot be achieved because of other considerations.

Noninverting Amplifier with Gain

Figure 7-11 shows the basic *buffer amplifier with gain capability added*. V_s, an ideal voltage source, is connected to the (+) terminal of the

OPERATIONAL AMPLIFIER APPLICATIONS

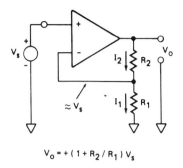

$V_o = +(1 + R_2/R_1) V_s$

Figure 7-11. Noninverting amplifier circuit with gain

amplifier. The output will drive the R_1–R_2 divider until the (−) terminal of the amplifier is at a value of V_s volts. The voltage drop across R_1 will be V_s, and the current through it will be V_s/R_1 amperes. This current can only come from the output. Thus, I_2 will be equal to I_1, and

$$V_o = V_s + I_2 R_2 = V_s + \left(\frac{V_s}{R_1}\right) R_2$$

$$= V_s \left(1 + \frac{R_2}{R_1}\right)$$

For this circuit, the closed-loop gain is

$$A_{CL} = 1 + \frac{R_2}{R_1}$$

The circuit can be used to (1) provide voltage gain, and (2) buffer or interface two circuits.

Example

If $V_s = -3$ V, $R_1 = 2$ kΩ, and $R_2 = 6$ kΩ, then the gain is

$$A_{CL} = \frac{V_o}{V_s} = 1 + \frac{6 \text{ k}\Omega}{2 \text{ k}\Omega} = 1 + 3 = 4$$

and the output voltage is

$$V_o = 4 V_s = 4(-3 \text{ V})$$
$$= -12 \text{ V}$$

Bandwidth

An equation or mathematical model describing the closed-loop or circuit bandwidth is available. However, *a Bode plot more clearly identifies the open-loop and closed-loop gain-frequency characteristics.*

The open-loop or amplifier frequency response is shown in Figure 7-12. Superimposed on this graph is the closed-loop response. The closed-loop circuit has a voltage gain, at dc, of

$$A_{CL} \text{ (dB)} = 20 \log \frac{V_o}{V_s} = 20 \log \left(1 + \frac{R_2}{R_1}\right)$$

The circuit gain remains constant until the closed-loop response intersects the open-loop response and then rolls off as the amplifier's gain decreases with frequency. The circuit's rolloff will also be −20 dB per decade. The closed-loop bandwidth can also be controlled through several circuit techniques using discrete resistors and capacitors.

Single-Supply Amplifier Circuit

Not all op amp circuits use two supplies. The *noninverting ac amplifier* in Figure 7-13 *uses a single supply* connected between the amplifier's V^+ and V^- pins. The negative terminal

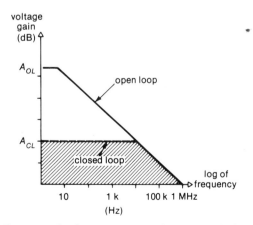

Figure 7-12. Open- and closed-loop bandwidths

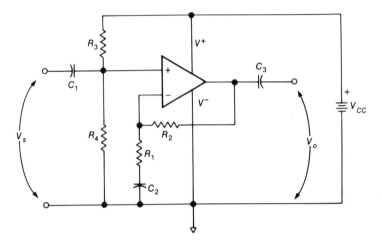

Figure 7-13. Single-supply amplifier circuit.

of the supply is a designated common or ground. For single-supply circuits, the amplifier inputs and outputs must be dc biased above ground to allow ac signals to swing above and below this value. The inputs will be at a dc voltage of

$$V_{NI}(\text{dc}) \cong V_I(\text{dc}) = \frac{R_4}{R_3 + R_4}(+V_{cc})$$

Capacitor C_2 is equivalent to an open at dc, and hence the amplifier is a dc follower. Thus,

$$V_o(\text{dc}) \cong V_{NI}(\text{dc}) = \frac{R_4}{R_3 + R_4}(+V_{cc})$$

C_1, C_2, and C_3 are normally large-valued capacitors, and within the frequency range of interest of the circuit, they are equivalent to short circuits. C_1 and C_2 are input and output coupling capacitors whose value of reactance is small compared to their associated circuit resistances. The ac voltage gain of the circuit is

$$A_v(\text{ac}) = \frac{R_1 + R_2}{R_1} \quad \text{for } X_c(C_2) \ll R_1$$

Example

If $+V_{cc} = 12$ V, $R_3 = R_4 = 10$ kΩ, $R_1 = R_2 = 5$ kΩ, and C_1, C_2, and C_3 are large, then

$$V_o(\text{dc}) = \frac{1}{2}(+12 \text{ V}) = +6 \text{ V}$$

and $A_{CL}(\text{ac}) = \dfrac{10 \text{ k}\Omega}{5 \text{ k}\Omega} = 2$

Inverting Amplifier Circuit

A second basic application of the operational amplifier is the *inverting amplifier circuit*. This circuit is characterized by a feedback network and an input network that can be any two-terminal circuit. The noninverting amplifier input is grounded. Two methods will be used to develop the input–output relationship.

Analysis Using a Circuit Model

The circuit model for the inverting amplifier is shown in Figure 7-14. Z_f and Z_i represent

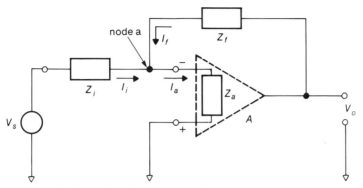

Figure 7-14. Inverting-amplifier circuit model.

the impedances of the feedback and input networks, and Z_a models the *nonideal* amplifier's input impedance. The real amplifier has a voltage gain of A. The inverting input of the amplifier is identified as node a, and the voltage associated with this terminal is V_a. If KCL is applied to this node,

$$I_f + I_i = I_a$$

These currents are associated with their respective impedances and are determined by using Ohm's law; that is,

$$I_f = \frac{V_o - V_a}{Z_f}$$

$$I_i = \frac{V_s - V_a}{Z_i}$$

and $I_a = \dfrac{V_a}{Z_a} = I_f + I_i$

V_a is also related to the output voltage by

$$V_o = -AV_a$$

The unknowns in the above equations are I_i, I_f, and V_a. Solving these equations simultaneously provides the desired input–output relationship.

$$\frac{V_o}{V_s} = -\frac{Z_f}{Z_i} \frac{1}{1 + (1/A)[1 + (Z_f/Z_i \| Z_a)]}$$

For $A \gg 1$, $1/A$ approaches zero, and

$$\frac{V_o}{V_s} \cong -\frac{Z_f}{Z_i}$$

This expression states that the output voltage is equal to the inverted (−) input times the ratio of the feedback network impedance divided by the input network impedance. Impedance is proper in describing the two networks because resistor–capacitor networks are frequently used. For Z_i and Z_f comprised of a single resistor each,

$$\frac{V_o}{V_s} \cong -\frac{R_f}{R_i}$$

Analysis by Inspection

The noninverting terminal of the amplifier in the inverting circuit of Figure 7-15 is grounded or at 0 V. The inverting terminal must track the noninverting, and hence it will be electrically zero volts (not ground). The voltage across R_1 will be V_s volts, and hence a current will flow through R_1 of value V_s/R_1 amperes. This current cannot enter the inverting input because of point 2 and must be sunk by the amplifier output via R_2. The output will go to a voltage of $-IR_2$, or

$$V_o = -IR_2 = \left(\frac{-V_s}{R_1}\right) R_2 = \left(\frac{-R_2}{R_1}\right) V_s$$

AMPLIFIER INTEGRATED CIRCUITS AND APPLICATIONS

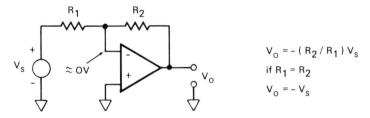

Figure 7-15. Inverting amplifier circuit.

The closed-loop gain of the circuit is determined by the ratio $-R_2/R_1$. Since the gain is the ratio of the output over the input,

$$A_{CL} = \text{closed-loop gain} = \frac{-R_2}{R_1}$$

It can be less than 1 ($R_1 > R_2$) or greater than 1 ($R_1 < R_2$). To achieve reasonable accuracy and stability, the closed-loop gain is usually made 500 times less than the amplifier open-loop gain. This restricts the circuit closed-loop gain to a practical maximum of 400 at dc.

The circuit can be used to

1. Provide voltage gain.
2. Generate a signal of the opposite voltage polarity.
3. Generate an output signal 180° out of phase with respect to its input.

Example

If $V_s = -3$ V, $R_1 = 10$ kΩ, and $R_2 = 5$ kΩ, then the gain is

$$A_{CL} = \frac{-R_2}{R_1} = \frac{-5\ \text{k}\Omega}{10\ \text{k}\Omega} = -0.5$$

and the output voltage is

$$V_o = -0.5 V_s = -0.5(-3\ \text{V}) = +1.5\ \text{V}$$

Input and Output Resistance and Bandwidth

The *input resistance* for the circuit is R_1. One end of R_1 is driven by the input source and the other end is at zero volts (not ground).

$$r_{in} \cong R_1$$

The *output resistance* for the circuit is low and is determined in the same manner as for the noninverting circuit; that is,

$$r_{out} \cong \frac{r_0}{AF}$$

The *closed-loop bandwidth* for an inverting amplifier is less than that for the equivalent noninverting circuit. The closed-loop and open-loop responses are illustrated in Figure 7-16.

The output resistances of both the noninverting and inverting circuits are very small. They are approximately equal to the output resistance of the amplifier ($\approx 100\ \Omega$) divided by the loop gain (AF). The exact value of the output resistance of the amplifier is generally not important because of the extremely low circuit output resistance.

The input resistance of the noninverting circuit is higher than that of the inverting version. The exact value of the amplifier's input resistance is generally not important because of the extremely high circuit input resistance (noninverting case).

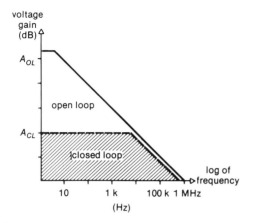

Figure 7-16. Open- and closed-loop bandwidths, inverting.

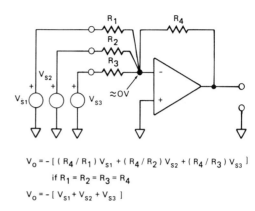

$V_o = -[(R_4/R_1)V_{s1} + (R_4/R_2)V_{s2} + (R_4/R_3)V_{s3}]$

if $R_1 = R_2 = R_3 = R_4$

$V_o = -[V_{s1} + V_{s2} + V_{s3}]$

Figure 7-17. Summing amplifier circuit.

The bandwidth of the noninverting circuit is slightly greater than the bandwidth of the equivalent inverting circuit. The key to making the circuit's characteristics approach the ideal is to ensure a high loop gain ($AF \gg 1$).

Summing Amplifier

The algebraic operation of summing means the addition of signed (+ and −) numbers. *The summing amplifier sums or algebraically adds signed (+ and −) voltages.* Figure 7-17 shows the summing amplifier circuit. Its behavior is an extension of the inverting circuit's. The circuit is characterized by a single feedback resistor and two or more input resistors. The voltages across R_1, R_2, and R_3 will be V_{s1}, V_{s2}, and V_{s3} because the inverting input of the amplifier is at zero volts. The inverting input for this configuration is called the *summing junction*, since it will sum the currents I_1, I_2, and I_3. These currents will flow through R_4 to the amplifier output. The amplifier output voltage

$V_o = -R_4(I_1 + I_2 + I_3) =$

$-\left[\left(\dfrac{V_{s1}}{R_1}\right)R_4 + \left(\dfrac{V_{s2}}{R_2}\right)R_4 + \left(\dfrac{V_{s3}}{R_3}\right)R_4\right]$

The gain for each input voltage can be different. However, for $R_1 = R_2 = R_3$, the closed-loop gain is

$A_{CL} = -\dfrac{R_4}{R_1}$

and if $R_1 = R_2 = R_3 = R_4$, then

$V_o = -(V_{s1} + V_{s2} + V_{s3})$

which more explicitly shows the summing operation. The circuit can be used to (1) sum signed voltages, and (2) amplify the sum of signed voltages.

Example

If $R_1 = R_2 = 2\ k\Omega$, $R_3 = 5\ k\Omega$, $R_4 = 10\ k\Omega$, $V_{s1} = -4$ V, $V_{s2} = +6$ V, and $V_{s3} = -2$ V, then the gains are

$$A_{CL1} = \frac{-10\ k\Omega}{2\ k\Omega} = -5$$

$$A_{CL2} = \frac{-10\ k\Omega}{2\ k\Omega} = -5$$

$$A_{CL3} = \frac{-10\ k\Omega}{5\ k\Omega} = -2$$

and the output voltage is

$$V_o = (-5)(-4\ V) + (-5)(+6\ V) + (-2)(-2\ V)$$
$$= -6\ V$$

Difference Amplifier

The *difference amplifier* (Figure 7-18) is the complement of the summing amplifier and *allows the subtraction of two signed voltages.* The voltage at the noninverting (*NI*) input is

$$V_{NI} = \frac{R_2}{R_1 + R_2} V_{s2}$$

Since the two amplifier inputs track each other with their net difference ideally equal to zero, the voltage at the inverting (*I*) input is

$$V_I \cong V_{NI} = \frac{R_2}{R_1 + R_2} V_{s2}$$

The current through R_1 is

$$I_1 = \frac{V_I - V_{s1}}{R_1} = \left(\frac{R_2}{R_1 + R_2}\frac{V_{s2}}{R_1} - \frac{V_{s1}}{R_1}\right)$$

Since no current enters the amplifier inputs,

$$I_1 \cong I_2$$

and the output voltage will be the sum of the voltage at the inverting input plus the product of I_2 and R_2.

$$V_o = \frac{R_2}{R_1 + R_2} V_{s2} + \left[\left(\frac{R_2}{R_1 + R_2}\right)\frac{V_{s2}}{R_1} - \frac{V_{s1}}{R_1}\right]R_2$$
$$= \frac{R_2}{R_1}(V_{s2} - V_{s1})$$

The closed-loop gain is

$$A_{CL} = \frac{R_2}{R_1}$$

and if $R_2 = R_1, A_{CL} = 1$, then

$$V_o = V_{s2} - V_{s1}$$

which more explicitly shows the difference operation. The circuit can be used to

1. Find the difference between two signed voltages.
2. Amplify the difference of two voltages.
3. Convert a potential difference to a voltage whose reference is ground.

Example

If $R_1 = R_2 = 2\ k\Omega$, $V_{s1} = -4\ V$, and $V_{s2} = -8\ V$, then the gain is

$$A_{CL} = 1$$

and the output voltage is

$$V_o = 1[-8\ V - (-4\ V)] = 1(-8\ V + 4\ V)$$
$$= -4\ V$$

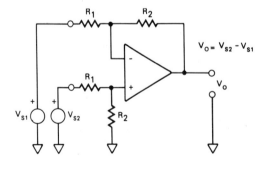

Figure 7-18. Difference amplifier circuit.

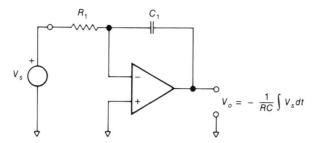

Figure 7-19. Integrating amplifier circuit.

Integrating Amplifier

The mathematical operation of integration is performed by the circuit of Figure 7-19 and is called an *integrator*. *In an integrator, the input voltage is converted to a higher order function at the output.* If the input is a dc voltage, the circuit will convert it to a voltage ramp at the output. A voltage ramp is a potential that increases linearly with time. If the input is a voltage ramp, the output will follow the time square (t^2) function. Mathematically,

k_1 is converted to $k_2 t$, or
$k_3 t$ is converted to $k_4 t^2$, or
$k_5 t^2$ is converted to $k_6 t^3$, and so on,

where k_x are constants (circuit) and t is time. The constant k_1 is analogous to a dc voltage; that is, its value is constant and independent of time. The function $k_2 t$ is analogous to a voltage ramp; that is, the value of voltage increases, in time, in a straight-line manner.

In the circuit of Figure 7-20, the input source V_s is a dc voltage. With V_s constant,

$$I_1 = \frac{V_s}{R_1}$$

When the switch S_1 is closed, this dc current will charge the capacitor C_1. For a positive V_s, the output voltage will go negative, since the capacitor terminal at the inverting (−) input is at zero volts and the capacitor is being charged plus (inverting input) to minus (output).

Prior to the closure of the switch, the amplifier output, ideally, is at zero volts. When the switch is closed ($t = 0$),

$$V_o = -\frac{V_s}{R_1 C_1} t$$

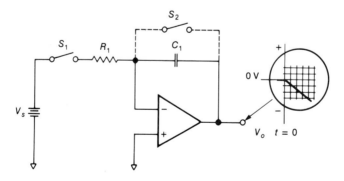

Figure 7-20. Integrator application: ramp generator.

where the slope of the line or ramp is

$$m = \frac{V_s}{R_1 C_1}$$

The switch S_2, in shunt with C_1, is used to discharge the capacitor. This particular amplifier circuit has many variations and is extensively used to generate signals with ramplike features. The circuit can be used to (1) integrate an input voltage signal, and (2) provide a voltage ramp for timing and sweep circuits.

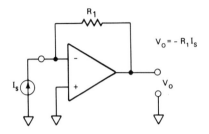

Figure 7-21. Current-to-voltage converter.

Example

If $R_1 = 1$ kΩ, $C_1 = 1$ μF, $V_s = 1$ V and the switch is closed at some time zero (t_0), then the output will go from zero volts (at t_0) in a negative direction with a slope

$$m = -\frac{1 \text{ V}}{(1 \text{ k}\Omega)(1 \text{ }\mu\text{F})} = \frac{-1 \text{ V}}{1 \text{ ms}}$$

The current charging the capacitor will be

$$I_1 = \frac{1 \text{ V}}{1 \text{ k}\Omega} = 1 \text{ mA}$$

At the end of 5 ms,

$$V_o = \left(\frac{-1 \text{ V}}{1 \text{ ms}}\right) 5 \text{ ms} = -5 \text{ V}$$

Current-to-Voltage Converter

The output variable for many devices, sensors, and transducers is current. This current can be converted to a voltage with the *current-to-voltage converter* circuit of Figure 7-21. The current from the source I_s flows through the feedback resistor R_1 because, ideally, the amplifier inputs can neither source nor sink any current. The amplifier output voltage is

$$V_o = -I_s R_1$$

The current source output is voltage clamped to zero volts because the inverting (−) input tracks the inverting (+) input, which is at ground potential. The circuit can be used to (1) convert current to voltage, and (2) amplify photocell, photodiode, and photovoltaic current signals.

Example

If $R_1 = 1$ MΩ and $I_s = 100$ nA, then

$$V_o = -(10^7 \text{ }\Omega)(10^{-7} \text{ A})$$
$$= -1 \text{ V}$$

7.5 ADVANCED AMPLIFIER APPLICATIONS

Operational amplifier circuits are used to

1. *Amplify* ac and dc voltages.
2. Perform mathematical *operations*.
3. *Convert* a nonvoltage input variable to an output voltage.
4. *Filter* ac voltages.
5. *Generate* voltage signals.
6. *Regulate* dc voltages.

The examples given in the preceding section were circuits that used only one amplifier. Single-amplifier circuits were used to illustrate amplification, conversion, and mathematical operations. These fundamental configurations are repeatedly used, intact or modified, to build more sophisticated linear circuits and systems. Other noteworthy one-amplifier circuits are

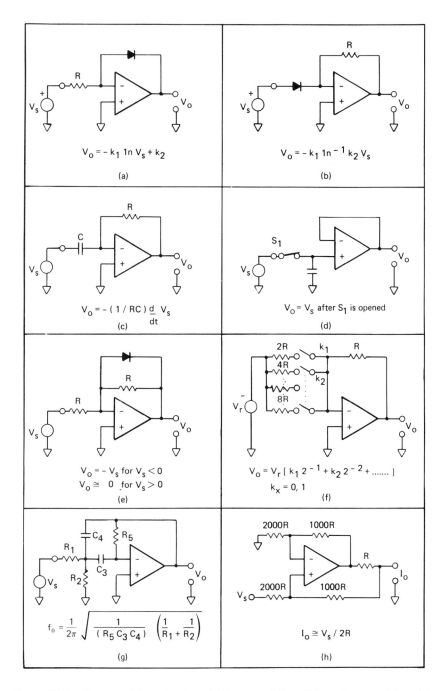

Figure 7-22. One-amplifier circuits: (a) log amplifier; (b) antilog amplifier; (c) differentiator; (d) sample-and-hold; (e) half-wave rectifier; (f) digital-to-analog converter; (g) bandpass filter; (h) current source circuit.

shown in Figure 7-22. The operational amplifier is *the* linear building block. Single- and multiple-amplifier circuits will be used to illustrate signal generation, filtering, and regulating under the topics of oscillation, filtering, and regulation.

Most amplifier applications use two or more amplifiers. The applications vary from highly specialized circuits designed to do a unique task, to circuits that have widespread application. This section looks at a few of the more common multiple-amplifier circuits.

Simulated Inductor

The reactance of an inductor for the ac steady-state condition is given as

$$\frac{V(t)}{I(t)} = X_L = 2\pi f L$$

The variables in this equation are reactance and frequency, and they are directly proportional to each other, since 2, π, and L are fixed constants. The equation can be rewritten as

$$X_L = kf$$

where k is a constant or number whose unit of measurement is the henry or ohm-second. Then it must be true that any two terminals that generate this relationship will also be an inductor. The device or circuit may not be an inductor in the traditional sense of wire wound on a bobbin, but it will be a simulated inductor. What a component is made of is immaterial, as long as we know its current and voltage relationship. *The simulated inductor circuit in Figure 7-23 is called a gyrator, and its function is to rotate or gyrate a capacitor into an inductor.* This circuit can be monolithically fabricated, and it provides the IC designer with inductors to use in circuit designs. The circuit also illustrates the use of modified basic amplifier configurations. The analysis of the circuit is not mathematically complex, but it is tedious. The step-by-step procedure is as follows:

Step 1: $V_{A1} = 2V_1$ (noninverting amplifier with a gain of 2)

Step 2: $V_{A2} = -2V_1 + 2V_2$ (use superposition for the V_2 and V_{A1} circuit voltages)

Step 3: $I_A = \dfrac{V_1 - V_{A1}}{R} = \dfrac{V_1 - 2V_1}{R} = \dfrac{-V_1}{R}$

$I_B = \dfrac{V_1 - V_2}{R}$

$I_C = \dfrac{V_{A2} - V_2}{R} = \dfrac{2V_2 - 2V_1 - V_2}{R}$

$= \dfrac{V_2 - 2V_1}{R}$

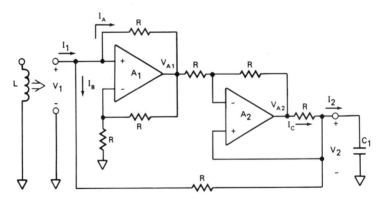

Figure 7-23. Simulated inductor circuit.

Step 4: $I_2 = I_B + I_C = \dfrac{V_1 - V_2}{R}$

$= \dfrac{V_2 - 2V_1}{R} = \dfrac{-V_1}{R}$

Step 5: $I_1 = I_A + I_B = \dfrac{-V_1}{R} + \dfrac{V_1 - V_2}{R} = \dfrac{-V_2}{R}$

The ratio of V_1/I_1 defines the impedance, or in the ac steady state the reactance, of the *input* terminals. From steps 4 and 5,

$$X_{in} = \dfrac{V_1}{I_1} = \dfrac{-RI_2}{-V_2/R} = \dfrac{R^2}{V_2/I_2}$$

However, the ratio of V_2/I_2 defines the impedance, or in the ac steady state the reactance, of the *output* terminals. Hence,

$$X_{in} = \dfrac{R^2}{X_{out}}$$

$$X_L = \dfrac{R^2}{X_C}$$

$$= \dfrac{R^2}{1/(2\pi RC)}$$

$$= 2\pi f C R^2 = 2\pi f (CR^2) = 2\pi f L_{eq} = kf$$

The equivalent value of the inductance, L_{eq}, is equal to CR^2. Its unit of measurement is ohm-seconds or henries (H) and can be verified by noting that the unit of measurement, or dimension, for RC is seconds.

$$R^2 C = R(RC) = [\text{ohm-seconds}] = [\text{henries}]$$

All discrete filter circuits that use a grounded inductor can now be implemented with monolithic techniques. The restriction of having one inductor terminal grounded is a disadvantage; however, other circuits are available that allow the simulated inductor to "float." This amplifier circuit, in its simplest sense, simulates the two-terminal passive device called an inductor.

Example

If $C = 0.1 \ \mu F$ and $R = 10 \ k\Omega$, then

$L_{eq} = CR^2 = (10^{-7} \text{ F})(10^4 \ \Omega)^2$

$L_{eq} = 10 \text{ H}$

Current Source

There are two electronic energy sources, voltage and current. The *ideal* voltage (dc) source maintains a constant potential between its terminals independent of the current that is being drawn from them. *The ideal current (dc) source maintains a constant current between its terminals (through a load) independent of the potential difference developed across them.* The *real* voltage and current sources will maintain a constant voltage and current within certain restraints determined by the circuits that implement them. In practice, constant voltage sources greatly outnumber the constant

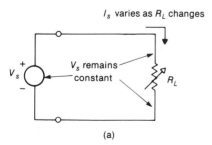
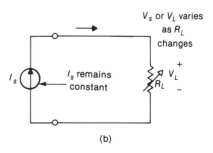

Figure 7-24. Electronic energy sources: (a) constant voltage source; (b) constant current source.

current sources; however, current sources in certain applications play an important role. Figure 7-24 highlights the differences and similarities of the two sources.

The current source in Figure 7-25 illustrates the use of three basic one-amplifier circuits. A_1 is a summing amplifier, A_2 a unity-gain buffer, and A_3 a unity-gain inverting amplifier. This current source is also called a voltage-controlled current source (VCCS) because the value of the current can be controlled, or programmed, by the voltage source, V_c.

The objective in the design of a current source is to make the load current constant and the load voltage independent of what establishes the constant current. A constant current is developed if a constant voltage is established across a fixed resistor. This is the key point in the VCCS circuit.

Assume a current of value I_s into the load R_L. This current times R_L produces the potential V_L. The current I_s must be sourced, through R, by the output of the amplifier A_1. The current at the input (+) of A_2 is ideally zero. The input and hence the output of the buffer A_2 is V_L. A_2's high input impedance prevents any shunting of the load current that is sourced by A_1. A_3 is an inverting amplifier with a gain of -1. Its output is $-V_L$. A_1 is a summing amplifier; that is, it sums the source voltage, $-V_c$, and the voltage, $-V_L$. A_1's output is

$$V_{oA1} = V_L + V_c$$

The potential difference across R is

$$V_R = (V_L + V_c) - (V_L)$$
$$= V_c$$

and the current is

$$I_s = \frac{V_R}{R} = \frac{V_c}{R}$$

If V_c is fixed, then I_s will be a constant current determined by V_c/R. It is independent of the voltage developed at the load. The maximum load voltage is determined by the maximum output voltage of A_1, that is,

$$V_{oA1}(\text{max.}) \geq V_L + V_c$$

and the maximum input voltage of A_2.

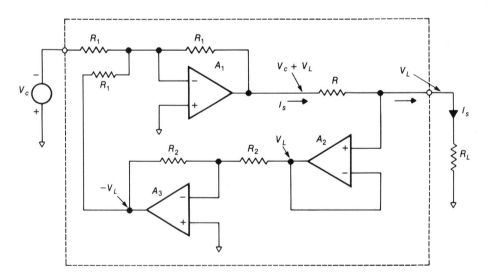

Figure 7-25. Operational amplifier constant-current source.

DATA SHEET PARAMETERS

The current source can be a programmable, computer-controlled source if V_c is a digital-to-analog voltage converter (D/A). If V_c is an ac voltage source, the output current will be a constant ac current. The polarity of V_c establishes the direction of I_s.

7.6 DATA SHEET PARAMETERS

The real amplifier deviates from the ideal because of the finite values associated with the primary characteristics of voltage gain, input and output resistance, and bandwidth. These finite values introduce error terms in the input–output relationship. *Data sheet parameters also introduce amplifier error terms.* They are specified for a given set of test conditions and can be found in the manufacturer's list of specifications. *These parameters are the result of the manufacturing process and the circuit devices of the amplifier.* A composite data sheet for the 201A IC operational amplifier is shown in Figure 7-26. The following is a description and definition of the key IC amplifier parameter terms.

Input Offset Voltage (V_{OS})

The difference of potential between the amplifier inputs with the output at zero volts. Physically, this potential difference represents a mismatch, or difference, in the base–emitter voltages of the bipolar transistors in the input stage's differential amplifier (Figure 7-27a). For this stage,

$$V_{OS} = \Delta V_{BE} = V_{BE1} - V_{BE2}$$

The typical range of values for V_{OS} is

$$0.5 \text{ mv} \lesssim V_{OS} \lesssim 15 \text{ mV}$$

V_{OS} can be modeled as a voltage source (low value) in series with either of the amplifier inputs. It can be either polarity.

In an actual circuit, V_{OS} is reflected to the output and sums, as an error, with the reflected input signal. In the inverting amplifier circuit in Figure 7-27b, V_O will be a function of the signal source V_S and the error voltage V_{OS}. The output voltage is determined using superposition.

$$V_O = -\left(\frac{R_2}{R_1}\right)V_S + \left(1 + \frac{R_2}{R_1}\right)V_{OS}$$

The offset error voltage is reflected to the output with a magnitude of $(1 + R_2/R_1)V_{OS}$. To minimize this error term, the magnitude of the signal source should be kept significantly larger than V_{OS}, or V_{OS} can be externally adjusted or nulled out to zero volts using an external network. Manufacturer's data sheets specify the proper offset null circuit for each type of amplifier.

Input Bias Current (I_B)

The magnitude of the average of the dc biasing currents at the inputs of an operational amplifier at zero volts output. Mathematically, I_B is defined as follows:

$$I_B = \left|\frac{I_B^+ + I_B^-}{2}\right|$$

The range of values (typical) of I_B is

$$100 \text{ pA} \lesssim I_B \lesssim 1.5 \text{ μA}$$

I_B^+ and I_B^- represent the currents of the noninverting (+) and inverting (−) inputs. Physically, I_B represents the average of the base currents (Figure 7-28a) of the differential amplifier's bipolar transistors. For monolithic op amps with *npn* transistors in the differential amplifier stage, the direction of the current will be *into* the amplifier.

The bias currents are modeled by two current sources. For *npn* differential amplifier transistors, the current sources will *sink* current.

In an actual circuit, I_B is reflected to the output and sums, as an error, with the reflected

LM101A/LM201A

absolute maximum ratings

Supply Voltage	±22V
Power Dissipation (Note 1)	500 mW
Differential Input Voltage	±30V
Input Voltage (Note 2)	±15V
Output Short Circuit Duration (Note 3)	Indefinite
Operating Temperature Range LM101A	−55°C to 125°C
LM201A	−25°C to 85°C
Storage Temperature Range	−65°C to 150°C
Lead Temperature (Soldering, 10 sec)	300°C

electrical characteristics (Note 4)

PARAMETER	CONDITIONS	MIN	TYP	MAX	UNITS
Input Offset Voltage	$T_A = 25°C$, $R_S \leq 50 k\Omega$		0.7	2.0	mV
Input Offset Current	$T_A = 25°C$		1.5	10	nA
Input Bias Current	$T_A = 25°C$		30	75	nA
Input Resistance	$T_A = 25°C$	1.5	4		MΩ
Supply Current	$T_A = 25°C$, $V_S = \pm 20V$		1.8	3.0	mA
Large Signal Voltage Gain	$T_A = 25°C$, $V_S = \pm 15V$ $V_{OUT} = \pm 10V$, $R_L \geq 2 k\Omega$	50	160		V/mV
Input Offset Voltage	$R_S \leq 50 k\Omega$			3.0	mV
Average Temperature Coefficient of Input Offset Voltage	$25°C \leq T_A \leq 125°C$ $-55°C \leq T_A \leq 25°C$		3.0	15 20	μV/°C
Average Temperature Coefficient of Input Offset Current			0.01 0.02	0.1 0.2	nA/°C nA/°C
Input Bias Current				100	nA
Supply Current	$T_A = +125°C$, $V_S = \pm 20V$		1.2	2.5	mA
Large Signal Voltage Gain	$V_S = \pm 15V$, $V_{OUT} = \pm 10V$ $R_L \geq 2 k\Omega$	25			V/mV
Output Voltage Swing	$V_S = \pm 15V$, $R_L = 10 k\Omega$ $R_L = 2 k\Omega$	±12 ±10	±14 ±13		V V
Input Voltage Range	$V_S = \pm 20V$	±15			V

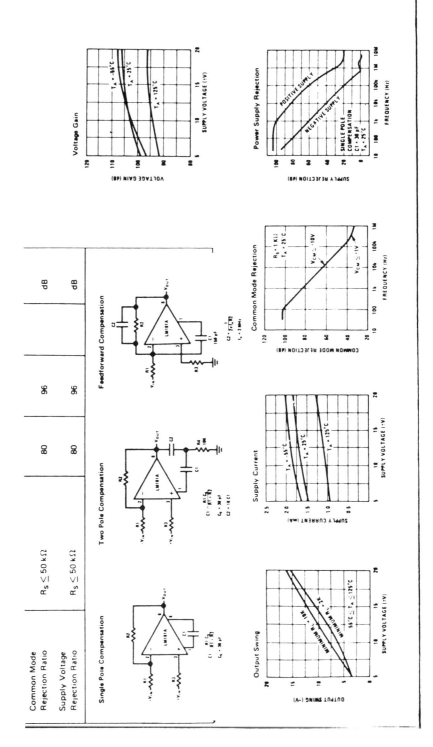

Figure 7-26.

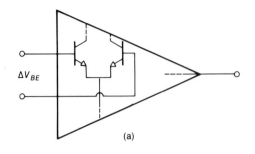

(a)

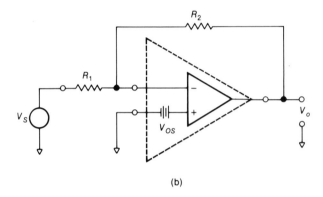

(b)

Figure 7-27. Input offset voltage (V_{OS}): (a) ΔV_{BE} of the differential amplifier stage; (b) circuit model in an inverting amplifier circuit.

input signal. In the inverting amplifier circuit in Figure 7-28b, V_O will be a function of the signal source V_S and the error currents I_B^+ and I_B^-. The output voltage is found using superposition.

$$V_O = -\frac{R_2}{R_1} V_S + I_B^- R_2 - I_B^+ R_3 \left(1 + \frac{R_2}{R_1}\right)$$

If R_3 is made equal to the parallel combination of R_1 and R_2,

$$R_3 = R_1 \| R_2 = \frac{R_1 R_2}{R_1 + R_2}$$

then

$$V_O = -\left(\frac{R_2}{R_1}\right) V_S + (I_B^- - I_B^+) R_2$$

The *difference* in the bias currents (called offset current) is a smaller error than the *average* of the bias currents. The purpose of a resistor in series with the noninverting amplifier input in an inverting amplifier circuit is to reduce the error from the bias current (I_B) to the offset current (I_{OS}).

Input Offset Current (I_{OS})

The magnitude of the difference between the amplifier input currents of an amplifier at zero volts output.

$$I_{OS} = |I_B^+ - I_B^-|$$

Normally, I_{OS} is much smaller than I_B. The typical range of values for I_{OS} is

$$10 \text{ pA} \lesssim I_{OS} \lesssim 0.5 \text{ μA}$$

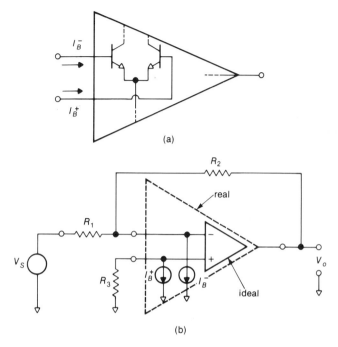

Figure 7-28. Input bias current: (a) base currents of the differential amplifier stage; (b) circuit representation and an inverting amplifier circuit.

Common-Mode Rejection Ratio (CMRR)

The ratio of the input voltage range over the peak-to-peak change in the offset voltage over this range.

$$\text{CMRR (dB)} = 20 \log \frac{A_{vd}}{A_{vc}}$$

Common-mode rejection ratio is mathematically defined as 20 times the log of the differential voltage gain to the common-mode voltage gain. The circuit in Figure 7-29 will be used to illustrate the meaning of CMRR and common mode. In this circuit, V_1 is a dc voltage. In series with *each* amplifier input is a dc voltage of V_1 volts. In the inverting input, v_2 is summed with V_1, and in the noninverting input the ac signal v_3 is summed with V_1. The com-

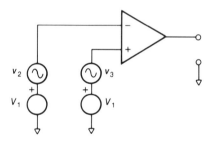

Figure 7-29. Common- and differential-mode voltages.

binations of $V_1 - v_2$ and $V_1 - v_3$ represent ac signals riding on dc voltages. The difference in potential between the amplifier inputs is

$$v_{\text{diff}} = v_3 - v_2$$

The differential input voltage is amplified by the differential voltage gain A_{vd}.

The common-mode voltage, or the voltage common to both input terminals, is V_1. There is also a common-mode voltage gain (very small) that will amplify the common-mode voltage and reflect it to the output. The common-mode voltage gain A_{vc} introduces an error into the circuit and is specified through the common-mode rejection ratio (CMRR) parameter. This error can also be reflected back to the input terminals as a change in the offset voltage. The typical range of values for CMRR is

$$60 \text{ dB} \lesssim \text{CMRR} \lesssim 120 \text{ dB}$$

Output Voltage Swing (V_O Swing)

The peak output voltage swing, referred to zero, that can be obtained without clipping. This parameter is measured for a given load resistor and defines the maximum usable output voltage. For $^{\pm}V_{CC} = 15$ V and $R_L = 10$ kΩ,

$$12 \text{ V} \lesssim V_O \text{ (swing)} \lesssim 14 \text{ V}$$

Supply Current (I_S)

The current required from the power supply to operate the amplifier with no load and the output at zero volts. The typical range of values for I_S is

$$0.2 \text{ mA} \lesssim I_S \lesssim 6 \text{ mA}$$

Measuring Amplifier Parameters

Operational amplifier parameters are not measured directly. Rather, *they are measured by connecting the amplifier under test (AUT) in series with a buffer amplifier (A_2) in a closed-loop configuration* (Figure 7-30a) that has a certain gain. The AUT parameters, multiplied by the loop gain, are reflected to the output of the buffer amplifier. The closed-loop gain of the circuit is

$$A_v = 1 + \frac{R_2}{R_1} = 1 + \frac{99.9 \text{ k}\Omega}{100 \text{ }\Omega} = 1000$$

A signal source is not present, but the high closed-loop gain will reflect the parameters to the circuit's output (A_2). The output of A_2 is equal to

$$V_O(A_2) = 1000 \, V_{in}$$

If the switches S_1 and S_2 are closed, the resistors R_S will be shorted and V_{in} will be equal to V_{OS}. Thus,

$$V_O = 1000 V_{in} = 1000 V_{OS}$$

The resistors in series with the input terminals convert the input bias currents to voltages that are reflected to the output along with the offset voltage. Normally, V_{OS} is the first parameter measured, and its value is stored in memory and subtracted from subsequent measurements. If both R_S resistors are *not* shorted, then the voltage V_{in} is

$$V_{in} = I_B^- R_S + V_{OS} - I_B^+ R_S$$

and

$$V_O = 1000 V_{OS} + 1000(I_B^- - I_B^+) R_S$$

The output of the AUT is held to near zero volts (large signal) by the inputs of A_2. The inverting (−) input of A_2 is grounded, and the noninverting (+) input will be a 0 V or virtual ground. Since A_2's input current is very small, the large-signal output of the AUT is 0 V.

The supply terminals of the amplifier under test (AUT) are connected to programmable voltage sources. The equivalent of an ammeter in series with one of the V_{CC} supplies can be used to measure the I_S or the supply current parameter.

Other dc parameters can be measured by adding R_4 and a programmable voltage source V_p. The R_3, R_4, and V_p circuit will cause the output of the AUT to go to minus the value of V_p.

$$V_o (\text{AUT}) = -V_p$$

With the addition of the above circuit, common-mode, gain, and output swing parameters

PRACTICAL CONSIDERATIONS IN AN AMPLIFIER CIRCUIT

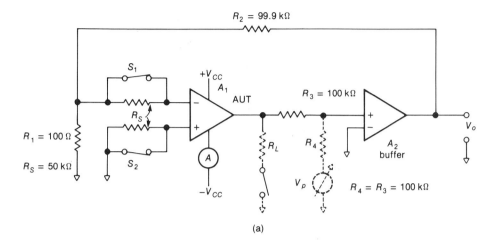

Parameter	S_1	S_2	Measure
V_{OS}	close	close	$V_o = 1000\, V_{OS}$
I_B^-	open	close	$V_o = 1000(V_{OS} + I_B^- \cdot 50\,k\Omega)$
I_B^+	close	open	$V_o = 1000(V_{OS} - I_B^+ \cdot 50\,k\Omega)$
I_{OS}	open	open	$V_o = 1000(V_{OS} + I_{OS} \cdot 50\,k\Omega)$
I_S	close	close	Ammeter output

(b)

Figure 7-30. Measuring op-amp parameters: (a) test circuit; (b) test circuit conditions.

can be measured. Various load resistors can be connected to the output of the AUT. They present a load if their series relay is closed.

The test conditions to measure the basic input parameters are presented in Figure 7-30b. This test circuit is used commercially to measure op amp parameters and, with a few modifications, it can measure all the dc parameters, slewing rate, and the gain–bandwidth product (GBW).

7.7 PRACTICAL CONSIDERATIONS IN AN AMPLIFIER CIRCUIT

An *actual schematic of an inverting amplifier circuit* may be difficult to recognize. The additional components tend to obscure the fundamental nature of the circuit, even though the extra components do not directly enter into the input–output relationship. The circuit in Figure 7-31a illustrates the use of additional components to compensate for data sheet parameter errors and to maintain frequency stability. The additional components in the circuit of Figure 7-31b provide protection for the amplifier against supply- and output-related problems. Most practical amplifier circuits will utilize some, but not necessarily all, of the protective devices.

The function of the components R_3, R_5, and R_6 is to reduce the data sheet parameter errors, and components C_f, C_b, R_b, and R_4 are primarily used to prevent circuit oscillations. The addition of C_f in parallel with R_2 forms a low-pass filter. This combination reduces the closed-

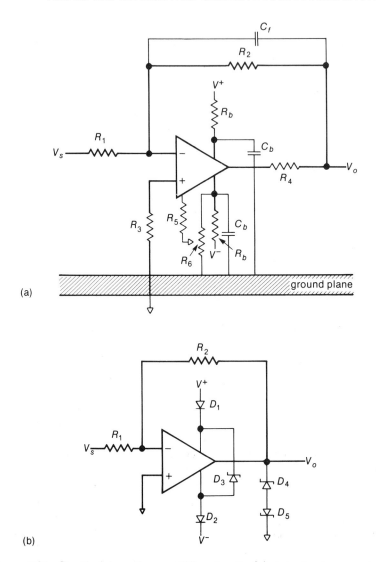

Figure 7-31. Practical inverting amplifier circuit: (a) data-sheet parameters and stability considerations; (b) supply and output problem considerations.

loop circuit bandwidth and increases the phase margin. If the application does not require the amplifier's full bandwidth, reducing the bandwidth will decrease the amplifier's phase shift and reduce the chances of circuit oscillation. The corner or cutoff frequency for the feedback RC network is

$$f_c = \frac{1}{2\pi R_2 C_f}$$

The components R_b and C_b form supply decoupling networks. R_b is usually small (<10 Ω, because of current considerations) and C_b large (>5 μF), and together they form a very low pass filter. The networks should be physically close to the amplifier's V^+ and V^- terminals to reduce supply transients and to present a low impedance.

R_4 is used in low-frequency applications where the amplifier may have to drive reactive (capacitive or inductive) loads. This resistor (<100 Ω) reduces the peak currents that the amplifier has to supply under transient conditions.

The proper handling of grounds in linear systems is an art and a science. The drawing depicts a heavy trace (for pc boards) for the ground to minimize the resistance, capacitance, and inductance associated with the circuit's reference. In the layout of the printed circuit, care should be exercised to ensure that the leads of the input and output networks do not cross or parallel each other. Otherwise, the distributed capacitance between the leads will provide unwanted feedback paths.

Resistors R_5 and R_6 are used to reduce the amplifier's initial offset voltage (V_{os}) error. The specific values and circuit arrangement are specified for each type of op amp in the manufacturer's data sheets. R_3 reduces the error due to the amplifier's input currents from I_b to I_{os}. It is equal to the parallel combination of R_1 and R_2; that is,

$$R_3 = R_1 \| R_2$$

The diodes D_1 through D_5 in Figure 7-31b protect the amplifier IC against supply and loading problems. D_1 and D_2 guard against the accidental application of the wrong polarity of the supplies. The zener voltage of D_3 ensures that the supplies' differential voltage does not exceed a certain limit. The back-to-back zener diodes, D_4 and D_5, limit the maximum output voltage, in both directions, that is applied to the load.

7.8 INVERTING AMPLIFIER CIRCUIT DESIGN

The circuit of Figure 7-32 is an *inverting amplifier with a gain of 5*. The amplifier is designed to meet the following specifications:

Input voltage range: $0.01 V_{pk} \leqslant V_s \leqslant 2 V_{pk}$

Input frequency range: $0 \text{ Hz} \leqslant f_s \leqslant 10 \text{ kHz}$

Input–output relationship: $V_o = -5 V_s$

Accuracy: 0.5% ± 1 mV

Input resistance: $\geqslant 1.5$ kΩ

Output resistance: $\leqslant 100$ Ω

To ensure that the open-loop gain is much greater than the closed-loop gain, the frequency compensation capacitor for the 301A is specified at 3 pF. This is a small-signal consideration and is determined by examining the open-loop frequency response graph in Figure 7-26. Worst cast for this circuit is at 10 kHz, where the 3 pF capacitor ensures a 46 dB separation between the open- and closed-loop gains.

The worst-case consideration for large-signal conditions is when the output must swing $10 V_{pk}$ at 10 kHz. For this case,

$$S_R = 2\pi f_{max} V_{pk}$$
$$= (6.28)(10^4 \text{ Hz})(10 V_{pk})$$
$$= .628 \text{ V}/\mu\text{s}$$

Hence, an amplifier is chosen whose slew rate is greater than 0.63 V/μs to minimize the distortion.

The 3 pF compensation capacitor and the 10 MΩ resistor are the manufacturer's standard frequency compensation circuit for the 301A. It will allow a nominal gain–bandwidth product near 10 MHz.

The 100 Ω resistor in series with the amplifier output isolates the amplifier from large capacitive loads. It forms a high-frequency lag network along with the 0.001 μF feedback capacitor.

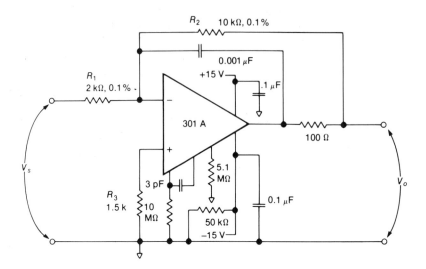

Figure 7-32. Detailed practical design of inverting amplifier.

However, its effect is not felt because of the feedback filter circuit comprised of the precision 10 kΩ resistor and the 0.001 μF capacitor. f_c for the feedback circuit is about 16 kHz, which is above the maximum input signal frequency of 10 kHz. The primary function of this filter circuit is to increase the circuit's phase margin. It reduces the circuit's phase shift and minimizes the chances for instability.

R_1 and R_2 are related by the circuit gain requirement; that is, $R_2 = 5R_1$. The lower limit of the input resistor's (R_1) value is determined by the specified input resistance and amplifier bias current considerations. The signal current of R_1 must be greater than that of the amplifier's bias current (approximately 100 nA). The maximum output current (20 mA) and the load current requirements define R_1 and R_2's upper limit. A range of values is possible for R_1 and R_2.

The 50 kΩ potentiometer is used to null out the input offset voltage error (V_{os}). For V_s = 0 V, it is adjusted until V_o = 0 V. The 1.5 kΩ resistor (R_3) in series with the noninverting input reduces the bias current error to that of the offset current. For inverting circuits,

$R_3 = R_1 \| R_2$

The output resistance of the circuit is guaranteed to be significantly less than 100 Ω (see Appendix B).

7.9 CURRENT-MODE OPERATIONAL AMPLIFIER

Fundamentals

The conventional operational amplifier is a voltage-mode device. The input and output variables are voltage, and they are related by the amplifier's gain, A. Another class of monolithic amplifiers exists called *current-mode operational amplifiers. The output variable is voltage, but the input variable is current.* In a conventional amplifier, the input differential amplifier senses the difference of two voltages. In the current-mode amplifier, the input current-mirror circuit senses the difference of two currents. The name *Norton amplifier* is used to indicate this new type of operation. It was specifically designed to operate from a single power supply voltage, a requirement of many industrial control systems, including automobiles.

The *uniqueness* of this amplifier is in the form of its input circuit, which is called a *current mirror*. This circuit (Figure 7-33) operates in the current mode, since input currents are compared or differenced. It can be thought of as a Norton differential amplifier. The voltages at the input terminals are fixed at one diode drop above ground. The current-mode amplifier is used in circuits similar to the op amp, where external components are connected to provide feedback, or the components interconnect the amplifier to signal sources and dc voltages. Signal voltages are converted to currents by using input resistors.

The input circuit or current mirror senses current. A current I_{NI} will flow into the noninverting (+) terminal of the current-mirror circuit of Figure 7-33. This current will turn D_1 on, and the (+) terminal will go to approximately +0.6 V. This voltage is impressed across the base to the emitter of Q_1 and will cause a collector current of the same value as I_{NI}. This occurs because the *pn* junctions associated with the diode and *B-E* of the transistor are monolithic. They are made at the same time, in the same die area, and under the same conditions; hence, they will exhibit the same voltage-versus-current characteristics. The collector current of Q_1 can be sourced only from the inverting (-) terminal. The amplifier output, through the feedback components, forces the condition of I_I equals I_{NI}. The base currents of Q_1 and Q_2 are much smaller than the I_{NI} and I_I currents and represent error currents. However, the true signal current of the amplifier is the base current of Q_2, which is relatively small. With Q_2 on, the voltage at the inverting (-) input will equal its base-emitter voltage and is labeled V_I. The amplifier output and input are related through the voltage gain from V_I to V_O.

A new *symbol* is used for the current-mode amplifier. The current arrow (Figure 7-34a) on the (+) input is used to indicate that this functions as the current input. The current source symbol between the inputs implies the current mode of operation and that current is removed from the (-) input. Included in this figure is the

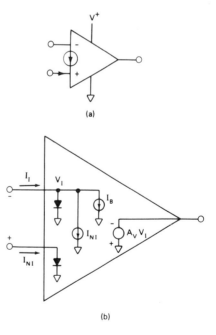

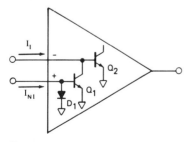

Figure 7-33. Current mirror circuit.

Figure 7-34. Current-mode amplifier: (a) schematic symbol; (b) circuit model.

first-order circuit model. In the model, the two diodes fix the voltage of the inputs to approximately 0.6 V. The current I_{NI}, determined by external sources and resistors, is mirrored by the current source connected to the (−) input. Its value is established by the current in the noninverting input. I_B represents the base current of Q_2. The amplifier output voltage equals the voltage gain A_V times the inverting input voltage V_I and is symbolized by a dependent voltage source.

Integrated Current-Mode Amplifier

The *schematic for a single amplifier of the quad 3900* is shown in Figure 7-35. Because of the circuit simplicity, four of these amplifiers are usually fabricated on a single chip. One common biasing circuit is used for all the individual amplifiers. All the voltage gain is provided by the common-emitter circuit of Q_2. This stage achieves a large voltage gain (70 dB) through the use of an active load implemented by Q_5, which is a current source, and the gain of Q_4. Q_4 also functions as an interstage buffer to reduce the loading at the high-impedance collector of Q_2. Transistor Q_7 is an output emitter follower and serves as a driver for the load currents. The collector–base junction of Q_4 becomes forward biased under a large negative output voltage swing condition. This transistor converts to a vertical *pnp* during this mode of operation, which causes the output to change from class A bias to class B. This allows the amplifier to sink more current than that provided by the Q_6 current source. The transistor Q_3 provides the class B action, which exists under large-signal operating conditions.

The *complete schematic* (Figure 7-36) of the current-mode 3900 amplifier includes the four amplifiers, input clamps, turn-on circuitry, and current source biasing. Q_1, Q_2, Q_4, and Q_5 are the *pnp*, 200 µA current sources. These multiple-collector devices have one collector tied to the base to fix its current gain. The base current for these transistors is predetermined through Q_{28}. The base and emitter voltages of this transistor are fixed; hence, the emitter and collector currents are established. One-fourth of this current is distributed to each *pnp* current source base. The emitter current of Q_{28} voltage biases, through Q_{33}, the 1.3 mA *npn* current sources of Q_{31}, Q_{32}, Q_{34}, and Q_{35}. Components Q_{20}, Q_{30}, and D_6 "start up" the biasing circuits. The input clamps are provided by the multiple-emitter transistor Q_{21}. These clamps protect the amplifier inputs against damage from negative voltages.

Current-Mode Amplifier Applications

AC Amplifier: Inverting

The Norton amplifier readily lends itself to use as an *ac amplifier*, because the output can be biased to any desired dc level within the range of the output voltage swing. The (+) input terminal current (Figure 7-37) is

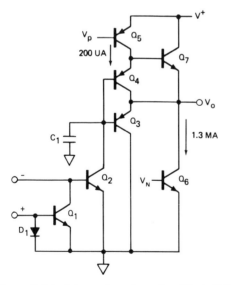

Figure 7-35. Schematic of a simplified 3900 amplifier.

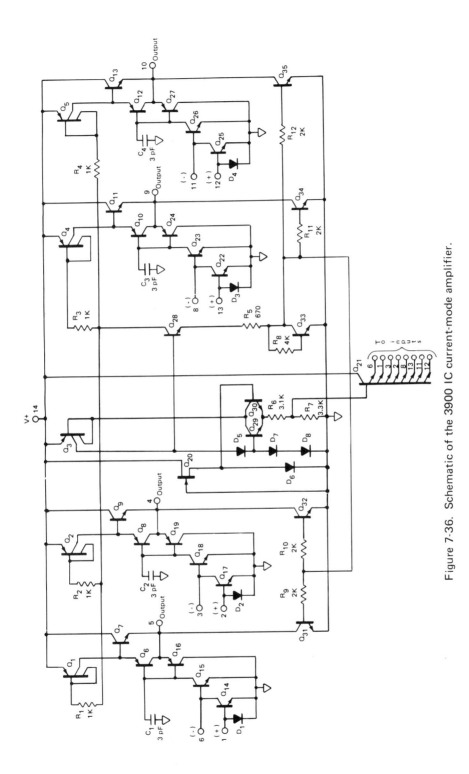

Figure 7-36. Schematic of the 3900 IC current-mode amplifier.

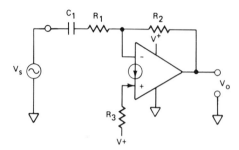

Figure 7-37. AC amplifier, inverting.

$$A_{CL} \text{ (ac)} = -\frac{12 \text{ k}\Omega}{3 \text{ k}\Omega}$$
$$= -4$$

For a 1 V peak ac input voltage, the output voltage will swing 4 V peak from +2 V to +10 V.

The negative sign in the gain equation indicates a phase reversal at the output. The value of C_1 is such that the capacitive reactance is extremely small compared to R_1 in the frequency range of the input signal.

$$I_{NI} = \frac{V^+ - 0.60 \text{ V}}{R_3}$$

This current is mirrored at the (−) terminal and must be sourced by the amplifier output. The coupling capacitor blocks the dc signal current. The output dc voltage is

$$V_O \text{ (dc)} = +0.6 \text{ V} + I_{NI} R_2$$
$$= 0.6 \text{ V} + \frac{V^+ - 0.6 \text{ V}}{R_3} R_2$$

If $2R_2 = R_3$, then

$$V_O \text{ (dc)} = 0.6 \text{ V} + \frac{V^+}{2} - 0.3 \text{ V}$$
$$\cong \frac{V^+}{2}, \text{ since } V^+ \gg 0.3 \text{ V}$$

Thus, the amplifier output is biased to a dc voltage between ground and the positive supply and will allow maximum ac signal transfer.

The (−) terminal is the equivalent of an ac common, or ground, and thus the closed-loop gain is determined in the same manner as for an operational amplifier.

$$A_{VCL} \text{ (ac)} = \frac{-R_2}{R_1}$$

Example

If $V^+ = +12$ V, $R_1 = 3$ kΩ, $R_2 = 12$ kΩ, and $R_3 = 24$ kΩ, then $V_O \text{ (dc)} \cong +6$ V and the gain

AC Amplifier: Noninverting

The amplifier in Figure 7-38 shows *both a noninverting ac amplifier* and a second method for dc biasing. By making $R_2 = R_3$, V_O (dc) will be equal to the reference voltage that is applied to the resistor R_3. By making $R_4 = R_5$ and much smaller in value than R_2 and R_3, the reference voltage will be equal to $V^+/2$. The filtered $V^+/2$ reference shown can also be used for other amplifiers (remember, Norton amplifiers come four to a package). The (−) terminal current is

$$I_I = I_{NI} \cong \frac{V_{REF} - 0.6 \text{ V}}{R_3}$$

and

$$V_O \text{ (dc)} = 0.6 \text{ V} + I_{NI} \cdot R_2$$
$$= 0.6 \text{ V} + \frac{(V_{REF} - 0.6 \text{ V})}{R_3} R_2$$

For $R_2 = R_3$

$$V_O \text{ (dc)} = V_{REF} = \frac{V^+}{2}$$

V_{REF} is established by the lower-valued resistor divider of R_4 and R_5, and can be any value.

The ac current through R_1 is summed with the I_{NI} dc current and reflected to the output through R_2. Hence,

$$A_{CL} \text{ (ac)} \cong \frac{R_2}{R_1}$$

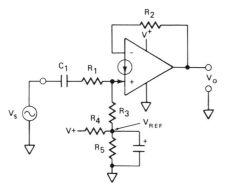

Figure 7-38. AC amplifier, noninverting.

The ac resistance of the amplifier input diode is a potential source of error and can be large for small currents. The following ac gain equation accounts for this diode resistance, r_d.

$$A_V \text{ (ac)} = \frac{R_2}{R_1 + r_d}$$

where $r_d = \dfrac{0.026}{I_{NI}}\ \Omega$

Of course, for any amplifier, the amplifier open-loop gain must be significantly greater than the closed-loop gain.

Example

If $V^+ = +12$ V, $R_1 = 6$ kΩ, $R_2 = R_3 = 12$ kΩ, and $R_4 = R_5 = 1$ kΩ, then

$$V_O \text{ (dc)} = \frac{V^+}{2} = +6 \text{ V}$$

and $A_{CL} \text{ (ac)} \cong \dfrac{R_2}{R_1} = 2$

The current $I_{NI} = 6$ V/12 kΩ = 0.5 mA, and the diode resistance is

$$r_d = \frac{0.026 \text{ V}}{0.0005 \text{ A}} = 52\ \Omega$$

which is small compared to R_1 but represents an error of approximately 1%.

Comparator

A comparator with hysteresis is called a Schmitt trigger. This type of comparator has different trip points for its high and low outputs, thus displaying a hysteresis effect. The inverting and noninverting configurations using Norton

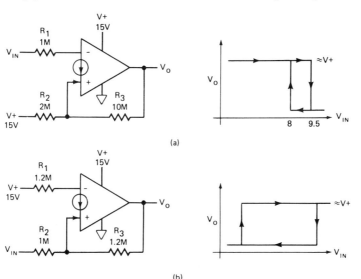

Figure 7-39. Norton comparators with hysteresis: (a) inverting; (b) noninverting.

amplifiers are shown with their corresponding transfer (V_O versus V_{in}) characteristics in Figure 7-39. In each case, positive feedback is provided by R_3.

In the inverting circuit, the comparator output voltage is high as V_{in} increases from 0 to +9.5 V. For this range of input voltage, the (+) input current is

$$I_{NI} = I_2 + I_3$$
$$= \frac{V^+ - V_{BE}}{R_2} + \frac{V_{SAT} - V_{BE}}{R_3}$$

For the component and source values given, I_{NI} is approximately 9 μA and the (−) input current is

$$I_I = \frac{V_{in} - V_{BE}}{R_1}$$

This current will be less than I_{NI} for V_{in} less than 9.5 V. At 9.5 V the current I_I is greater than I_{NI}, and V_O switches to the low state.

The comparator voltage is low as V_{in} is decreased from some high positive value to +8 V. For this range of input voltage, the (+) input current is

$$I_{NI} = I_2 + I_3$$
$$= \frac{V^+ - V_{BE}}{R_2} + 0$$
$$= 7.5 \text{ μA}$$

The (−) input current will be greater than the (+) input current for V_{in} greater than 8 V. At 8 V, I_I is less than I_{NI}, and V_O switches to the high state.

The noninverting circuit operates in a similar manner.

7.10 COMPARATORS

Most electronic systems contain analog and digital circuits. Both types do not exist as separate entities but are eventually joined together by specialized circuits called *interface circuits*. These circuits are not purely digital or linear but contain both functions. They convert one form to the other. Modern electronic systems are computer or microprocessor controlled. Coded digital information is transmitted to and from the computer to a portion of the system where the test, measurement, detection, and control occur. This portion is usually implemented with linear devices. The system-level interface devices that directly convert digital information to analog information and analog to digital are called D/A and A/D converters. These are usually modular or hybrid, but a few are monolithic. At the circuit level, a comparator is used to convert analog to digital.

Comparator Fundamentals

The comparator is a two-input, one-output voltage-amplifying linear device that is capable of high gain, high input resistance, and low output resistance. This definition is the same as that for an operational amplifier. In fact, the operational amplifier can be used as a comparator. However, a comparator performs a different function. *It detects two input voltages and provides an output that has two discrete states.* The two states indicate whether one input is greater in magnitude than the other, or vice versa. The comparator *differs* from the operational amplifier in four ways:

1. It performs a different function.
2. It is operated open loop.
3. It is significantly faster.
4. It has two output states whose voltage levels are digitally compatible.

The *operational amplifier in linear circuits operates closed loop*. The two inputs track each other, and the output can take on any value within its operational (saturation voltages) limits. The *comparator operates open loop*. It has differential amplifier inputs, but normally one input is connected to a voltage reference. The other input is connected to a voltage that varies with time. The two comparator inputs do not track each other. In fact, the two are the same only at the time the comparator changes its output state. The output states are based on whether one input is greater or less than the other input. Two output voltage levels represent the two states of the decision. Since the saturation voltage levels of op amps are not compatible with digital circuits, the output stage of most comparators is specifically designed to provide +5 V and 0 V. The values between these two voltages carry no information and occur only during the brief transition from one state to the other.

The response time for op amps is in the tens of microseconds. This is extremely slow compared to the operating time of digital circuits and systems. The comparator is designed for minimum response time, and although it is not quite as fast as a digital device, it is many orders of magnitude faster than operational amplifiers.

The *schematic symbol* for a comparator (Figure 7-40) is the same as for the amplifier. Both have two inputs, an output, and two power supply pins. However, most comparators also have ground and strobe pins. The ground pin is used to bias the output stage for digital logic compatibility. The strobe pin is used to enable the comparator to make a decision based on its inputs or to disable the comparator and have its output go to a predetermined permanent state. Two more pins complete the typical eight-pin package, and they are used to null out (balance) the inherent comparator input error. Specifically, this is called the offset voltage, and it represents the mismatch of the V_{BE}'s of the differential amplifier stage transistors.

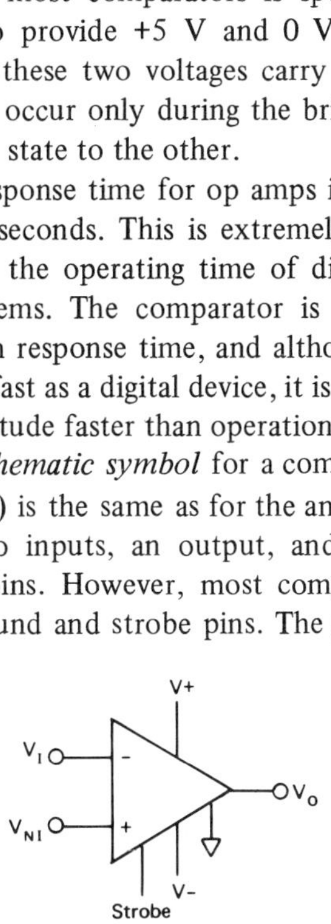

Integrated-Circuit Comparator

The simplified *schematic of the 311 comparator* is shown in Figure 7-41. The input differential amplifier is implemented with the Q_1-Q_3 and Q_2-Q_4 transistor pairs. This stage differences two voltages, provides gain, and minimizes the input error current. The output of this stage is further amplified by the Q_5-Q_6 pair. This stage feeds Q_9, which provides additional gain and drives the output stage. The circuit current sources are used to determine the biasing so that performance is not greatly affected by supply voltage changes. The output transistor is Q_{11}, and it is protected by Q_{10} and Q_9, which limit the peak output current. Q_{10}, normally off, will turn on when the voltage drop across R_6 approaches 0.6 V. With Q_{10} on, the base current of Q_{11} is shunted to the output, where current limiting takes place. The output lead, since it is not connected to any other point, can be returned to a positive supply through a pull-up resistor. This supply is usually +5 V, but is not limited to this value. The output stage can also be biased to interface with circuits that operate from 0 to a negative voltage. This can be accomplished by grounding the output pin through a pull-up resistor, and then returning the ground pin to a negative voltage.

The complete schematic of the comparator is shown in Figure 7-42.

(a)

$V_I > V_{NI}$ V_o = LOW
$V_I < V_{NI}$ V_o = HIGH

(b)

Figure 7-40. IC comparator: (a) schematic symbol; (b) mathematical model.

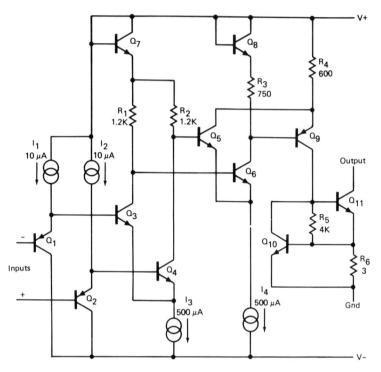

Figure 7-41. Simplified schematic of the 311 IC comparator.

Comparator Applications

Detector with Strobe

The voltage sources, V_{s1} and V_{s2}, in Figure 7-43a can represent time-varying signal sources or dc voltages. If one is used as a basis for comparison, it is called a reference voltage. When the comparator is enabled through its strobe, the output

$$V_o = +5 \text{ V} \quad \text{for } V_{s1} < V_{s2}$$

and $V_o = 0$ V for $V_{s1} > V_{s2}$

Resistor, R_1, is the pull-up resistor for the open-collector transistor in the output stage of the comparator. Q_1 and R_2 are external components necessary to interface a digital logic signal to the device's strobe pin. If the comparator is disabled or inhibited via the strobe signal, the output will go to a predetermined permanent state (the output transistor is off for the 311).

When one of the input sources is at zero volts or an input pin is grounded, the circuit is referred to as a zero crossing detector. The output will change state each time the signal source crosses zero volts to the opposite voltage polarity.

The transfer characteristics of Figure 7-43b illustrate the use of this circuit as a *reference detector*. The source V_{s2} is a fixed dc reference source. The comparator output will be low for any value of V_{s1} greater than V_{s2} and high for the opposite condition.

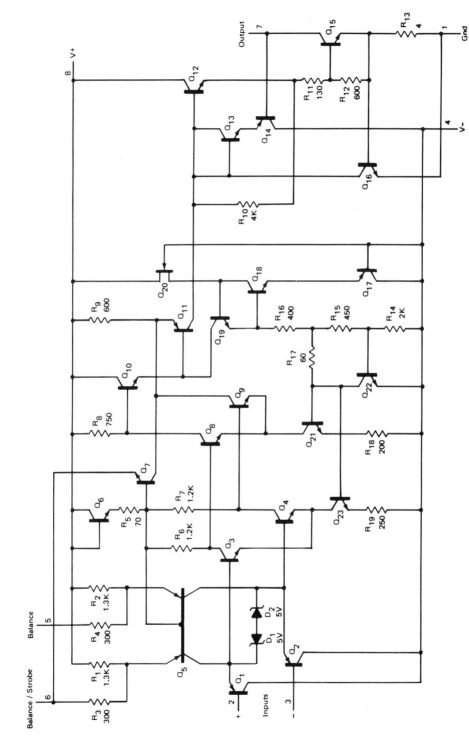

Figure 7-42. Schematic of the 311 IC comparator.

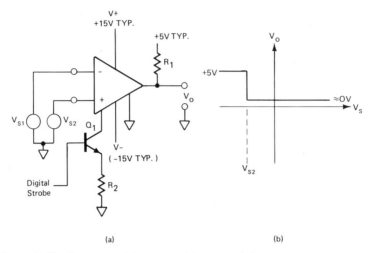

Figure 7-43. Detector with strobe: (a) circuit; (b) transfer characteristics.

Window Comparator

The use of two devices, connected as shown in Figure 7-44a, produces a *window comparator*. This circuit provides a low output voltage if the signal source V_s is more positive than the lower limit (LL) voltage and less positive than the upper limit (UL) voltage. The circuit output is high if the signal source is less positive than the lower limit or more positive than the upper limit. The circuit detects a range or window of input voltage and is shown in Figure 7-44b.

Mathematically,

$$V_o = \text{LOW} \quad \text{for } V_{LL} < V_S < V_{UL}$$

and $V_o = \text{HIGH}$ for $V_s > V_{UL}$ or $V_s < V_{LL}$

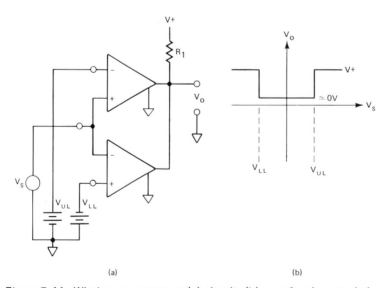

Figure 7-44. Window comparator: (a) circuit; (b) transfer characteristics.

The open collectors of each comparator are tied together to form a wire ORed configuration. They are pulled up to V^+ through R_1.

Digital Interface

The voltage levels of bipolar and unipolar logic families differ. Typically, TTL device outputs swing from 0 to +5 V, and pMOS swings from 0 to -10 V. The circuit of Figure 7-45 joins or *interfaces the two logic families*. V_s models the output of a TTL device. When V_s is low or near 0 V, the more positive voltage (+2.5 V) on the inverting input will cause the comparator output to go to its low state, which is near -10 V. The comparator ground pin, which essentially is the emitter of the output stage transistor, is connected to -10 V. When V_s is high or near +5 V, the more positive voltage or the noninverting input will cause the comparator output to go to its high state, 0 V, through the pull-up resistor R_1.

This type of output stage increases the comparator's versatility in interfacing with other devices and circuits, notably solid-state switches.

Free-Running Multivibrator

Comparators, occasionally, are operated closed loop. The circuit of Figure 7-46 is a *square-wave oscillator* with positive (R_4) and negative (R_3) feedback. The comparator does not operate in its linear region, but switches

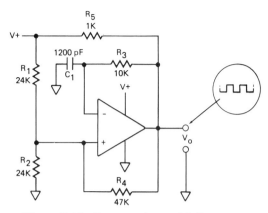

Figure 7-46. Free-running multivibrator.

from one saturation voltage level to the other. The time at which the output switches is a function of the two input voltages. The noninverting (+) input voltage is dc and is one of two values depending on the comparator output voltage. The inverting (−) input is a time-varying voltage due to the charging and discharging of the capacitor C_1 through R_3. If the output is at the high level, V^+,

$$V_{NI} = \frac{R_2}{R_2 + R_1 \| R_4} \cdot V^+$$

and if the output is at the low level, $\simeq 0$ V,

$$V_{NI} = \frac{R_2 \| R_4}{R_2 \| R_4 + R_1} \cdot V^+$$

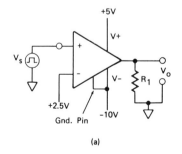

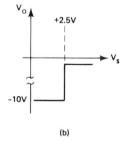

(a)   (b)

Figure 7-45. Digital interface: (a) circuit; (b) transfer characteristics.

For V^+ equal to +5 V and the component values given,

$$V_{NI} \simeq +3 \text{ V}, +2 \text{ V}$$

The capacitor is either being charged through R_3 and the comparator output to V^+, or it is being discharged because the comparator output is at zero volts.

When the output is at +5 V (through R_5), the noninverting (+) input is at +3 V. The inverting (−) input is charging toward +3 V. When it reaches +3 V, the comparator output switches to near zero volts, and the noninverting input voltage goes to +2 V. The capacitor then discharges toward 0 V; but when it reaches +2 V, the output switches back to +5 V and the process repeats itself.

Components C_1 and R_3 determine the frequency of oscillation. The time period of the oscillating frequency is approximately

$$T = \frac{1.1}{\tau}, \quad \text{where } \tau = R_3 C_1$$

The ratio of the resistors R_1 and R_2, with the proper R_4, determines the symmetry or duty cycle of the waveform. Equal values produce a symmetric waveform with the high and low voltage times equal. Resistor R_5 is the pull-up resistor and must be small compared to R_3 and R_4.

Problems

1. In Figure 7-11, $V_s = +1$ Vdc, $R_1 = 1$ kΩ, $R_2 = 9$ kΩ, and the open-loop dc gain of the amplifier is 100 dB. For the above conditions,
   (a) Calculate the *exact* closed-loop gain.
   (b) Calculate the *exact* value of the output voltage.
   (c) What is the actual difference in potential between the amplifier inputs?
   (d) What is the percent of difference between the ideal and actual closed-loop gain?

2. In Figure 7-11, $R_1 = 1$ kΩ, $R_2 = 9$ kΩ, and the GBW product of the amplifier is 1 MHz. The amplifier's gain–frequency response follows that of a single lag network.
   (a) Determine the amplifier's open-loop gain at 1 kHz.
   (b) Calculate the *actual* closed-loop gain if the frequency of V_s is 1 kHz.
   (c) What is the upper cutoff frequency of the closed-loop gain?
   (d) What is the *exact* value of the output voltage if V_s is a 1 V peak signal at 1 kHz?
   (e) If S_R of the amplifier is 0.5 V/μs and the maximum peak voltage of V_s is 1 V, what is the highest distortion-free frequency that the circuit can amplify?

3. The input and output resistances of the amplifier in the circuit shown in Figure 7-11 are 10 MΩ and 100 Ω, respectively. For this circuit $R_1 = 1$ kΩ, $R_2 = 9$ kΩ, and the open-loop dc gain of the amplifier is 100 dB.
   (a) What is the approximate input and output resistance of the circuit?
   (b) Does the circuit input and output resistance increase or decrease for a higher closed-loop gain?
   (c) Are these values achievable?

4. In Figure 7-15, $V_s = +1$ Vdc, $R_1 = 1$ kΩ, $R_2 = 9$ kΩ, $R_L = 10$ kΩ, and the open loop dc gain of the amplifier is 100 dB. For the above conditions,
   (a) Determine the *actual* closed-loop gain.
   (b) What is the *actual* difference in potential between the amplifier inputs?
   (c) Calculate the *exact* output voltage.
   (d) Estimate the circuit's input and output resistances if r_i and r_o of the amplifier are 10 MΩ and 100 Ω, respectively.
   (e) Calculate I_{R1} and I_{R2}.

5. In Figure 7-17, $V_{s1} = -1$ Vdc, $V_{s2} = +2$ Vdc, $V_{s3} = -3$ Vdc, $R_1 = 2$ kΩ, $R_2 = 4$ kΩ, $R_3 = 6$ kΩ, and $R_4 = 8$ kΩ.
   (a) Calculate V_o, I_{R2}, and I_{R4}.

(b) If a load resistor is connected at the circuit output (to ground) of value 5 kΩ, what is the amplifier output current direction and value?

6. In Figure 7-18, $V_{s1} = +1$ Vdc, $V_{s2} = -2$ Vdc, $R_1 = 5.6$ kΩ, and $R_2 = 10$ kΩ. Calculate V_o, I_{s1}, and I_{s2}.

7. In Figure 7-20, the switch S_1 is closed at t_o. What is the output voltage at the end of
   (a) $t = 5$ ms
   (b) $t = 7.5$ ms
   (c) $t = 10$ ms
   if $R_1 = 1$ kΩ, $C_1 = 1$ μF, and $V_s = +1$ Vdc.

8. Estimate the input resistance of the circuit in Figure 7-21.

9. In Figure 7-22f, specify a set of values for V_r and the circuit resistances for a 4-bit D/A convertor whose full-scale output is +7.5 V.

10. What factors affect the decay rate of the sample-and-hold circuit in the hold mode (S_1 open) in Figure 7-22d?

11. Specify a set of values for V_S, V_{CC}, and the circuit resistances in Figure 7-25 for a current source value of 1 mA and a maximum load voltage of 10 V.

12. An analog multiplier is comprised of two log amplifiers, a summing amplifier, and an antilog amplifier. In block diagram form, show how two analog voltages are multiplied.

13. What is the input impedance of the gyrator circuit in Figure 7-23 if it is terminated with an (a) inductor; an (b) open circuit?

14. In Figure 7-25, $V_c = -1$ V, $R = R_1 = R_2 = 2$ KΩ. $\pm V_{CC}$ for all amplifiers is 15 V.
    (a) Calculate the output current, I_o.
    (b) Estimate R_L (max).

15. Using Figure 7-26, at $V_{CC} = \pm 15$ Vdc,
    (a) Estimate the percent difference in open-loop gain between 25° and 125°C.
    (b) Estimate the S_R value of the amplifier if it is compensated with a 3 pF capacitor.
    (c) Estimate the power dissipated by the amplifier for $T_A = 25$°C under no-load conditions.
    (d) Estimate the maximum output swing for a load resistor of 10 kΩ.

16. For the circuit in Figure 7-27, $R_1 = R_2 = 10$ kΩ, $V_S = +1.00$ V, and $V_{OS} = +5$ mV. Find V_O and determine the percent of the V_{OS} error at the output.

17. In Figure 7-28, $R_1 = R_2 = 10$ kΩ, $R_3 = 0$ Ω, $I_B^+ = 110$ nA, $I_B^- = 90$ nA, and $V_S = 1$ V. Except for I_B, assume the other amplifier parameters are ideal. Find V_O and determine the percent of the I_B error at the output. Let $R_3 = 5$ kΩ, and recalculate V_O and the percent of the input current's error at the output.

18. An IC op amp is biased with two 15 V symmetric supplies. The amplifier's no-signal current drain is 5 mA. How much power is the amplifier dissipating under no-signal conditions? What is the maximum power that the amplifier can dissipate?

19. For the circuit in Figure 7-30, $V_{OS} = +5$ mV, $I_B^+ = 110$ nA, $I_B^- = 90$ nA, and S_1 and S_2 are opened. Find V_O.

20. What are the dc and ac components of V_O in Figure 7-37 if $V^+ = 12$ V, $C_1 = 0.05$ μF, $R_1 = 100$ kΩ, $R_2 = 1$ MΩ, $R_3 = 2$ MΩ, and $V_s = 0.2$ V (pk - pk)?

21. In Figure 7-38, what should the ratio of R_3 to R_4 and R_5 be to minimize the loading of V_{REF}?

22. Calculate the trip points of the noninverting Norton comparator in Figure 7-39b.

23. The typical slewing rate of an op amp is 0.5 V/μs and for a comparator is 100 V/μs. If both devices swing 10 V, how much faster is the comparator?

24. What is the frequency of oscillation of the circuit in Figure 7-46?

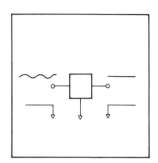

8 REGULATION CONCEPTS AND PRINCIPLES

Regulation is the concept of controlling or adjusting to some standard or requirement. Voltage regulators are circuits that maintain a constant output voltage for a varying load, input, and so on. This chapter discusses the concepts and principles of regulation and basic regulator circuits.

The most frequency application of voltage regulators is in dc power supplies. The basic circuits in a power supply are the transformer, *rectifier, filter, and regulator. The voltage regulator is described by its output voltage, input-voltage range, output-current range, and its line and load regulation. The zener diode or a voltage reference circuit provides the basis for the value of the output voltage in a regulator. The basic regulator circuits are the zener, series, and shunt regulator, but the series regulator with feedback is the model for the IC regulators.*

8.1 THE CONCEPT OF REGULATION

Regulation is the act of regulating or the state of being regulated. *To regulate means to control*, or direct by a rule, principle, or method. *To regulate is to adjust to some standard or requirement.*

A regulator is a device, or a person, that regulates. Examples of mechanical regulators are the governor and the valve. A governor is a mechanism for regulating the flow of steam, fuel, and the like, to an engine in order to maintain a constant speed under varying load. A valve is a device for regulating the pressure of a flowing gas or liquid. Electronically, a regulator is a component, circuit, or system that maintains a specific characteristic at a designated value or varies it according to a predetermined plan.

The most prominent electronic regulator is the voltage regulator. It is a component, circuit, or a system that maintains a constant output voltage for a varying load (current). A voltage source qualifies as a voltage regulator, but the name voltage regulator has come to be identified with the portion of the source that actually performs the regulation function. The dual of the voltage regulator is the current regulator or constant current source. It, however, exists in far fewer numbers than the voltage regulator.

Components that function as voltage regulators are the diode and the zener diode. These components maintain a constant potential across their terminals (to some degree, of course) independent of the current through them.

A large number of circuits function as voltage regulators. They vary from the simple (a resistor and diode) to the complex (switching regulators). The need for and widespread usage of voltage regulators have prompted IC manufacturers to develop integrated-circuit versions of these circuits.

The sophistication of these circuits reaches the system level when extraordinary demands are placed on the specifications and requirements for the regulator.

The real voltage regulator establishes and maintains a constant dc output voltage independent of input voltage, output current, temperature, and other variations. Of course, it maintains this constant voltage to a certain degree and under a certain range of conditions, which are identified in its specifications.

The most frequent application of voltage regulators is in dc power supplies. A power supply generally implies a constant voltage source with high output current capability. This is an electrical potential energy source, and it must be remembered that its dual exists, that is, the electrical kinetic energy source or constant current source. The latter exists in far fewer numbers than the former.

The design and choice of circuits used in electronic equipment vary greatly, depending on the equipment's ultimate application or usage. Primarily, digital circuits are used in computers, RF circuits in communications equipment, and analog circuits in linear control systems. However, as diverse as the equipment designs are, they have a common requirement. They must all have a minimum of one power source. A small number of the systems may have their circuitry powered by batteries or an ac source, but the greater requirement is for well-regulated dc voltage supplies. The supplies must provide a constant voltage to the electronic circuits and hold their value independent of changes in the input line voltage, load current, or even temperature. The portion of the power supply that maintains this constant voltage is called a voltage regulator.

A brief review of the power supply is necessary to illustrate the role and importance of the voltage regulator.

8.2 POWER SUPPLY FUNDAMENTALS

Figure 8-1 is a block diagram of a general-case power supply. Power supplies are necessary to convert the ac voltage of our electric utility to the specific dc voltage required in electronics equipment. The power supply, including the load, is comprised of five functional blocks:

1. Transformer
2. Rectifier
3. Filter
4. Regulator
5. Load

Transformer

The *ideal transformer* (Figure 8-2) is a lossless, two-terminal pair device that transforms an input ac voltage to an output ac voltage of the same frequency but with the same or a different magnitude. The transformer is an ac-voltage-in, ac-voltage-out device and transfers energy from one circuit to another without physical connection. The transformer is an application of magnetic induction. An alternating voltage applied to the input or primary winding generates an alternating magnetic flux that links with a secondary winding and induces an alternating voltage in the secondary. The magnitude of the output voltage, relative to the input, is determined by the ratio of the turns of the windings.

POWER SUPPLY FUNDAMENTALS

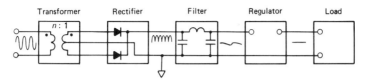

Figure 8-1. General-case power supply.

$$e_s = e_p \left(\frac{N_s}{N_p}\right) = e_p \frac{1}{a}$$

N_s and N_p identify the number of mutually coupled turns in the secondary and primary windings. Phase dots are placed on the primary and secondary windings to show the input and output terminals whose signals have the same potential slope or phase.

The transformer is called a step-up transformer for $N_s > N_p$ and step-down for $N_s < N_p$. The transformer is referred to as 1:1 when $N_s = N_p$ and, for this case, is used for its input–output isolation feature. The ideal transformer is a lossless device; hence,

$$p_{out} = p_{in}$$

Other ideal transformer relationships include

$$i_s = \left(\frac{N_p}{N_s}\right) i_p = a i_p$$

$$z_{in} = \left(\frac{N_p}{N_s}\right)^2 z_L$$

where z_{in} is the reflected load impedance.

The transformer steps down or steps up the ac line voltage to accommodate the output voltage (dc) requirement. The power supply, in converting ac to dc, requires the peak magnitude of the input ac to be only moderately greater than the dc output. Probably, a more

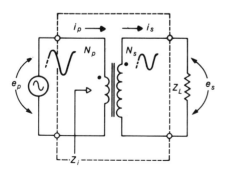

Figure 8-2. The transformer.

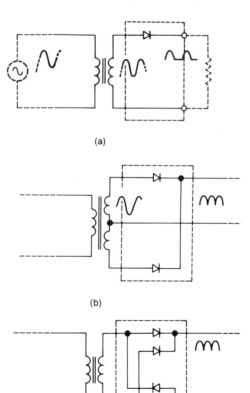

Figure 8-3. Rectifier circuits: (a) half-wave; (b) full-wave, center tapped; (c) full-wave bridge.

important function of the transformer in power supplies is to provide isolation between the utility ac and the equipment. This is necessary for safety and protection reasons.

Rectifier

Rectifier circuits rectify or change the ac voltage, at the transformer secondary, to a pulsating direct current. The most common rectifier circuits (Figure 8-3) are

1. Half-wave
2. Full-wave center tapped
3. Full-wave bridge

The full-wave bridge and full-wave center-tapped circuits are the ones most frequently employed. For these circuits, secondary current surges occur twice per cycle so that they are of smaller amplitude than that found in the half-wave circuit. The fundamental ripple frequency is double the supply, that is, 120 Hz.

The advantages of the half-wave rectifier are its simplicity and the savings in cost of one diode. Disadvantages include higher current peaks and the need for a transformer with a larger core.

Filter

The pulsating dc of the rectifier is smoothed by a capacitor or a passive π or L *filter* (Figure 8-4).

For low output current supplies (<1 A), a capacitor will sufficiently reduce the ripple voltage. At this point, a load could be connected. In fact, in simple applications with minimal load requirements, many power supply designs stop here. The obvious advantages are simplicity and economy. The time constant associated with the capacitor and load resistance is made much longer than the period of the input ac signal.

Combining the action of the inductor and the

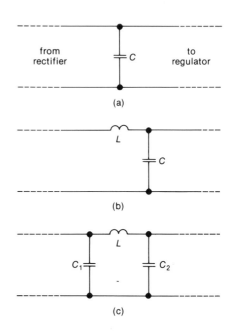

Figure 8-4. Power supply filters: (a) capacitor; (b) L-section; (c) π-section.

capacitor in a simple filter improves the filter's performance in reducing the ripple voltage. Power supply inductors and capacitors are quite large in value and, therefore, bulky and expensive. The inductors in power supplies are called chokes. Two filters in common use are the π and L-section filters. In both types, the choke is employed in series with the load to smooth out the current, and the capacitor (or capacitors) is placed in parallel to smooth out the voltage. In some special applications, where the ripple must be reduced further, the filter is made up of multiple π or L-sections.

8.3 VOLTAGE REGULATOR CHARACTERISTICS

The primary characteristics of the voltage regulator are

1. Output voltage

2. Input-voltage range
3. Output-current range
4. Load regulation
5. Line regulation

Secondary characteristics include ripple rejection, quiescent current, output noise voltage, stability, thermal resistance, maximum power dissipation, and those characteristics unique to the components and circuit implementing the regulator.

The regulator is a variable-voltage input, constant-voltage output device. The extent to which the input voltage can vary is specified, and is called the input voltage range. The regulator will maintain a constant output voltage for a specified output current range. The range is from no-load (0 A) to some maximum value (I_L MAX), or what is called full load. The degree to which the output voltage is maintained (in percent) for a change in input voltage is called line regulation. The input voltage change can be reflected back to the utility change in voltage. The degree to which the output voltage is maintained (in percent) for a change in output current is called load regulation.

Ripple rejection is the ratio of rms input ripple voltage to rms output ripple voltage. It is expressed in decibels. Quiescent current is that part of the input current that is not delivered to the load. The output noise voltage is the rms output noise voltage generated by the regulator.

Maximum power dissipation is the maximum device (IC) dissipation for which the regulator will operate within specifications.

8.4 BASIC REGULATOR CIRCUITS

Regulator Elements

The diode and the zener diode are simple voltage-regulating components. Both components will maintain a reasonably constant voltage across their terminals independent of the current through them. The I-V characteristics of these two devices are shown in Figure 8-5. The forward voltage or barrier potential of the silicon diode is approximately 0.6 V, but the zener voltage (more accurately called the breakdown voltage) can be of almost any value. The ordinary diode is more commonly used as a control device. The zener diode is primarily used as a voltage-regulating device. The change in voltage for a change in current is much smaller for the zener. Integrated-circuit versions of the zener diode are available that even further minimize the change in voltage. Two IC versions of reference diodes are shown in Figure 8-6.

Zener Regulator Circuit

The most basic and inexpensive form of voltage regulator uses a zener diode. The input of this *basic regulator circuit* (Figure 8-7a) is the

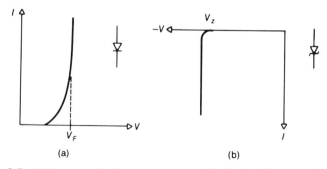

Figure 8-5. Voltage-regulating components: (a) diode; (b) Zener diode.

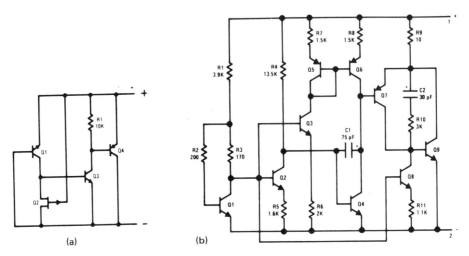

Figure 8-6. Integrated-circuit reference diodes: (a) 103; (b) 113.

unregulated output voltage of the filter, V_i. This voltage is applied to the current-limiting resistor R, and the regulated output is taken across the zener diode. The unregulated dc voltage at the input reverse biases the zener diode and causes it to operate in its breakdown region. The input voltage must be greater than the zener voltage.

The output voltage of this regulator is essentially equal to the zener voltage. It does, however, change somewhat when a load is connected because of the nonzero resistance of the diode. We can see this change if we replace the zener diode by its equivalent circuit, as indicated in Figure 8-7b. The resistance value of r_z is between 10 and 30 Ω, but is significantly reduced in the IC zener. The slope of the zener I–V characteristic in the breakdown region is the reciprocal of the zener resistance. The current through the resistor R is

$$I_R = \frac{V_i - V_z}{R}$$

For the no-load condition ($R_L = \infty$),

$$I_Z = I_R$$

With a load,

$$I_Z = I_R - I_L$$

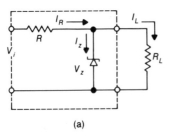
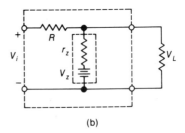

Figure 8-7. Zener regulator circuit: (a) Zener regulator circuit; (b) regulator with Zener equivalent circuit.

The load current can never be allowed to exceed I_R minus the minimum required zener current.

$$I_L = I_R - I_Z \text{ (min)}$$

Temperature and *power dissipation* play an extremely important role in regulators. All components are temperature sensitive. For zener or voltage-reference diodes, their voltage will change with temperature. This characteristic is reflected in the diode's temperature coefficient (TC) specification, usually expressed in millivolts per degree Celsius. Zener diodes or circuits used as voltage references are optimized for low TCs. Most component failures are due to excessive power dissipation. Because of the high values of current and voltage in regulator circuits, care must be exercised to ensure that the components do not dissipate more power than their ratings allow.

Example

If $V_1 = 15$ V, $V_Z = 10$ V, $R = 100$ Ω, $r_z = 10$ Ω, and I_z (min) = 1 mA, then as a first approximation,

$V_o \cong +10$ V at $R_L = \infty$ and r_z assumed small

$$I_R \cong \frac{15 \text{ V} - 10 \text{ V}}{100 \text{ Ω}} = 50 \text{ mA}$$

Hence,

$$I_L \text{ (max)} = 49 \text{ mA}$$
and $\quad R_L > 204$ Ω

As a second approximation (at no load),

$$V_o = 10 \text{ V} - (10 \text{ Ω})(0.049 \text{ A}) = 9.5 \text{ V}$$

Line Regulation

Line regulation is the change in output voltage for a given change in input voltage. It is expressed as a percentage and is defined as follows:

$$\text{Line regulation [\%]} \triangleq \frac{\Delta V_o}{\Delta V_i} \cdot 100$$

where $\Delta V_o = V_{o1} - V_{o2}$ and $\Delta V_i = V_{i1} - V_{i2}$.
V_{o1} is the output voltage resulting from the input voltage V_{i1}, and V_{o2} is the output voltage for the input voltage V_{i2}.

In the zener regulator circuit of Figure 8-7b,

$$V_O = \left(\frac{R}{R+r_z}\right)V_Z + \left(\frac{r_z}{R+r_z}\right)V_i$$

(use superposition)

If the input voltage is changed by some small amount, it will cause a change in the output voltage.

$$\Delta V_O = \left(\frac{R}{R+r_z}\right)V_Z + \left(\frac{r_z}{R+r_z}\right)\Delta V_i$$

V_Z is the ideal zener voltage and does not change when V_i changes. Hence,

$$\frac{\Delta V_O}{\Delta V_i} = \frac{r_z}{R+r_z}$$

and $\quad$ Line regulation [%] $= \left(\frac{r_z}{R+r_z}\right) 100$

If the zener resistance, r_z, is 10 Ω, and the resistor, R, is 1 kΩ, then

$$\text{Line regulation} = \frac{10 \text{ Ω}}{1010 \text{ Ω}} \cdot 100 = 0.99\% \cong 1\%$$

The output voltage will change approximately 1% of the amount that the input voltage changes. This number is typical of moderate-performance regulators. This specification measures the degree to which the output voltage will change if the input voltage changes.

The above relationship for line regulation assumes a no-load condition. In practice, line regulation is measured for a fixed, constant load. To measure the effect of V_i on V_O, all other conditions must remain constant. It is usually a pulsed test so that the temperature of the components remain the same.

Load Regulation

Load regulation is the change in output voltage for a given change in load current. It is expressed as a percentage and is defined as follows:

$$\text{Load regulation [\%]} = \frac{V_O(\text{NL}) - V_O(\text{FL})}{V_O(\text{NL})} \cdot 100$$

where $V_O(\text{NL})$ is the output voltage under the no-load ($R_L = \infty$) condition, and $V_O(\text{FL})$ is the output voltage where the value of R_L will result in the maximum load current. To determine or measure the regulator's load-regulation capability (only), all other conditions must be held constant, including the input voltage and temperature.

To find the load regulation of the zener regulator circuit in Figure 8-8a, the circuit must be Thevenized. The Thevenin equivalent circuit is shown in Figure 8-8b. V_{TH} is V_O for the no-load condition and R_{TH} is redefined as r_o or the output resistance of the circuit.

A resistor R_L is added in Figure 8-8c and represents the R_L that provides the full-load condition. Thus,

$$V_O(\text{FL}) = \left(\frac{R_L}{R_L + r_o}\right) V_O(\text{NL})$$

$$\frac{V_O(\text{NL}) - V_O(\text{FL})}{V_O(\text{NL})} = \frac{V_O - \left(\frac{R_L}{R_L + r_o}\right) V_O}{V_O}$$

$$= \frac{r_o}{R_L + r_o}$$

and $\quad \text{Load regulation [\%]} = \left(\frac{r_o}{R_L + r_o}\right) \cdot 100$

where $r_o = r_z \| R$.

If the zener resistance is 10 Ω, R is 1 kΩ, and the full-current load resistance is 500 Ω,

$$r_o = 10\ \Omega \| 1\ \text{k}\Omega = 9.9\ \Omega$$

$$\text{Load regulation} = \left(\frac{9.9\ \Omega}{509.9\ \Omega}\right) \cdot 100$$

$$= 1.94\% \cong 2\%$$

For this circuit, the output voltage will change approximately 2% when the load varies from $R_L = \infty$ to $R_L = 500$ Ω. This specification measures the degree to which the output voltage will change as the output current changes. This number is typical of regulators with moderate performance.

Basic Series Regulator

Regulation over a wider range of loads is possible with the *series regulator circuit* illustrated in Figure 8-9. Q_1, the regulating element, is in series with the load. The output voltage of this circuit is

$$V_o = V_Z - V_{BE}$$

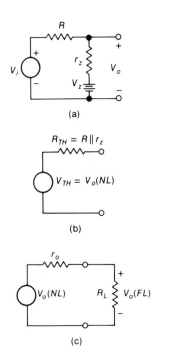

Figure 8-8. Load regulation and the Zener regulator circuit: (a) regulator circuit; (b) Thevenin equivalent, no load; (c) Thevenin equivalent, full load.

BASIC REGULATOR CIRCUITS

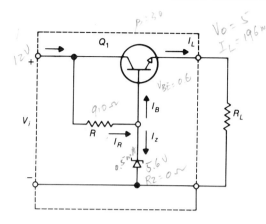

Figure 8-9. Series regulator.

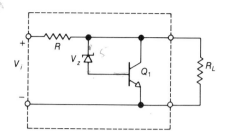

Figure 8-10. Basic shunt regulator.

The unregulated dc input voltage must exceed the desired output voltage by at least 1 V.

The current through R is

$$I_R = \frac{V_i - V_z}{R} = I_B + I_Z$$

The current supplied to the load is essentially the same as the collector current of the transistor. As I_L (and hence I_C) changes, the transistor base current I_B changes. However, the current gain of the transistor makes the values of I_B small, and hence I_Z remains relatively constant if $I_Z \gg I_B$. A Darlington transistor pair is often used to keep I_B small.

Basic Shunt Regulator

The *shunt regulator circuit* is shown in Figure 8-10. Q_1, the regulating element, is in shunt with the load. For this circuit, the output voltage

$$V_o = V_z + V_{BE}$$

For the no-load condition ($R_L = \infty$),

$$I_C = I_R - I_B \cong I_R$$

and $$I_R = \frac{V_i - (V_z + V_{BE})}{R}$$

With a load,

$$I_C \cong I_R - I_L$$

The maximum current that the load can draw must be slightly less than I_R.

$$I_L \lesssim I_R$$

If R_L is made small enough such that $I_L > I_R$, the transistor and zener are cut off and the circuit ceases to regulate.

Transistor Q_1 dissipates the maximum power when the load current is minimum. For this condition ($R_L = \infty$),

$$P_{D\,max}(Q_1) = (I_{C\,max})(V_{CE}) = (I_R)(V_o)$$

Example

If $V_z = 5.6$ V, $V_{BE} = 0.6$ V, $V_i = 10$ V,

$R = 100\ \Omega$

$\beta = 100$, then

$V_o = V_z - V_{BE} = +5.0$ V

$I_R = \dfrac{10\text{ V} - 5.6\text{ V}}{0.1\text{ k}\Omega} = 44$ mA

For $I_L = 0$ A, $\quad I_Z = I_R = 44$ mA

For $I_L = 250$ mA, $\quad I_B = 2.50$ mA and

$I_Z = 41.5$ mA

Example

If $V_z = 5.4$ V, $V_{BE} = 0.6$ V, $V_i = 10$ V, and $R = 0.2$ kΩ, then

$$V_o = V_z + V_{BE} = +6.0 \text{ V}$$

$$I_R = \frac{4 \text{ V}}{0.2 \text{ k}\Omega} = 20 \text{ mA}$$

and $I_L \text{ (max)} < 20$ mA

Series Regulator with Feedback

The *series regulator circuit, with feedback*, shown in Figure 8-11, is the model for IC regulators. Q_1 is called the regulator pass transistor. This transistor passes the required output current. Its collector voltage is the unregulated dc input, and its emitter voltage is the regulated dc output.

Q_2 functions as a voltage amplifier. Its output is its collector voltage, which drives the pass transistor Q_1. Q_2's inputs are the zener voltage at the emitter and the output voltage, divided by R_1 and R_2, at the base. To illustrate the negative feedback configuration, assume the output voltage momentarily decreases. When V_o decreases, the base voltage of Q_2 decreases, and hence V_{BE} decreases since V_Z is constant. If V_{BE} decreases, I_C decreases, and the collector voltage of Q_1 rises, bringing V_O back up. Resistor R_2 is the collector load resistor for Q_2, and R_1 biases the zener diode on.

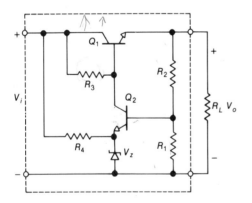

Figure 8-11. Series regulator with feedback.

The output voltage is a function of V_Z, $V_{BE}(Q_2)$, R_1, and R_2. Q_2's base voltage is

$$V_B(Q_2) = V_Z + V_{BE}$$

This voltage establishes the current through R_1; that is,

$$I_{R1} = \frac{V_B(Q_2)}{R_1} = \frac{V_Z + V_{BE}}{R_1}$$

If $I_B(Q_2) \ll I_{R1}$, then

$$I_{R2} \cong I_{R1}$$

and $\quad V_o = V_B + (I_{R2})(R_2)$

$$= (V_Z + V_{BE}) + (V_Z + V_{BE})\frac{R_2}{R_1}$$

Hence,

$$V_O = (V_Z + V_{BE})\left(1 + \frac{R_2}{R_1}\right)$$

Problems

1. In Figure 8-2, $e_p = 115$ V (rms) at 60 Hz, $R_L = 100 \; \Omega$, and $a = 10$.
   (a) Calculate $e_s, i_s,$ and i_p.
   (b) Calculate $p_i, p_o,$ and z_{in}.

2. A 5 V power supply with a full-wave bridge requires a transformer secondary voltage of 10 V (rms). Calculate the transformer windings turns ratio to provide the required secondary voltage.

3. In Figure 8-7, $V_i = 12$ V, $V_z = 5$ V, $R = 100 \; \Omega, r_z = 5 \; \Omega$, and $I_z \text{ (min)} = 0.5$ mA.
   (a) Calculate $V_o, I_R, R_L \text{ (min)},$ and r_o.
   (b) Calculate line and load regulation.
   (c) Calculate the maximum power dissipated by the zener.

4. For the series regulator circuit in Figure 8-9, $V_i = 12$ V, $V_z = 5.6$ V, $R = 910 \; \Omega, r_z = 0 \; \Omega$, $I_z \text{ (min)} = 0.5$ mA, $V_{BE} = 0.6$ V, and $\beta \text{ (min)} = 30$. Calculate V_o and $I_L \text{ (max)}$.

PROBLEMS

5 In Figure 8-9, $R = 910\ \Omega$, $r_z = 30\ \Omega$, and $\beta = 30$. Estimate r_o.

6 For the basic shunt regulator in Figure 8-10, $V_z = 5.4\ \text{V}$, $V_{BE} = 0.6\ \text{V}$, $V_i = 10\ \text{V}$, and $R = 200\ \Omega$. Calculate $P_D(Q_1)$ if $R_L = \infty$.

7 In Figure 8-11, $V_o = 18\ \text{V}$, $V_{BE} = 0.6\ \text{V}$, and $V_z = 2.4\ \text{V}$. Find a set of values for R_1 and R_2 to satisfy these requirements.

8 Estimate the voltage gain of Q_2 in Figure 8-11 if $r_z = 10\ \Omega$, $R_3 = 510\ \Omega$, $I_C(Q_2) = 4$ mA, $\beta(Q_2) = 50$, $R_1 = 2.2\ \text{k}\Omega$, and $R_2 = 10$ kΩ.

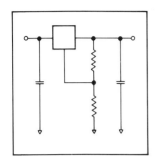

9 REGULATOR INTEGRATED CIRCUITS AND APPLICATIONS

Voltage regulator ICs are monolithic circuits that regulate. This chapter looks at several popular regulator ICs and illustrates their applications.

Three-terminal IC regulators consist of an error amplifier, pass transistor, voltage reference, and a number of self-protection features. The self-protection features include current limiting, excessive power dissipation protection, and thermal shutdown. The regulator comes complete and ready to use. Fixed-value regulators can be made variable with a minimum number of external parts. The output current of the regulator can be increased by using an external transistor circuit. The design and applications of negative voltage regulators are similar to those of the positive regulator.

9.1 HISTORY OF INTEGRATED REGULATORS

The frequency and widespread usage of regulator circuits has prompted semiconductor manufacturers to develop IC regulator product lines. Early (1960s) IC regulators required a reasonably large number of external components to complete the circuit. These circuits were versatile, and they represented a step toward the fully integrated regulators we find today. Two popular IC regulators of that era were the 104 and the 723.

Today, monolithic voltage regulators come complete and are one of the easiest class of devices to use. They have an input that accepts a nonregulated voltage and an output that provides a well-regulated dc voltage. They are typically three-pin devices, with the third pin used as a ground or reference for the input and output voltages. Most three-terminal regulators have fixed dc output voltages but can be made adjustable by using external resistors. After the operational amplifier, the regulator is the most popular linear IC. Our attention will be devoted exclusively to *three-terminal regulators*.

9.2 INTEGRATED POSITIVE REGULATOR

Block Diagram

The simplified diagram for a positive voltage regulator is shown in Figure 9-1. This series-type regulator is a basic proven design and lends itself well to integration. The *regulator consists of a pass transistor Q_1, an error amplifier A_1, and a reference voltage V_Z*. The reference establishes a value for the regulator's output voltage. The output voltage will be proportional to the reference voltage, with R_1 and R_2 establishing the constant of proportionality.

$$V_o = V_Z \left(\frac{R_1 + R_2}{R_1} \right) = kV_Z$$

The voltage regulator, in a way, represents an amplifier application. The signal voltage V_Z is fixed, with the bias voltage, V_{in}, being the input variable. The amplifier is a noninverting negative feedback circuit with a closed-loop gain of $1 + (R_2/R_1)$. Transistor Q_1 is within the loop and represents the output stage of the amplifier. It is an emitter follower that is used for its current and power gain capabilities.

Error Amplifier and Pass Transistor

For fixed-output, three-terminal regulators, resistors R_1 and R_2 are part of the device. *The error amplifier is a simplified voltage amplifier.* It compares a portion of the output voltage, divided by R_1 and R_2, with the reference voltage. When an error or difference exists, it drives the base of Q_1 to compensate for the difference. The emitter of Q_1, which is the output, follows the base. If the output voltage drops below the established value (kV_Z) because of load conditions, A_1's output will go more positive, raising the output voltage. If the output voltage rises above the established value, A_1 will go less positive, lowering the output voltage.

Q_1 *is the regulator pass transistor.* This transistor passes the required output current. Its collector voltage is the nonregulated dc input, and its emitter voltage is the regulated dc output. The transistor's V_{CE} makes up the difference between the two and can be large. The load current is supplied by V_{in} through Q_1 and along with V_{CE} determines the power dissipated by the pass element. Q_1 is designed as a power transistor and occupies over half of the IC chip in the higher current regulators.

Voltage Reference

The performance of the voltage reference is crucial to the overall performance of the regulator. It is the *basis for comparison by the error amplifier in establishing and regulating the output voltage.* The noise, stability, and thermal effects of the reference are directly reflected to the regulator's output. The simplest voltage reference is a zener diode. Monolithic voltage regulators use either a zener diode or a complex circuit that can be modeled as a zener diode. Both can produce a precise and stable reference voltage.

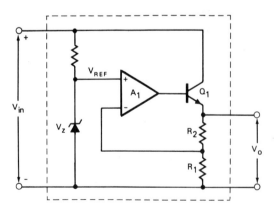

Figure 9-1. Simplified positive voltage regulator.

Protection Features

The proper functioning of the power supply is vital to the operation of electronics equipment and systems. If it is "down," the entire system is inoperable, since all other circuits require the supply's voltage and current to operate. Troubleshooting any part of the system cannot begin until the power supplies are "up." It is also a function that is subjected to the adverse conditions of high temperature, high power dissipation, and excessive loading. Included in the regulator design are features to safeguard the device against these conditions. Figure 9-2 is a simplified regulator circuit and includes the main self-protection features. The three features illustrated are

1. *Thermal shutdown* (Q_4).
2. *Excessive power dissipation protection* (D_1, R_3, and R_4).
3. *Current limiting* (Q_3 and R_{CL}).

The basic regulator circuit has been modified to more closely represent the actual IC design. The output of the error amplifier drives the pass transistor, but its pull-up is from a current source, not previously shown. V_{REF} is now shown as an independent source. Components Q_3, Q_4, D_1, R_4, and R_{CL} have been added to show the protective features of the regulator. Most monolithic regulators today include in their design current limiting, excessive power dissipation protection, and thermal shutdown.

Current limiting is a method of limiting output current to avoid regulator failure resulting from large output current loads or dead shorts. The current-limiting transistor, Q_3, is normally off. Under a heavy load or short condition, the voltage drop across R_{CL} increases enough to turn on Q_3, which, in turn, sinks the base drive current that is being supplied to Q_1. The amount of base drive is limited. With less Q_1 base current, Q_1 cannot conduct as much collector current and is limited until the adverse output condition has been corrected. This technique is similar to that used in the output stage of operational amplifiers.

Excessive power dissipation protection is intended to protect the output pass transistor from breakdown caused by excessive V_{CE} at

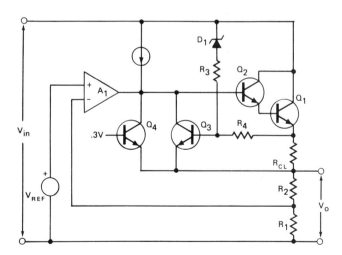

Figure 9-2. Simplified positive voltage regulator with self-protection features.

high currents. If the input voltage, V_{in}, becomes too great, D_1 breaks down. This will cause current to be drawn through R_4, lowering the amount of current it takes through R_{CL} to turn on the current-limiting transistor Q_3. The voltage drops of R_4 and R_{CL} add. The pass transistor has been converted to a Darlington configuration. The addition of Q_2 decreases the loading of the error amplifier.

Thermal shutdown prevents the IC chip from overheating in case of excessive load, momentary short, or increase in ambient temperature. When the chip reaches a temperature of 165°C to 180°C, the output current is limited to prevent further heating. This protective feature uses the breakdown voltage temperature coefficient of the forward-biased junction of Q_4 (about -2.0 mV/°C). The base of Q_4 is biased to about 0.3 V, too low to turn it on at room temperature. As the die temperature exceeds the prespecified value, the turn-on voltage of Q_4 approaches 0.3 V, and it begins to conduct. This again shunts the base drive to Q_1 and limits output current.

The 340 IC Regulator

The complete schematic for the *monolithic 340 regulator* is shown in Figure 9-3. In design and performance, it is similar to another positive regulator, the 78XX. This device contains 16 transistors, 2 diodes, and 18 resistors. Included in the above components are *npn*s, *pnp*s, multiple-collector lateral *pnp*s, a capacitor, and diffused resistors. The functions and their associated components follow.

1. Pass transistor (Darlington): Q_{16}, Q_{15}, R_{15}

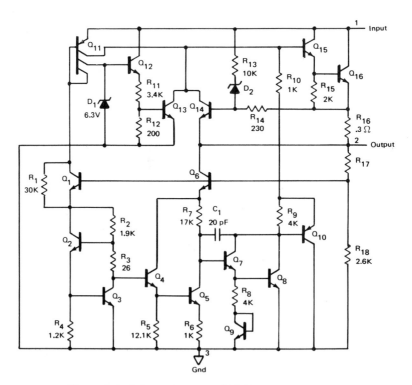

Figure 9-3. Schematic of the 340 IC regulator.

2. Error amplifier: $Q_{11}, R_{10}, R_9, Q_{10}, Q_8,$ Q_7, R_8, Q_9, C_1, Q_6
3. Voltage reference: $Q_{11}, R_1, Q_1, R_2, Q_2,$ $R_3, Q_3, R_4, Q_4, R_5, Q_5, R_6, R_7$
4. Current limiting: Q_{14}, R_{16}
5. Excessive power dissipation: D_2, R_{13}, R_{14}
6. Thermal shutdown: $D_1, Q_{12}, R_{11}, R_{12}, Q_{13}$

The voltage gain for the regulating loop, symbolized in the simplified diagram by an *amplifier* symbol, is provided by the *common emitter stage of* Q_8. It is configured as a Darlington pair with Q_7, which buffers its input and increases the stage's gain. The collector load for Q_8 is a current source. Q_8's collector current is sourced from the multiple-collector pnp, Q_{11}, through R_{10}. The voltage at the junction of R_{10} and the collector of Q_{11} drives the pass transistors Q_{15} and Q_{16}. The capacitor, C_1, is connected between the collector and base of the Q_7-Q_8 gain stage. It provides negative feedback and frequency stability for the circuit.

Resistors R_{17} and R_{18} are the gain setting resistors used in establishing the value of output voltage. They are the equivalent of R_2 and R_1 in the regulator simplified diagram. The junction of the two resistors is connected to the base of Q_6, whose emitter voltage is the temperature-stable V_{REF}.

The collectors of Q_{13} and Q_{14} are connected to the base of the pass transistor pair. Q_{13} and Q_{14} are normally off and will turn on and shunt the base current of the pass transistors under fault conditions. R_{16} is the current-limiting resistor (R_{CL}). When the voltage drop across this resistor approaches 0.6 V, Q_{14} will turn on and limit the output current of the regulator.

Components D_2, R_{13}, and R_{14} implement the excessive power dissipation protection circuit. When the input voltage to the regulator becomes excessive, D_2 will break down and cause current through R_{13} and R_{14}. The voltage drop across R_{14} will cause less voltage drop across R_{16} than necessary to turn the current-limiting transistor Q_{14} on.

Transistor Q_{13} will turn on when the chip temperature exceeds a predetermined value, approximately 175°C. The reference voltage for the base of this transistor is developed with a conventional zener diode, D_1. This voltage is buffered by Q_{12} and voltage divided with R_{10} and R_{11}. The divided voltage, about 300 mV, is the base bias voltage of Q_{13}. The emitter–base voltage of Q_{13} is the actual temperature sensor. As temperature increases, the turn-on base-to-emitter voltage of the transistor decreases at a rate of about -2 mV/°C. As V_{BE} approaches the base bias voltage, it will turn on and shunt the pass transistor's base current.

Q_1, Q_2, Q_3, Q_4, Q_5, and associated components establish the *reference voltage* at the emitter of Q_6. The reference voltage must be temperature stable and in actuality is the sum of two voltages with opposing temperature coefficients. This circuit technique of generating a voltage reference is called a *band gap* or ΔV_{BE} reference. In the simplified diagram, it is modeled as a zener diode. The circuit operation is somewhat complex and is redrawn in the simplified form in Figure 9-4. In this circuit, Q_1 is connected as a diode whose forward voltage is V_{BE} and whose current is high compared to Q_2. Becuase of the different emitter currents of Q_1 and Q_2, the emitter voltage of Q_2 will be equal to their differences ($V_{BE2} - V_{BE1}$) or ΔV_{BE}. The voltage drop across R_2 will be R_2/R_3 times ΔV_{BE}. The reference voltage at the collector of Q_3 will be the sum of $V_{BE}(Q_3)$ plus $(R_2/R_3)\Delta V_{BE}$. If the gain R_2/R_3 is properly chosen, the negative temperature coefficient (TC) of $V_{BE}(Q_3)$ can be made to cancel the positive TC of ΔV_{BE}, producing nearly zero temperature drift for the reference.

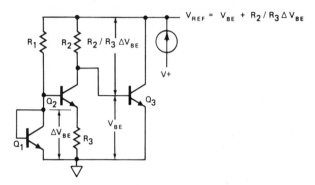

Figure 9-4. Simplified schematic of the band gap reference.

Three-terminal regulators also use zener diode references. These diodes are subject to adverse electrical effects within the chip and are placed below the die surface with a technology known as ion implantation. Figure 9-5 shows the integrated structure of a reference zener diode. Band-gap references are generally chosen for higher current devices (0.5 to 3 A), where they offer low noise without significantly increasing the die area, whereas zeners are chosen for small-die, lower current (0.1 to 0.25 A) devices.

Q_{11} is a multiple-collector *pnp* transistor.

Figure 9-5. Regulator Zener reference: (a) integrated structure; (b) schematic symbol.

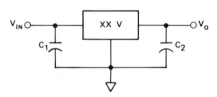

Figure 9-6. Fixed output regulator.

9.3 POSITIVE REGULATOR APPLICATIONS

Fixed-Output Regulator

The monolithic, three-terminal regulator is complete and ready to use. Apply a nonregulated input voltage, and the device produces a fixed, regulated output voltage. Figure 9-6 illustrates the simplicity of its application. An input and output capacitor are added to aid frequency stability and transient response of the circuit.

The two most common positive regulator values are +5 V and +15 V. Five volts is used to power digital circuitry, and +15 V is used for analog circuitry. Typical values of the input and output capacitors are 0.22 µF and 0.1 µF, respectively.

Adjustable-Output Regulator

Although positive regulators that have various output voltage values can be purchased, many

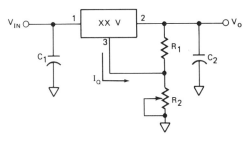

Figure 9-7. Adjustable output regulator.

times odd voltages or variable voltages are required. *The addition of two resistors, R_1 and R_2 in Figure 9-7, converts the fixed output regulator to a variable output.* The output voltage of the regulator circuit is

$$V_o = V_{REG} + \left(\frac{V_{REG}}{R_1} + I_Q\right) R_2$$

where V_{REG} is the regulator voltage and I_Q is the device's quiescent current. Quiescent current is a parameter of monolithic voltage regulators and is defined as that part of the input current that is not delivered to the load. Its value can be found in the data sheet and is moderately constant. In this circuit, the output voltage is the sum of the voltage drop across R_2 and the regulator voltage (pins 2 to 3). The voltage drop across R_2 is the sum of the I_1 and I_Q currents times R_2. The value of the current I_1 is equal to V_{REG}/R_1, and I_Q, from terminal 3, is a function of the particular type of regulator. If R_2 is a potentiometer, then V_o can be adjusted over a range of values. Capacitors C_1 and C_2 are added for stability.

Example

The typical quiescent current for an 8 V 340 regulator is 8 mA. If R_1 is a fixed 1 kΩ resistor and R_2 is a 250 Ω potentiometer, then

$$V_o = 8 \text{ V} + \left(\frac{8 \text{ V}}{1 \text{ k}\Omega} + 8 \text{ mA}\right) \cdot R_2$$

and the output voltage can be adjusted from +8 to +12 V.

Current Regulator

The regulator is a voltage-in, voltage-out device. However, it can be configured as a *voltage-in, current-out device*. The voltage between pins 2 and 3 of the regulator (Figure 9-8) is a fixed value. If a resistor R_1 is added, then the constant voltage across the resistor will develop a constant current. The only major error is the quiescent current of pin 3, but it is a known constant value. Hence,

$$I_o = \frac{V_{REG}}{R_1} + I_Q$$

Example

For a +12 V regulator, $I_Q = 8$ mA, $R_1 = 1$ kΩ, and the output current is

$$I_o = \frac{12 \text{ V}}{1 \text{ k}\Omega} + 8 \text{ mA} = 20 \text{ mA}$$

This circuit will deliver a constant 20 mA to the load. The maximum load voltage will

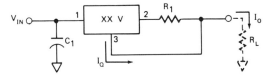

Figure 9-8. Current regulator.

depend on the maximum unregulated input voltage to the regulator.

High-Current Voltage Regulator

Large electronic stysems use power supplies that deliver extremely *heavy load currents*. The voltage regulator must also be capable of handling this heavy current. Monolithic voltage regulators are usually not capable of delivering more than 5 A because of the power dissipation of the pass transistor and its lower gain at the higher current level. These problems are circumvented by using an external transistor, as shown in the 10 A regulator of Figure 9-9. At lower current levels, Q_1 is off. The current is sourced by V_{in} through R_1 and the regulator to the load. When the current through R_1 develops a voltage drop of about 0.6 V, Q_1 will turn on and shunt the excess current to the load. Voltage regulation at the output is still performed by the regulator.

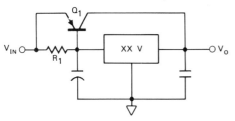

Figure 9-9. High-current voltage regulator.

High-Current Regulator with Short-Circuit Protection

The circuit in Figure 9-10 provides a *high-output current* and takes advantage of the internal current limiting of the regulator to provide *short-circuit (SC) current protection* for the booster transistor as well. If the diode voltage V_{D1} equals the V_{BE} of Q_1, the regulator and Q_1 share load current according to

$$I_1 = \frac{R_2}{R_1} I_{REG}$$

Similarly, during output short circuits

$$I_1 \text{ (SC)} = \frac{R_2}{R_1} I_{REG} \text{ (SC)}$$

If the thermal characteristics of the regulator and Q_1 are similar, the thermal protection of the regulator will be extended to Q_1. Resistor R_3 decreases the response time of Q_1 to load changes.

9.4 NEGATIVE REGULATOR APPLICATIONS

Fixed-Output Regulator

The application of negative three-terminal regulators is similar to that of positive three-terminal regulators. *Apply a negative nonregulated input voltage, and the device produces*

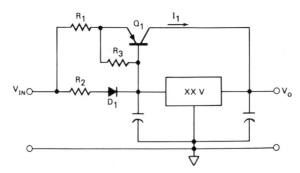

Figure 9-10. High-current regulator with short-circuit protection.

NEGATIVE REGULATOR APPLICATIONS

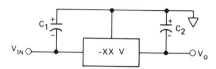

Figure 9-11. Fixed-output regulator.

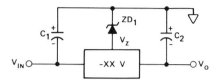

Figure 9-12. Fixed-output regulator of odd value.

a fixed, negative, regulated output voltage. Figure 9-11 illustrates the simplicity of its application. An input and an output capacitor are added to aid frequency stability and transient response.

The most common negative regulator values are -5 and -15 V. Negative five volts is used to power digital circuitry (ECL), and -15 V is used to power analog circuitry. Typical values of the input and output solid tantalum capacitors are 2.2 and 1 μF, respectively.

The 320 and 79XX are two of several popular negative regulators. All their designs, in principle, are similar.

Adjustable-Output Regulator

Although negative regulators that have various output voltage values can be purchased, many times *odd voltages or variable voltages* are required. The technique of using a resistor divider to increase the output voltage in the positive regulator circuit can also be used with negative regulators.

The output voltage may also be raised by simply placing a *zener diode in series with the ground pin*, as indicated in Figure 9-12. The ground or quiescent current, 1 mA for the 320, biases the zener diode on. The zener diode must have a low temperature coefficient where low output voltage drift is required. The output voltage is the sum of the zener and regulator voltages.

Basic Dual Supplies

Figure 9-13 is a schematic diagram of a simple *dual output supply* with the output voltage equal to the preset value of the regulators. Diodes D_2 and D_3 protect the supply against polarity reversal of the outputs during overloads. Diode D_1 and resistor R_1 allow the positive regulator to start up when $+V_{in}$ is delayed relative to $-V_{in}$ with a heavy load drawn between the outputs. Most positive regulators will latch up during start-up, with a heavy load current flowing to the negative regulator.

The addition of three resistors and two trim potentiometers (Figure 9-14) converts the fixed

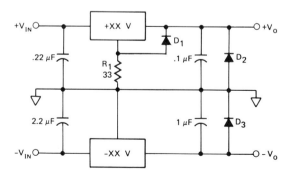

Figure 9-13. Fixed dual supply.

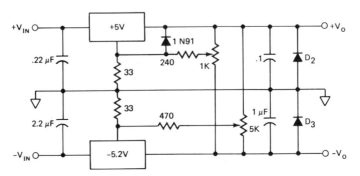

Figure 9-14. Trimmable dual supply.

dual supply to one whose output can be *adjusted to a tighter tolerance*.

Variable Output Lab Supply

The circuit in Figure 9-15 will provide 0 to −20 V output with up to 1 A load current. It uses the 120 as a pass element, a 301A as an external error amplifier, and a 113 as the voltage reference. The regulator is short-circuit proof and, together with the amplifier, features superior regulation characteristics at all voltages.

Transformer T_1 steps down the line ac voltage and has a tapped secondary to provide the positive voltage for the op amp. D_1 and D_2 are half-wave rectifiers. The positive-rectified output of D_1 is filtered by R_1 and C_1 and applied to the V^+ pin of the op amp. The negative-rectified output of D_3 is filtered and connected to the nonregulated input of the regulator. The ground pin of the regulator is driven by the error amplifier, whose voltage reference at the noninverting input is established by R_6 and the 113. The error amplifier will voltage drive the regulator until the voltage at the amplifier in-

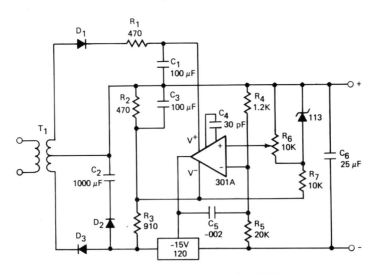

Figure 9-15. Lab supply 0--20 V.

verting input is approximately the same as the voltage established at the noninverting input. The voltage at the inverting input is the output voltage divided by R_4 and R_5. Capacitors C_4 and C_5 provide frequency compensation and stability for the amplifier. C_6 prevents overshoot and minimizes ripple and noise.

Problems

1. In the simplified positive regulator circuit of Figure 9-1, determine the output voltage if $V_z = 1.5$ V, $R_1 = 2$ kΩ, and $R_2 = 18$ kΩ. What is the smallest load resistor that the regulator can drive if its rated output is 200 mA maximum?

2. The thermal shutdown transistor Q_4 shown in Figure 9-2 is biased to 0.350 V at 25°C, its B-E barrier potential is 0.600 V, and the B-E temperature coefficient is -2.2 mV/°C. At what temperature will the device turn on if the die temperature is 75°C in an environment of 25°C?

3. The current-limiting resistor R_{CL} in Figure 9-2 is 3 Ω. The barrier potential of Q_3 is 0.660 V. If Q_3's base current is neglected, at what value of load current will Q_3 turn on and begin to cause the regulator to limit current?

4. In Figure 9-2, $V_O = 5$ V, $R_3 = 10$ kΩ, $R_4 = 130$ Ω, $R_{CL} = 3$ Ω, $R_L = \infty\Omega$, $V_Z(D_3) = 6.3$ V, $V_{BE}(Q_3) = 0.600$ V, and $V_i = 18$ V. At what load current will the regulator begin to current limit? What is the voltage drop across the 130 Ω resistor?

5. What components in Figure 9-3 correspond to $D_1, R_3, Q_1, Q_2, R_{CL}, R_1, R_2, R_4, Q_3$, and Q_4 in Figure 9-2?

6. The reference voltage in Figure 9-4 is 1.8 V. Find the value of ΔV_{BE} if $V_{BE} = 0.550$ V, $R_2 = 6$ kΩ, and $R_3 = 600$ Ω.

7. A fixed 12 V regulator, Figure 9-7, must be made adjustable from 12 V to 18 V. Its quiescent current is 5 mA. If $R_1 = 1.5$ kΩ, find R_2.

8. Determine the value of R_1 in the circuit of Figure 9-8 if I_O must be 20 mA, I_Q is 5 mA, and an 18 V regulator is used.

9. At what value of R_L will Q_1 in Figure 9-9 begin to conduct if R_1 is 3.5 Ω, the barrier potential of Q_1 is .660 V, and an 18 V regulator is used?

10. Modify the block diagram of the positive regulator in Figure 9-1 to make it a negative voltage regulator.

11. Modify the circuit in Figure 9-7 to make it an adjustable negative voltage regulator. What is the output voltage range if the regulator voltage is -5.2 V, R_2 is a 500 Ω potentiometer, and R_1 is 510 Ω. Assume the quiescent current is very, very small. What is the maximum no-load regulator output current?

12. What is the output voltage of the circuit in Figure 9-12 if $C_1 = 2.2$ μF, $C_2 = 1$ μF, $V_Z = 1.8$ V, and the regulator voltage is -5.2 V?

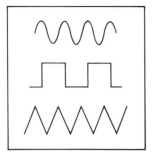

10 OSCILLATION CONCEPTS AND PRINCIPLES

Oscillation is an effect that repeatedly and regularly fluctuates about some mean value. Oscillators are circuits that produce oscillations. This chapter discusses the concepts and principles of oscillation and basic oscillator circuits.

The oscillator is basically described by its waveshape, frequency, and amplitude. The most common oscillator waveshapes are sinusoidal and rectangular wave. Frequency describes the number of cycles occurring per second, and amplitude is a measurement of the oscillator signal's magnitude. The positive feedback system model serves as a sound conceptual basis for numerous oscillator types, and it demonstrates the conditions necessary for oscillation. The Wien bridge is a common low-frequency sinusoidal oscillator, with the LC tuned tank oscillator found more in the high-frequency cases. A single op amp circuit illustrates the rectangular oscillator features.

10.1 THE CONCEPT OF OSCILLATION

Oscillation is the act or state of oscillating. It is an effect expressed as a quantity that repeatedly and regularly fluctuates above or below some mean value. *To oscillate means to swing or move to and fro; to vibrate.*

An oscillator is a device or machine that produces oscillations. An example of a simple mechanical oscillator is the pendulum. The motion of the pendulum's arm is cyclic and periodic and is called simple harmonic motion. It swings from one limit (maximum height in one direction) to the other limit (maximum height in the other direction). Its motion or cycle is averaged about a mean, which, for the pendulum, is the lowest height of the pendulum's arm. The amplitude of each cycle will be reduced because of the loss (energy conversion) due to friction, wind resistance, and the like. To maintain a constant amplitude, energy would have to be introduced into the system in a prescribed manner to make up for the loss.

In any oscillator, there is an *interchange of energy forms.* For the pendulum, the potential energy (the energy of position) is maximum and the kinetic energy (the energy of motion) is zero when the arm is at its maximum height. The amount of kinetic and potential energy shifts until the kinetic energy is maximum and the potential energy is zero. This occurs when

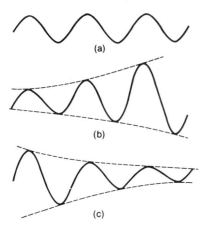

Figure 10-1. Oscillations: (a) critically damped; (b) underdamped; (c) overdamped.

into the system in excess of that lost, the oscillator's amplitude will increase in an exponential manner. The oscillations are referred to as underdamped. Figure 10-1 illustrates the three cases.

An electronic circuit that produces oscillations is also called an oscillator. It is a circuit whose output voltage is a cyclic and periodic voltage with a constant amplitude. The circuit does not have a signal input, but energy is introduced into the system through the dc bias supply to maintain a constant amplitude output. The most common output voltage waveforms of electronic oscillators are sinusoidal, rectangular, and sawtooth. Nonsinusoidal oscillators are called relaxation oscillators.

the pendulum's arm is at its lowest position. An analogous situation occurs in electronic oscillators where the kinetic energy (current) and potential energy (voltage) forms are interchanged.

If energy is not introduced into an oscillating system to sustain it or to compensate for the system's energy loss, the oscillations will decay. The oscillations or oscillatory system is then referred to as overdamped. The oscillations remain periodic, but the amplitude of oscillation decreases to zero at an exponential rate. If an amount of energy is introduced to exactly replenish the loss, the amplitude of oscillation will be sustained at the same magnitude. The oscillations or system is referred to as critically damped. If an amount of energy is introduced

10.2 OSCILLATOR CHARACTERISTICS

The *primary characteristics* (Figure 10-2) of the oscillator are (1) *waveshape*, (2) *frequency*, and (3) *amplitude.* Secondary characteristics include distortion, stability, programmability, range, and those characteristics unique to the components and circuit implementing the oscillator. For certain applications, the secondary characteristics can be of prime importance.

Waveshape

Waveshape is the characteristic of the oscillator that describes the *outline or form* of the output signal. The most common output voltage waveforms are sinusoidal, rectangular, and

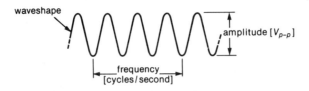

Figure 10-2. Primary oscillator characteristics.

OSCILLATOR CHARACTERISTICS

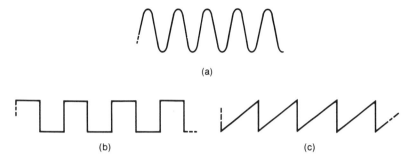

Figure 10-3. Common oscillator output voltage waveforms: (a) sinusoidal; (b) square wave; (c) sawtooth.

sawtooth. These basic output waveforms are shown in Figure 10-3. A sinusoidally varying voltage follows the sinusoidal trigonometric function. Its mathematical model is

$$v(t) = V_p \sin \omega t = V_p \sin 2\pi f t$$

The peak amplitude of the sine wave is V_p volts. Its frequency f is $1/T$ cycles per second (hertz), where T is the time period of one cycle.

$$f = \frac{1}{T}$$

The variable time or t varies from 0 to T seconds, which forces the voltage $v(t)$ to vary from 0 V (at $t = 0$) to V_p (at $t = T/H$) in a sinusoidal manner. The variable $v(t)$ means "the voltage as a function of time." The value of v depends on the specific t or the value of t. The mathematical modeling of the rectangular and sawtooth waves is mathematically complex.

Frequency

The output of an oscillator is cyclic, repetitive, and continuous. *Frequency* refers to the *number of cycles* that occur *in unit time* or one second. A cycle is one revolution or complete iteration. It is one of a succession of periodically recurring events. Frequency can be described in cycles per second or radians per second. Cyclic frequency f is measured in hertz, or cycles per second, and radian frequency ω is measured in radians per second. The two are related by

$$\omega = 2\pi f$$

There are 2π radians per cycle.

The specific frequency that an oscillator is designed for depends on its application. For timers, clocks, test generators, and some signal sources, the frequency is low, typically less than 10 MHz. For these applications, the oscillator is frequently implemented with an operational amplifier circuit. For high-frequency applications (e.g., communications equipment) oscillators are implemented with transistors and discrete parts. In these circuits, the distributed and device inductance and capacitance and the physical characteristics of the circuit play as important a role as the discrete components.

Amplitude

Amplitude refers to the *magnitude* or *size* of the output voltage of an oscillator. For sinusoidal oscillators, amplitude is measured in rms volts or peak-to-peak volts. The two are related by

$$V_{rms} = 0.707 V_p = 0.3535 V_{p-p}$$

For nonsinusoidal voltages, the amplitude is measured in peak-to-peak volts. Sinusoidal output voltages are always symmetric about a reference (dc voltage), but nonsinusoidal outputs are usually not.

Secondary Characteristics

Distortion is the characteristic that describes the *purity* of a sinusoidal signal. An actual sinusoidal voltage of a given frequency will contain components of other frequencies called harmonics. A harmonic is an integral multiple of the base or fundamental frequency. Harmonic distortion describes or measures the amount of a harmonic relative to the fundamental, and it is expressed as a percentage. If V_f is the magnitude of the voltage of the fundamental frequency and V_n is the magnitude of the nth harmonic, the harmonic distortion is

$$\% \ (n\text{th harmonic}) = \frac{V_n}{V_f} \, 100$$

Test equipment and sinusoidal signals used as standards must have low distortion specifications. A typical low harmonic distortion number is 0.05%.

Stability describes the ability of a circuit to maintain one or more of its characteristics constant for some given condition or conditions. The conditions can be time, temperature, environment, or any operating criteria. The oscillator that establishes the frequency of a communications transmitter must be ultrastable.

Programmability and range become very important characteristics for oscillators in test equipment. Programmability allows the equipment operator to *select different values* of one or more of the oscillator's characteristics. *Range* refers to the *degree to which one can select* the various values. A circuit or an instrument can be programmed using front panel controls or by a computer.

10.3 THEORY OF OSCILLATION

RC Feedback Oscillators

An amplifier circuit is an example of a negative feedback control system. The typical *oscillator* circuit is an example of a *positive feedback control system*. The amplifier circuit has an input signal whose effect, at the output, is reduced because of the circuit's negative feedback. Negative feedback is introduced by connecting a component or circuit from the output of an amplifier back to the inverting (−) input. The oscillator does not have an input signal, and its feedback is positive or regenerative. Positive feedback is introduced by connecting a component or circuit from the output of an amplifier back to the noninverting (+) input. Feedback oscillators and amplifiers must have at least one active device or gain-producing element. Transistors and operational amplifiers fall into this category.

The *block diagram* of a positive feedback control system model is shown in Figure 10-4. The technique of finding the gain for the system is similar to that developed for a negative feedback control system. A is the amplifier gain and F the feedback factor. The summation point illustrates positive feedback by algebraically summing the signal source V_s and the feedback signal V_f. The input voltage of the amplifier is called the error voltage V_e.

$$V_e = V_s + V_f$$

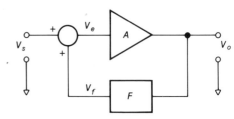

Figure 10-4. Positive feedback control system model.

The output voltage of the feedback network is

$V_f = FV_o$

where F is the feedback factor. Thus,

$V_e = V_s + FV_o$

The amplifier multiplies V_e by A; that is,

$V_o = AV_e = A(V_s + FV_o)$

$V_o(1 - AF) = AV_s$

or $\dfrac{V_o}{V_s} = \dfrac{A}{1 - AF} = A_v(CL)$

The output of the oscillator is a time-varying voltage. This implies that the feedback network will contain inductance and/or capacitance. The impedance of these components will contain phase and magnitude information, and hence F will also contain phase and magnitude information.

An oscillator does not have a signal source. If $V_s = o$, then

$A_v(CL) = \infty$

For the above to be true,

$1 - AF = 0$

Since F or A, and hence AF, will be a complex quantity, then

$|AF| = 1$ (magnitude)

$\phi_{AF} = 0°$ (phase)

These are the required conditions for oscillations to occur.

If the magnitude of the loop gain is greater than 1,

$|AF| > 1$

the amplitude of the oscillation will increase exponentially until, in an actual system, something (the supply voltage) restricts the growth. If the magnitude of the loop gain is less than 1,

$|AF| < 1$

the amplitude of the oscillation will decrease exponentially to 0 V.

The *frequency of oscillation* is the frequency of the signal that travels around the loop. The signal at the input must return back to the input through the feedback network in phase and with the same amplitude. A specific set of values for the components in the feedback network F will establish this condition and hence the frequency of oscillation. If the loop gain AF is sufficient, oscillation is ensured.

The feedback network of an oscillator is typically an RC network or an LC network. The theory of oscillation for both types is nearly identical. However, oscillators with RC feedback networks are normally used in low-frequency applications, and LC oscillators are used in high-frequency cases. The common active device or gain element in low-frequency oscillators is the operational amplifier. In high-frequency circuits, the transistor is the only choice.

Negative Conductance (Resistance) Oscillators

A second circuit configuration that produces sinusoidal oscillations is the *negative conductance (resistance) oscillator*. This type of oscillator *uses an LC tank circuit and a device with a negative conductance (resistance) I-V characteristic*. Examples of devices with this type of characteristic are the tunnel diode, the four-layer switch or *pnpn* diode, and the unijunction (UJT) transistor.

The *circuit model* for the negative conductance oscillator is shown in Figure 10-5a. The inductance and capacitance form a tuned tank circuit that exchanges electrical energy forms and produces a sinusoidal voltage across its terminals. The conductance G_1 models the loss of the circuit. It is the sum effect of the resistances associated with the physical inductor and capacitor. For the G_1 LC circuit only, the

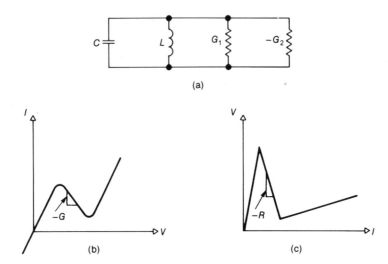

Figure 10-5. Negative conductance and negative resistance characteristics: (a) negative conductance oscillator model; (b) negative conductance, tunnel diode; (c) negative resistance, *pnpn* diode.

oscillations would decay, because any initial energy stored in the circuit would be converted to another energy form (heat). G_2 is a device with a negative conductance characteristic; an example is the tunnel diode of Figure 10-5b. For a certain area in the characteristic, the slope is negative. If this device is forced to operate in its negative conductance region and the slope is equal to G_1, the net oscillator circuit conductance is zero, that is, an open circuit. The tank circuit will resonate at a fixed frequency established by the capacitance and inductance values and with a fixed amplitude.

In terms of power, the *negative conductance device is supplying power to the circuit equal in amount to that lost by the circuit's positive conductance* ($|G_1| = |G_2|$).

LC Feedback Oscillators

Radio-frequency (RF) oscillators generally use an *LC* circuit to fix the frequency of oscillation. In the *LC* circuit, also called a tank, the form of the energy stored is changing from kinetic (current) to potential (voltage) in a cyclic manner. The maximum potential energy occurs when the capacitor is fully charged and the electric field is at its maximum value. At this time, the current and magnetic field are zero. The maximum kinetic energy occurs when the capacitor is fully discharged, and the magnetic field and the current of the inductor is at its maximum. The interchange of energy forms associated with the fields of the capacitor and the inductor is repeated over and over. The potential across the parallel *LC* circuit is a sinusoidal voltage. The *circuit is resonant* when the capacitive and inductive reactances are equal to each other, and the frequency of oscillation is determined by the constants of the circuit.

$$X_L = 2\pi f L = X_C = \frac{1}{2\pi f C}$$

$$f_o = \frac{1}{2\pi \sqrt{LC}}$$

The *practical LC tank circuit* (Figure 10-6) always includes one or more resistances. The

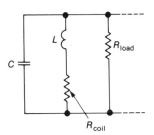

Figure 10-6. Practical *LC* tank circuit.

10.4 BASIC OSCILLATOR CIRCUITS

Sinusoidal: Wien-Bridge Oscillator

The *Wien-bridge oscillator* (Figure 10-8) is a popular low-frequency (<1 MHz) oscillator. It is characterized by two feedback networks. The R_3-R_4 negative feedback network establishes the voltage gain of the circuit, and the R_1-C_1 and R_2-C_2 positive feedback network establishes the phase shift.

Conditions for oscillation require that the signal, V_i, must be returned around the loop in phase and with the same amplitude or magnitude. The voltage gain of the noninverting amplifier circuit is determined by R_3 and R_4.

$$A_v = 1 + \frac{R_4}{R_3}$$

The output of the op amp will be greater in magnitude than V_i by A_v. The R_1-C_1 and R_2-C_2 network will attenuate the output signal as it proceeds from the output back to the noninverting (+) input. This network functions as an ac voltage divider. The degree of attenuation through the positive feedback network must equal the gain provided by the nega-

two most common ones are the coil resistance and the load. The tank resistance converts the electrical energy to another form (heat) and will cause the oscillations to dampen out or decay to zero. To sustain the oscillations, the *LC* tank is placed in a feedback control system with an amplifying device. When the conditions for oscillation are met, the losses are replaced and the oscillations will be sustained at a constant amplitude.

The *tuned-collector oscillator* in Figure 10-7 is a basic *LC* oscillator. The gain element is a transistor (Q_1) that is voltage divider (R_1-R_2) biased. The tuned *LC* tank circuit ($C_1 L_1$) is in the collector, and the feedback transformer provides the proper phase (transformer dot orientation) and magnitude (winding ratio) to ensure oscillation. The output of the *LC* tank is transformer coupled to the load.

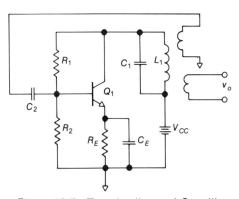

Figure 10-7. Tuned-collector *LC* oscillator.

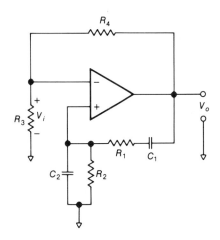

Figure 10-8. Wien-bridge oscillator.

tive feedback circuit. The phase of V_i relative to the amplifier output via the negative feedback circuit is 0° because of the noninverting mode. Hence, the phase through the positive feedback network must also be 0° to guarantee a net 0° phase shift. The signal at the input is

$$V_i = \left\{ \frac{Z_2}{Z_1 + Z_2} \right\} V_o$$

where

$$Z_1 = R_1 + \frac{1}{j\omega C_1}$$

and $Z_2 = \dfrac{R_2/j\omega C_2}{R_2 + (1/j\omega C_2)}$

Thus,

$$V_i = \frac{V_o}{\left\{ \dfrac{R_2 C_1 + R_1 C_1 + R_2 C_2}{R_2 C_1} \right\} + j \left\{ \omega R_1 C_2 - \dfrac{1}{\omega R_2 C_1} \right\}}$$

$$= \frac{V_o}{a + jb}$$

If the phase through this network is to be 0°, then the coefficient of j must be equal to 0, or

$$\phi = \tan^{-1} \frac{b}{a} = \tan^{-1} \frac{0}{a} = 0°$$

The frequency of oscillation is determined from the above condition; that is,

$$\omega_o C_2 R_1 - \frac{1}{\omega_o C_1 R_2} = 0$$

or $\omega_o = \sqrt{\dfrac{1}{C_1 C_2 R_1 R_2}}$

With the coefficient of j equal to zero, the gain function reduces to

$$\frac{V_o}{V_i} = \frac{1}{1 + (R_1/R_2) + (C_2/C_1)}$$

To ensure a loop gain of 1 the gain through the negative feedback circuit must equal the attenuation through the positive feedback circuit.

$$1 + \frac{R_4}{R_3} = 1 + \frac{R_1}{R_2} + \frac{C_2}{C_1}$$

For the special case where $R = R_1 = R_2$ and $C = C_1 = C_2$,

$$\omega_o = \frac{1}{RC}$$

and the minimum gain for oscillation is 3. Consequently,

$$R_4 \geq 2R_3$$

will ensure that the circuit will oscillate.

Tungsten Lamp Amplitude Stabilization Regulator

The *amplitude of the Wien-bridge oscillator* may be established by replacing R_3 with a tungsten lamp, as shown in Figure 10-9a. When power is first applied to the circuit, the *tungsten lamp* has a low resistance, and the amount of negative feedback is small. For this case, the loop gain AF is greater than 1, and the oscillations build up at the resonant frequency. As the oscillations increase in magnitude, the tungsten lamp heats up and its resistance also increases. When the loop gain is 1, the oscillator output is maintained constant. For the case where $R_1 = R_2$ and $C_1 = C_2$, the loop gain will be 1 when the resistance of the tungsten lamp is one-half of that of R_4. The gain through the negative feedback loop for this case will be 3. Data or specifications sheets for the tungsten lamp provide a graph of lamp resistance versus voltage (Figure 10-9b). The output voltage of the oscillator will be three times the lamp voltage whose corresponding value of lamp resistance satisfies the condition that the gain through the negative feedback loop is 3.

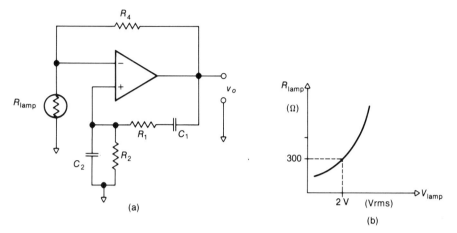

Figure 10-9. Establishing the amplitude of a Wien-bridge oscillator: (a) oscillator circuit; (b) graph of lamp resistance versus voltage.

Example

Design a Wien-bridge oscillator for a resonant frequency of 2 kHz and an output amplitude of 6 V (rms).

Let $R_1 = R_2$ and $C_1 = C_2$. Hence, $f_o = 1/(2\pi RC)$. Select $R_1 = 1$ kΩ. Thus,

$C = 1/(2\pi f_o R_1) = 0.0796$ μF

From Figure 10-9b, the lamp resistance will be 300 Ω at 2 V (rms).

$1 + \dfrac{R_4}{300 \; \Omega} = 3$

$R_4 = 600 \; \Omega$

JFET Amplitude Stabilization Regulator

The circuit in Figure 10-10 is a *Wien-bridge oscillator with a different amplitude stabilization scheme.* An important advantage of this circuit is that the traditional filament lamp amplitude regulator is eliminated, along with its time constant, linearity, and reliability problems.

The negative feedback circuit gain is basically established by R_4 and the drain-to-source resistance of the JFET transistor Q_1.

$A_v \cong 1 + \dfrac{R_4}{r_{ds}}$

The capacitor C_3 has a low-frequency rolloff, and its reactance is low at the frequency of oscillation. It prevents the dc offset voltage and offset current errors from being multiplied by the amplifier gain. The JFET drain-to-source or channel resistance is established by the gate voltage, which is the dc voltage stored in C_4. Negative output voltage peaks in excess of -8.25 V will cause D_1 and D_2 to conduct and thus charge C_4. The time constant of C_4 and R_5 is long compared to the oscillation time period. R_3 is chosen to adjust the negative feedback loop so that the JFET is operated at a small negative gate bias.

The frequency of oscillation is still determined by the positive feedback path provided by the R_1-C_1 and R_2-C_2 branches.

$f_o = \dfrac{1}{2\pi \sqrt{R_1 R_2 C_1 C_2}}$

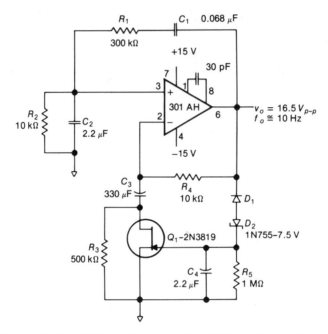

Figure 10-10. JFET amplitude stabilized Wien-bridge oscillator.

Rectangular Wave Oscillator: Op Amp

Rectangular wave oscillators are also referred to as astable or free-running multivibrators. The *output* of this class of oscillators *is a rectangular wave* whose up and down times may or may not be the same. The percentage of the up time in relation to the waveform's total period is called the *duty cycle*.

$$\text{Duty cycle (\%)} = \frac{\text{up time}}{\text{up time + down time}} \; 100$$

A square wave has a 50% duty cycle and is time symmetrical, but all others are not. Symmetric and nonsymmetric waveforms are illustrated in Figure 10-11. The ideal rectangular or square wave has a sharp rise and fall time or an infinite slope. In practice, this cannot be achieved; however, the finite rise and fall time should be significantly smaller ($<\frac{1}{20}$) than the waveform's time period.

The *low-frequency rectangular wave oscillator* in Figure 10-12 uses an op amp, four resistors, and a capacitor. The negative feed-

Figure 10-11. Rectangular-wave oscillator outputs: (a) symmetric, 50% duty cycle (square wave); (b) nonsymmetric, 25% duty cycle.

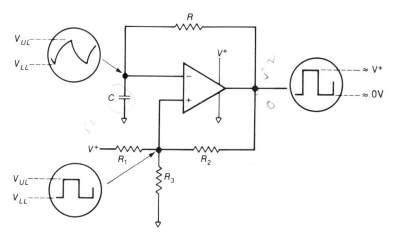

Figure 10-12. Rectangular wave oscillator.

back loop consists of a resistor R and capacitor C. The positive feedback loop consists of a three-resistor voltage divider. The amplifier is biased with a single supply connected to the amplifier supply terminals. The voltages at these pins establish the limits or amplitude of the oscillator output.

$$V_o \text{ (amplitude)} \cong V_{p-p}^+$$

The amplifier does not operate in its linear region, and the output is at one saturation limit or the other except for the brief transitional time.

The voltage at the inverting (−) amplifier input will be an exponentially growing and decaying waveform caused by the charging and discharging of the capacitor C. The charging and discharging of the capacitor is caused by the output of the oscillator (which drives R) changing from one output state to the other.

The voltage at the noninverting (+) amplifier input is a square wave whose limits are established by the R_1-R_2-R_3 voltage divider. The oscillator output drives one end of R_2, and since it switches between two values, so will the voltage at the noninverting input.

The *amplifier basically functions as a com-parator* whose comparison or trip limit alternates between two values. The reference voltages, or trip limits, for the (+) input are established by the R_1-R_2-R_3 divider. The time-varying exponential voltage at the (−) input charges or discharges toward these limits and causes the output to change states when these limits are reached. When the output changes state, a new reference voltage is established, and the exponential waveform changes direction toward the new reference. This process is continually repeated.

Frequency of Oscillation

To somewhat ease the analysis to determine the frequency of oscillation, R_1, R_2 and R_3 are made equal in value.

$$R_1 = R_2 = R_3 \quad \text{(square-wave output)}$$

Resistor R_2 will either shunt R_1 or R_3, depending on the output state of the oscillator. If the output is high, that is, at V^+ volts, then

$$V_{NI} = \frac{R_3}{R_3 + R_1 \| R_2} V^+ = \frac{R_3}{R_3 + (R_3/2)} V^+$$

$$= \frac{2}{3} V^+ \quad \text{(upper limit } V_{UL})$$

If the output voltage is low, that is, near 0 volts, then

$$V_{NI} = \frac{R_3 \| R_2}{R_3 \| R_2 + R_1} V^+$$

$$= \frac{1}{3} V^+ \quad \text{(lower limit } V_{LL})$$

Thus, the voltage waveform at the noninverting (+) input is a square wave whose maximum value is $\frac{2}{3} V^+$ and whose minimum value is $\frac{1}{3} V^+$.

The voltage at the inverting (−) input is an *exponentially growing or decaying waveform*. For $V_o = V^+$, this voltage is charging toward V^+, and when it slightly exceeds $\frac{2}{3} V^+$, it will cause the amplifier output to switch to the low state. The capacitor, at that time, is charged to $\frac{2}{3} V^+$ volts and then discharges toward 0 V. The discharge circuit is shown in Figure 10-13a. The mathematical model describing this relationship is

$$V_I = \frac{2}{3} V^+ e^{-t/RC} \quad \text{(discharge circuit)}$$

$$V_I = V_{UL} e^{-t/RC}$$

The capacitor voltage will discharge toward 0 V, but when it is slightly less than $\frac{1}{3} V^+$, it will cause the output to switch to the high (V^+) state. The time it takes to discharge from $\frac{2}{3} V^+$ to $\frac{1}{3} V^+$ is determined from the above relationship; that is,

$$\frac{1}{3} V^+ = \frac{2}{3} V^+ e^{-t/RC} \quad \text{(discharge circuit)}$$

Solving for $t_{discharge}$,

$$t_{discharge} = RC \ln 2$$

The exponentially growing voltage at the inverting input during the charge portion of the cycle (Figure 10-13b) is

$$V_I = V^+ - \frac{2}{3} V^+ e^{-t/RC} \quad \text{(charge circuit)}$$

$$V_I = V^+ (V^+ - V_{LL}) e^{t/RC}$$

When V_I reaches $\frac{2}{3} V^+$, the output will switch.

$$\frac{2}{3} V^+ = V^+ - \frac{2}{3} V^+ e^{-t/RC} \quad \text{(charge circuit)}$$

Solving for t_{charge},

$$t_{charge} = RC \ln 2$$

The *period of the oscillation* is

$$T = t_{charge} + t_{discharge}$$

or $T = 2RC \ln 2 = 1.386 \, RC$

The *frequency of oscillation* is the reciprocal of the time period.

$$f_o = \frac{1}{T}$$

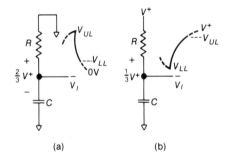

Figure 10-13. Determining the frequency of oscillation: (a) equivalent discharge circuit; (b) equivalent charge circuit.

Circuit Example

Find the frequency of oscillation, the peak-to-peak output voltage, and the duty cycle for the rectangular oscillator in Figure 10-14.

Solution

For high-frequency rectangular-wave oscillators, a comparator is often used in place of the operational amplifier. The circuit in Figure 10-14 uses the high-speed 311 whose bandwidth is higher than most op amps. Since the amplifying device in this type of circuit does not have to operate in its linear region, either a comparator or an amplifier may be used. This

BASIC OSCILLATOR CIRCUITS

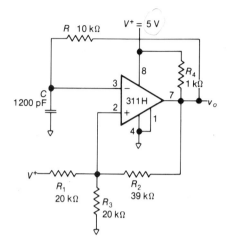

Figure 10-14. Circuit example of a rectangular wave oscillator.

comparator has an open collector output, and resistor R_4 is used as a collector pullup to V^+.

The voltage at the noninverting input V_{NI} is determined by the R_1-R_2-R_3 voltage divider. When $V_o = V^+$, V_{NI} is

$$V_{NI} = \frac{R_3}{R_3 + R_1 \| R_2} V^+ = \frac{3}{5} V^+ = 3.01 \text{ V}$$

When $V_o = 0$ V, V_{NI} is

$$V_{NI} = \frac{R_3 \| R_2}{R_3 \| R_2 + R_1} V^+ \cong \frac{2}{5} V^+ \cong 1.99 \text{ V}$$

When V_o switches from V^+ to 0 V, the capacitor voltage or V_I will discharge from the 3.01 V (upper trip limit) stored across it.

$$V_I = 3 \text{ V } e^{-t/RC}$$

When the capacitor voltage or V_I falls slightly below 2 V (the lower trip limit), the output will switch back to V^+. The time it takes to discharge from 3 V ($\frac{3}{5} V^+$) to 2 V ($\frac{2}{5} V^+$) is the discharge time.

$$2 \text{ V} = 3 \text{ V } e^{-t/RC}$$

$$t_{\text{discharge}} = RC \ln 1.5$$

The inverting input voltage V_I is, for the charge cycle, modeled by

$$V_I = V^+ - \tfrac{3}{5} V^+ e^{-t/RC} = 5 \text{ V} - 3 \text{ V } e^{-t/RC}$$

When V_I reaches 3 V, the output will switch from V^+ to 0 V. The time it takes to charge from 2 V to 3 V is the charge time.

$$3 \text{ V} = 5 \text{ V} - 3 \text{ V } e^{-t/RC}$$

$$t_{\text{charge}} = RC \ln 1.5$$

The oscillation period is

$$T = t_{\text{charge}} + t_{\text{discharge}} = 2RC \ln 1.5 = 0.81 \, RC$$

Since the charge and discharge times are the same, the output will be a square wave and the duty cycle is 50%. The ratio of R_1 to R_3 establishes the duty cycle. For the values given,

$$f_o = \frac{1}{T} = \frac{1}{(0.81)(10 \text{ k}\Omega)(1200 \text{ pF})}$$

$$= 102.8 \text{ kHz}$$

The peak-to-peak output voltage is approximately V^+ or 5 V.

Sinusoidal: LC Models

Oscillators using op amps or comparators are only used for low frequencies, typically less than 1 MHz. *High-frequency oscillators* (about 1 to 500 MHz) *use transistors and an LC resonant circuit.* In addition to the *tuned-collector oscillator*, other prominent members of the *LC* class include the *Hartley, Colpitts, Clapp*, and *crystal oscillators.*

The ac equivalent circuit of the common-emitter version of these oscillators is shown in Figure 10-15. The circuit characteristic of the *Hartley oscillator* (Figure 10-15a) is the split or tapped inductor in the *LC* tank circuit. At resonance ($X_L = X_C$), the tank in the collector appears resistive. The phase shift around the loop is 360° (or 0°), with 180° provided by

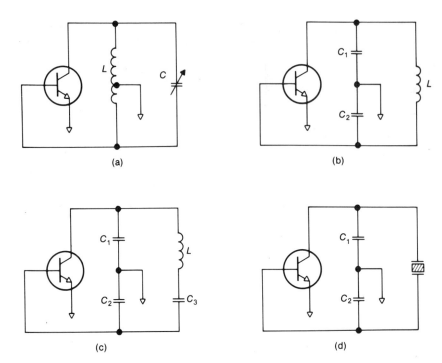

Figure 10-15. *LC* oscillators—ac equivalent circuits: (a) Hartley; (b) Colpitts; (c) Clapp; (d) crystal.

the base-to-collector signal inversion and 180° provided by the tapped inductor. The loop gain (AF) is established by the voltage gain of the transistor (A) and the turns ratio of the split inductor (F).

The circuit characteristic of the *Colpitts oscillator* is the split or tapped capacitor in the *LC* tank. The feedback technique for the Hartley and the Colpitts is the same except for which component is tapped. The amount of feedback in the Colpitts is dependent on the relative capacitance values of C_1 and C_2. The frequency of oscillation can be changed by gang tuning C_1 and C_2 so that their ratio remains fixed.

The *Clapp oscillator* in Figure 10-15c is a modified Colpitts. A third capacitor C_3 is added in series with the inductive branch. The feedback is still dependent on the relative values of C_1 and C_2. However, these capacitors are made much larger than C_3 so that the frequency of the tank circuit is essentially dependent on C_3 and L.

Crystal oscillators possess the important advantage of having an accurate and ultra-stable frequency of oscillation. Feedback is still provided by the capacitive tap in the circuit of Figure 10-15d, but the oscillator frequency is determined by the crystal. The crystal can be modeled as a large inductor in series with a small capacitor.

Problems

1 The mathematical model of a sinusoidal voltage is

$$v(t) = V_p \sin 2\pi ft$$

where the dimension of $2\pi ft$ is radians. What is $2\pi ft$ equal to when the sine function is maximum (one) at 90°?

2 A sinusoidal voltage is 10 V(rms). Find its peak and peak-to-peak values. Can a voltage square wave be measured in rms?

3 What is the only difference between the block diagrams of a positive feedback control system and a negative feedback control system?

4 If the feedback factor in a positive feedback control system is 0.05, what will be the value of the amplifier's gain if the system is to maintain a constant amplitude of oscillation? If the phase of the same feedback network is -90°, what will be the value of the amplifier's phase shift for a constant amplitude of oscillation?

5 The coordinates of two points on the negative resistance characteristic of the tunnel diode in Figure 10-5a are (10 mA, 0.04 V) and (1 mA, 0.4 V). What is the value of R?

6 In the circuit of Figure 10-8, $C_1 = C_2 = 0.01$ μF. Specify the value of $R_1 (= R_2)$ for the circuit to oscillate at a frequency of 5 kHz. If R_4 is 10 kΩ, what is the maximum value of R_3 to insure that the circuit will oscillate?

7 For the values listed in the circuit of Figure 10-10, find the exact (3 figures) value of the frequency of oscillation?

8 Specify the circuit change that is necessary to make the value of the peak-to-peak amplitude of the oscillator in Figure 10-10 equal to 12 V.

9 What is the recommended risetime of a rectangular voltage waveform if its period is 100 μSec?

10 Specify the new value of C in Figure 10-14 to make the frequency of oscillation 20 kHz. What change is necessary to make the peak amplitude of the output equal to 8 V?

11 In Figure 10-12, the resistor values are $R_3(R_3)$, $2R_3(R_2)$, and $4R_3(R_1)$. Find $t_{\text{discharge}}$ and t_{charge}.

12 If the circuit of Figure 10-15d is to oscillate, what must the equivalent circuit of the crystal include?

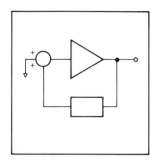

11 OSCILLATORS, INTEGRATED CIRCUITS, AND APPLICATIONS

Oscillators are circuits or monolithic ICs that oscillate. This chapter looks at several popular oscillator ICs and circuits and illustrates their applications.

The RC phase-shift and quadrature oscillators are additional examples of sinusoidal circuits that fall under the positive feedback system model. A function generator is an oscillator circuit with two outputs: a square-wave and a triangular-wave output. A technique of implementing this circuit uses two integrators. The timer IC (555) and the voltage-controlled oscillator or VCO (566) are two examples of monolithic oscillators. The VCO is a device whose output signal frequency is a function of the value of the input voltage. The 566 is a building block also used in the phase locked loop IC or the 565. The phase locked loop or PLL is a versatile servosystem IC whose applications include FM demodulation and motor-speed control. The local oscillator in an AM radio is a common example of a discrete circuit LC oscillator.

11.1 OTHER SINUSOIDAL OSCILLATORS

RC Phase-Shift Oscillator

The *RC phase-shift oscillator* in Figure 11-1 is another example (in addition to the Wien bridge) of an oscillator with a feedback network comprised of resistors and capacitors. The amplifier is in an inverting circuit with a gain magnitude of

$$|A_v| = \frac{R_2}{R_1}$$

The phase shift of the inverting amplifier circuit is 180°. The signal frequency that makes the phase shift through the RC feedback network 180° will be the frequency of oscillation. The feedback network is comprised of three identical, cascaded RC sections whose gain or feedback factor F is $\frac{1}{29}$.

$$F = \frac{1}{29}$$

To ensure oscillation, the loop gain must equal 1.

$$|AF| = 1$$

or $\quad |A_v| = \dfrac{R_2}{R_1} = 29$

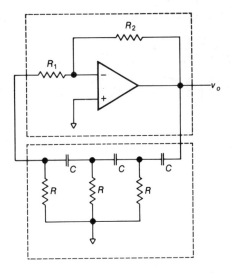

Figure 11-1. RC phase-shift oscillator—op amp.

The frequency of oscillation is given as

$$f_o = \frac{1}{2\pi(2.45)RC}$$

Two *transistorized versions* of the RC phase-shift oscillator are shown in Figure 11-2. The operation of both circuits is the same as that of the op amp version with the exception of the transistor gain. For the FET circuit, the gain is $g_m R_L$, where R_L is the total effective load resistance seen by the output of the transistor. It includes r_d, R_D, and the loading effect of the feedback network.

In the bipolar transistor version, R' is used to compensate for the low input resistance of the transistor. To obtain three matched RC sections, R' is added to the parallel combination of R_1, R_2, and r_i (transistor) to equal R of the other two sections. The bipolar transistor gain

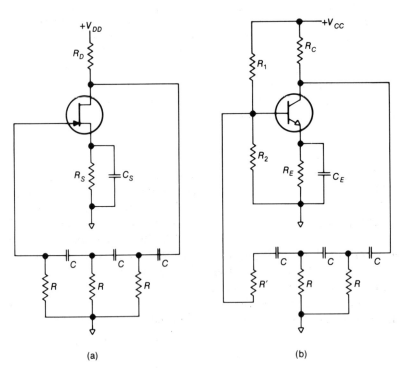

(a)   (b)

Figure 11-2. RC phase-shift oscillator—transistor: (a) JFET; (b) BJT.

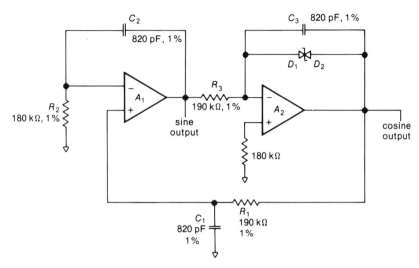

Figure 11-3. Quadrature oscillator.

is equal to r_C/r'_e, where r_C is the parallel combination of R_C, r_c, and the loading effect of the feedback network. The equations for the frequency of oscillation and minimum gain for the op amp version also apply to the transistor circuits.

Almost all oscillator types have amplifier and transistor versions. In addition, the transistor version comes in pairs because of the two transistor types: the bipolar junction transistor (BJT) and the junction field effect transistor (JFET).

Quadrature Oscillator

The *outputs of the quadrature oscillator* (Figure 11-3) *are two sinusoidal signals*. The output of A_1, called the sine output, is 90° out of phase with respect to the output of A_2, the cosine output. The circuit consists of a noninverting integrator A_1, an inverting integrator A_2, and an ac voltage-divider network comprised of R_1 and C_1.

The loop gain AF is the product of the gains of A_1, A_2, and the R_1-C_1 voltage divider. A_1 is a noninverting circuit whose gain is

$$A_{v1} = 1 + \frac{Z_{C2}}{Z_{R2}} = 1 - \frac{jX_{C2}}{R_2} = \frac{R_2 - jX_{C2}}{R_2}$$

The impedance of the capacitor is $-jX_C$. The j is required to convey the phase information present in the capacitor's behavior. A_2 is an inverting circuit whose gain is the ratio of the feedback impedance over the input impedance (minus).

$$A_{v2} = -\frac{Z_F}{Z_N} = -\frac{Z_{C3}}{Z_{R3}} = +\frac{jX_{C3}}{R_3}$$

The gain or the feedback factor F of the feedback network is

$$F = \frac{Z_{C1}}{Z_{C1} + Z_R} = \frac{-jX_{C1}}{-jX_{C1} + R_1}$$

The loop gain AF equals the product of A_{v1}, A_{v2}, and F.

$$AF = \left(\frac{R_2 - jX_{C2}}{R_2}\right)\left(\frac{jX_{C3}}{R_3}\right)\left(\frac{-jX_{C1}}{-jX_{C1} + R_1}\right)$$

The expression can be simplified by multiplying the complex quantities and simplifying the expression by letting $j^2 = -1$. The result is

$$AF = \frac{X_{C1}X_{C3}(R_2 - jX_{C2})}{R_2R_3(R_1 - jX_{C1})}$$

Oscillations will occur when the magnitude of AF equals 1 and the phase of AF equals $0°$. The loop gain is changed to polar coordinate form.

$$AF = \frac{X_{C1}X_{C3}\sqrt{R_2^2 + X_{C2}^2}\;\underline{/-\tan^{-1}(X_{C2}/R_2)}}{R_2R_3\sqrt{R_1^2 + X_{C1}^2}\;\underline{/-\tan^{-1}(X_{C1}/R_1)}}$$

The *magnitude of the loop gain* is

$$|AF| = \frac{X_{C1}X_{C3}\sqrt{R_2^2 + X_{C2}^2}}{R_2R_3\sqrt{R_1^2 + X_{C1}^2}}$$

The *phase of AF* is the angle in the numerator minus the angle in the denominator.

$$\phi = \tan^{-1}\left(\frac{X_{C2}}{R_2}\right) - \tan^{-1}\left(\frac{X_{C1}}{R_1}\right)$$

To simplify the circuit and the analysis, the following conditions are established.

$$R_1 = R_2 = R_3 = R$$
and $\quad C_1 = C_2 = C_3 = C$

The expression for the magnitude of the loop gain reduces to

$$AF = \frac{X_{C1}X_{C3}}{R_2R_3} = \frac{1/(4\pi^2 f_o^2 C^2)}{R^2}$$

If the magnitude of the loop gain is set equal to 1, the equation can be solved for the frequency of oscillation f_o.

$$f_o = \frac{1}{2\pi RC}$$

For the values given, f_o is approximately 1 kHz. The proper phase condition is met because

$$\phi = \tan^{-1}\frac{X_C}{R} - \tan^{-1}\frac{X_C}{R} = 0°$$

To ensure sufficient positive feedback, R_1 is slightly larger than R_2.

Zener diodes D_1 and D_2 limit the amplitude of the output signal without significantly affecting the waveform of the cosine. Because of the filtering the signal receives from the R_1-C_1 and R_2-C_2 networks, the sine output will be nearly distortionless. Measured distortion at both outputs is less than 1% at a 15 V peak-to-peak output level.

The frequency of oscillation and the values of the components are such that any of the popular IC op amps can be used. These include the 709, 741, and the 301A. The standard frequency compensation capacitor or network must be added for the 709 and 301A amplifiers.

Function Generator

Figure 11-4 is a schematic drawing of a *circuit that generates two voltage waveforms*. It is called a function generator, because one of its voltage outputs, A_2, is *triangular* in shape and the other output, A_1, is a *voltage square wave*. The generator consists of an integrator A_2 and a threshold detector or comparator A_1. The integrator implements the advanced mathematical function of integration. For a fixed dc voltage on its input, the output of the integrator will be a negative-going voltage ramp. The ramp will be positive going for a fixed negative dc voltage on its input. The model for the integrator is a constant current source charging a capacitor. The output of the comparator A_1 is a square wave whose polarity is a function of the polarity of the voltage at its noninverting input.

The complete circuit has three feedback paths: (1) C_1 around A_2, (2) R_3 around A_1, and (3) R_2 from A_2 to A_1. C_1 and R_2 provide negative feedback, and R_3 provides positive feedback. The net feedback for the circuit is positive and is a requirement for regenerative circuits or those that produce a repetitive output with no input required.

To understand the operation of the circuit, let the output of A_1 be at its positive saturation

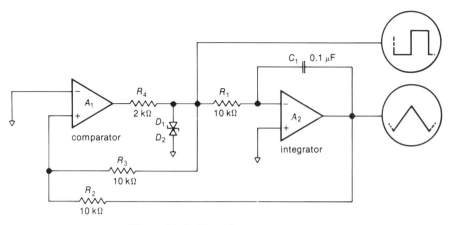

Figure 11-4. Function generator.

level. This fixed positive dc voltage is the input of the integrator. The integrator's output will be a negative-going voltage ramp. R_2 and R_3 form a voltage divider whose end points are the positive saturation voltage of A_1 and the steadily decreasing negative-going voltage of A_2. The junction of these two resistors is sensed by the noninverting input of A_1 with the inverting input at ground or 0 V. When the noninverting input senses a slightly negative voltage (originally it was positive), the output of A_1 will swing from its positive saturation voltage to its negative saturation voltage. The voltage ramp of A_2 will now stop going negatively and will begin to go positively. It will continue to increase until the junction of R_2 and R_3 goes to a slightly positive voltage with respect to ground. At this time A_1's output is switched back to the positive saturation level. This procedure is repeated continuously.

The triwave frequency is determined by R_1, C_1, and the positive and negative saturation voltages of the amplifier A_1. The frequency of the triangular and square voltage waveforms will be the same. The amplitude of the output voltage waveforms is established by the R_2/R_3 ratio and A_1's saturation voltages. The generator may be made independent of the saturation voltages (and hence the supply voltages) by clamping the output of A_1 with matched, back-to-back zener diodes.

Example

To simplify an analysis (and the design) of this circuit, resistors R_2 and R_3 are set equal to each other, and the back-to-back zener diodes are selected for a total voltage of 10 V. The output of A_1 will be a 10-V peak voltage square wave. The peak output of the triangular wave will also be 10 V. If $R_2 = R_3$, the noninverting input of A_1 will be at zero volts when the output of A_2 is of the same magnitude (10 V) and of the opposite polarity as the output of A_1. When the noninverting input of A_1 is at zero volts, the output of A_1 will switch to the opposite level or state.

The slope of the ramp of A_2 is determined by the dc current charging the capacitor and the capacitor's value. The current charging the capacitor is equal to the output voltage of A_1 divided by the input resistance R_1.

$$I_{C1} = \frac{V_o(A_1)}{R_1} = \frac{10 \text{ V}}{R_1}$$

230 OSCILLATORS, INTEGRATED CIRCUITS, AND APPLICATIONS

The slope of the voltage ramp is equal to the change in voltage (Δv) divided by the change in time (Δt).

$$m = \text{slope} = \frac{\Delta v}{\Delta t} = \frac{I_{C1}}{C_1} = \frac{10\text{ V}}{R_1 C_1}$$

In one oscillation period, the voltage ramp will go from -10 V to $+10$ V and back to -10 V. The total change in voltage is 40 V. The total change in time or the period of oscillation is

$$T = \frac{\Delta V}{m} = \frac{40\text{ V}}{10\text{ V}/(R_1 C_1)} = 4 R_1 C_1$$

The frequency of oscillation is equal to the reciprocal of the time period.

$$f_o = \frac{1}{T}$$

For the values given, $f_o = 1$ kHz.

Resistor R_4 is added to limit the output current of A_1 to some maximum value. The integrator's capacitor C_1 must be a high-quality, low-leakage type. Mylar, mica, or polystyrene capacitors are suitable. All capacitors used in integrators, sample-and-hold, and analog storage circuits require the same high quality. The comparator can be any general-purpose op amp, but the integrator amplifier should have a low bias current rating.

11.2 OSCILLATOR INTEGRATED CIRCUITS

Timer 555

The 555 is a linear integrated circuit that can be used as a *rectangular-wave oscillator* (astable operation), or it can generate a *single pulse* (monostable operation) upon application of an external trigger signal. External components (two resistors and a capacitor) establish the frequency of oscillation in the astable mode. Other applications include time-delay generation, pulse-width modulation, pulse-position modulation, and ramp generation. The schematic of the integrated circuit is shown in Figure 11-5.

A *functional block diagram of the 555* is shown in Figure 11-6. The device contains the

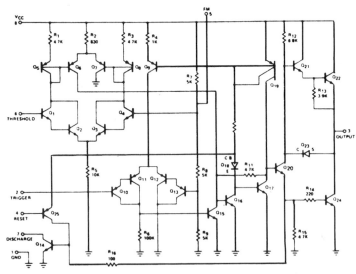

Figure 11-5. Schematic of the 555 integrated circuit.

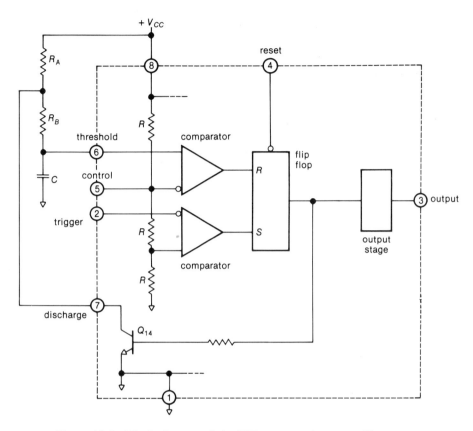

Figure 11-6. Block diagram of the 555 connected as an oscillator.

equivalent of two comparators, a flip flop, an output stage, a voltage divider, and discharge transistor. The threshold comparator is implemented by Q_1 through Q_4 and associated circuitry (refer to IC schematic). The trigger comparator is comprised of transistors Q_7 through Q_{10}, and the voltage divider is the 5 kΩ resistors labeled R_3, R_4, and R_5. The flip-flop section's (Q_{15}-Q_{23}) output drives the output stage (Q_{26}-Q_{28}) and the reset and discharge transistors numbered Q_{25} and Q_{14}.

The operation and function of the internal sections of the IC will be explained for the *astable condition*. If the timer is connected as shown in Figure 11-6, it will trigger itself and free run as an oscillator or multivibrator. The voltage at pin 6 (threshold) is an exponentially growing and decaying waveform. The capacitor is charged to $\frac{2}{3}V_{CC}$ through resistors R_A and R_B and discharged to $\frac{1}{3}V_{CC}$ through resistor R_A. The junction of R_A and R_B is connected to pin 7 (discharge). Internally, this pin is connected to the discharge transistor Q_{14}, which is on or conducting for the discharge portion of the cycle and off for the charge portion. The duty cycle is set by the ratio of the resistors R_A and R_B.

Pin 2 is the trigger input and is used for single-pulse or monostable applications. For this type of application, a falling signal applied to this pin will cause the output to go high for a predetermined length of time. For astable operation, the trigger pin or input is connected to pin 6.

Pin 4 is the reset input, and when it is brought low (below 0.4 V), the timing cycle is interrupted and the timer is returned to its non-triggered state. The timer cannot be triggered unless reset is released. For astable operation, pin 4 (reset) is returned to V_{CC}, disabling the reset function.

Pin 5 is the control voltage input. Internally, the control voltage input is connected to $\frac{2}{3}V_{CC}$ and to one input of the threshold comparator (I). A resistor to ground or an external voltage may be connected to this pin to change the comparator reference points. If the pin is not used, a capacitor to ground is added to filter power supply noise spikes, which could cause inconsistent timing.

The device is typically powered with a single supply (4.5 to 16 V) connected between pin 8 ($+V_{CC}$) and pin 1 (ground). Timing is relatively independent of this voltage.

The output of the device is at pin 3, which has sinking and sourcing current capability. For monostable operation, the output level is normally low and goes high during the timing interval.

Figure 11-7 shows actual waveforms for the output and the capacitor in the astable mode.

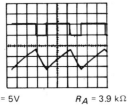

V_{CC} = 5V R_A = 3.9 kΩ
R_B = 3 kΩ C = 0.01 μf
Time 20 μs/Div. Top Trace Output 5V/div.
Bottom Trace Capacitor Voltage 1V/Div.

Figure 11-7. Output and capacitor voltage waveforms – 555.

The charge time (output high) is given by

$$t_{charge} = 0.693(R_A + R_B)C$$

and the discharge time (output low) by

$$t_{discharge} = 0.693(R_B)C$$

The total period is the sum of the two times.

$$T = 0.693(R_A + 2R_B)C$$

The frequency of oscillation is the reciprocal of the time period.

$$f_o = \frac{1}{T} = \frac{1.443}{(R_A + 2R_B)C}$$

The duty cycle is

$$D = \frac{R_B}{R_A + 2R_B}$$

Nomographs are graphs or charts used to show the relationship among several quantities. They provide a quick, visual solution to an equation, and they are a commonly used design and analysis tool. Figure 11-8 is a nomograph relating the RC values and frequency of oscillation for the 555. Nomographs are also extremely helpful in relating filter circuit characteristics and component values. Solving filter equations is usually a complex and tedious process.

Voltage-Controlled Oscillator 566

The 566 or *voltage-controlled oscillator* (VCO) *has two outputs: a square-wave and a triangular-wave output.* The VCO (Figure 11-9a) is made up of a precision current source and a Schmitt trigger. The triangular wave is generated by a current source alternately charging and discharging an external timing capacitor (C_1) between the two switching levels of the Schmitt trigger. The output of the Schmitt trigger, which is also the square-wave output, controls the direction of the current generated by the current source. The frequency of the device is programmed by (1) the voltage applied to the

OSCILLATOR INTEGRATED CIRCUITS

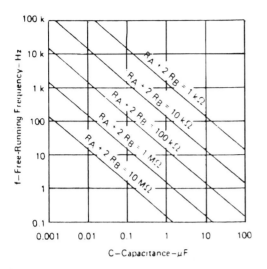

Figure 11-8. Nomograph—RC values and frequency.

control terminal or modulation input (pin 5), (2) the current injected into the timing resistor pin (pin 6), or (3) the value of the external resistor and capacitor (R_1 and C_1).

A *simplified diagram of the VCO* integrated circuit is shown in Figure 11-9b. I_1 is the charging current created by application of the control voltage V_C (to pin 6). In the initial state, transistor Q_3 is nonconducting or off, and the current I_1 charges the capacitor C_1 through the diode D_2. The voltage at the capacitor is a positive-sloped, linear voltage ramp. When the voltage on C_1 reaches the upper triggering threshold, the Schmitt trigger changes state (high) and turns on or activates Q_3. This transistor (previously off) now provides a current sink and essentially grounds the emitter of Q_1 and Q_2. The charging current I_1 (which previously flowed through D_2 and charged C_1) now flows through D_1, Q_1, and Q_3 to ground. Since the base-to-emitter voltage of Q_2 is the same as that of Q_1 (the devices are matched), an equal current (current-mirror action) flows through Q_2. This current removes the charge from C_1, or we say the current discharges the capacitor. This occurs until the lower threshold level of the Schmitt trigger is reached, at which point the output of the Schmitt trigger changes state, Q_3 is turned off, and the current I_1 begins to recharge C_1. The cycle is continuously repeated.

The schematic of the actual integrated circuit is shown in Figure 11-10. The precision current source is implemented with transistors Q_1-Q_7

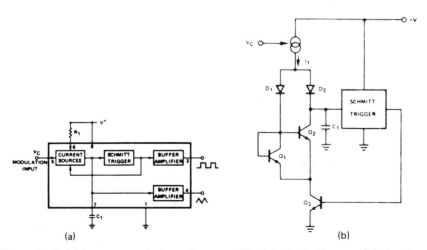

Figure 11-9. Voltage-controlled oscillator — 566: (a) block diagram; (b) simplified circuit diagram.

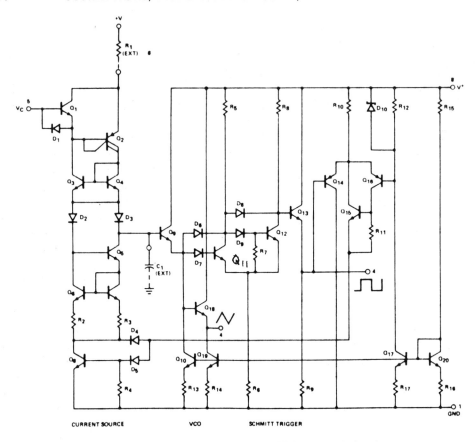

Figure 11-10. Schematic of the 566 integrated circuit.

and diodes D_1 through D_3. The base of Q_7 is the control voltage input. The external resistor R_1 and the control voltage V_C establish the value of the charging current I_1. In the actual IC, a duplicate current source (of opposite direction) discharges the capacitor. This duplicate current source is comprised of transistors Q_5 through Q_7 and resistors R_2 and R_3. Transistor Q_8, under the direction of the Schmitt trigger, determines which current source is applied to C_1. The capacitor voltage is buffered by Q_9 (emitter follower) and connected to the input of the Schmitt trigger (Q_{11} and Q_{12}). Transistor Q_9's output is buffered again by Q_{18} to provide the triangular output (pin 4). The Schmitt-trigger output is connected back to the current source through the differential amplifier (Q_{14}-Q_{16}).

Frequency-Modulated Function Generator

The circuit in Figure 11-11 illustrates the use of the 566 as a *function generator that can be frequency modulated*.

The dc control voltage at pin 5 is established by the R_2-R_3 resistor divider. It is a device requirement that the control voltage V_C be in the range from

$$0.75V^+ < V_C < V^+$$

For the R_2 and R_3 values given, the control voltage is in the middle of this range or $0.87V^+$. For a V^+ of 12 V, V_C equals 10.4 V.

OSCILLATOR INTEGRATED CIRCUITS

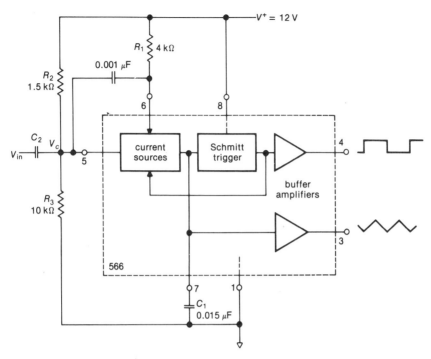

Figure 11-11. Frequency-modulated function generator.

The 0.001 μF capacitor between pins 5 and 6 is used to eliminate possible oscillation (unwanted) in the control current source. Capacitor C_2 is used to ac couple a modulating signal. This signal will algebraically add to V_C and change the VCO output frequency.

R_1, C_1, V_C, and V^+ establish the frequency of oscillation. The frequency is approximately

$$f_o = \frac{2(V^+ - V_C)}{R_1 C_1 V^+}$$

For the values given on the schematic,

$$f_o = \frac{2(12\text{ V} - 10.4\text{ V})}{(4\text{ k}\Omega)(0.015\text{ μF})(12\text{ V})} = 4.44\text{ kHz}$$

If the modulating input signal V_{in} causes V_C to rise to 11.9 V, the output frequency will be

$$f_o = \frac{2(12\text{ V} - 11.9\text{ V})}{(4\text{ k}\Omega)(0.015\text{ μF})(12\text{ V})} = 227\text{ Hz}$$

For $V^+ = 12$ V, the typical VCO output waveforms are shown in Figure 11-12.

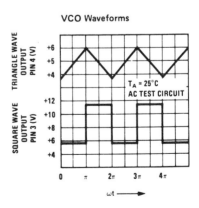

Figure 11-12. VCO output waveforms.

11.3 PHASE LOCKED LOOP

Principles of Operation

The basic concept of the phase locked loop or PLL has been around since the early 1930s; however, its current availability in a low-cost, self-contained, monolithic IC package has changed it from a specialized design technique to a general-purpose building block. The device represents an increase in the level of complexity in linear ICs because it contains four rather basic linear building blocks.

A phase locked loop is basically an electronic servosystem or loop consisting of a phase detector, a low-pass filter, an amplifier, and a voltage-controlled oscillator (VCO). The block diagram of the PLL is shown in Figure 11-13. The phase comparator, low-pass filter, and amplifier are in the signal forward path, and the VCO is in the feedback path. When in operation or in lock, the PLL can be approximated as a linear feedback system, except the signal variable is frequency and phase instead of voltage or current. The particular PLL illustrated (565) is a general-purpose circuit designed for highly linear FM demodulation. Other applications include frequency shift keying (FSK), frequency multiplication or translation, motor-speed control, and SCA (background music) decoder. The principles of operation of the PLL will be explained from two points of view.

With no system input signal $V_{in}(t)$ present, the voltage V_d is equal to zero. The VCO operates at a set frequency f_o, which is known as the free-running frequency. The time or frequency dependence of the system input signal is noted by the designation $V_{in}(t)$. If an input signal is applied to the system, the phase comparator compares the phase and frequency of the input with the VCO frequency and generates an error voltage $V_e(t)$ that is related to the phase and frequency difference between the two signals. This error voltage is filtered, amplified, and applied to the control terminal of the VCO. In this manner, the control voltage $V_{d(t)}$ forces the VCO frequency to vary in a direction that reduces the frequency difference between f_o and the input signal. If the input frequency f_{in} is sufficiently close to f_o, the feedback nature of the PLL will cause the VCO to synchronize or lock with the incoming signal. Once the two signals are in lock, the VCO and the input signal will be at the same frequency except for a finite phase difference. The net phase difference is necessary to shift the VCO from its free-running frequency to the input signal frequency f_{in} and, thus, keep the PLL in lock.

Another viewpoint of the PLL is to observe that the phase comparator is a multiplier or mixer circuit that mixes the input and VCO signals. This mix produces the sum and difference frequencies $f_{in} \pm f_o$. The loop will servo and lock such that the VCO duplicates the input frequency. The output of the multiplier contains ac components that are filtered, and a dc component that is a function of the phase angle between the VCO and the input signal. The dc component is amplified and fed back to the VCO.

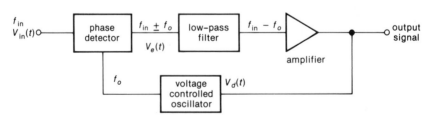

Figure 11-13. Block diagram of the phase locked loop.

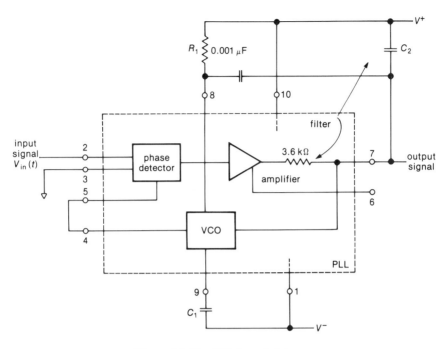

Figure 11-14. FM demodulator.

Lock and Capture Range

Two key parameters of a phase-locked-loop system are its lock and capture range. The *lock range* is the band of frequencies in the vicinity of f_o over which the PLL can *maintain lock* with an incoming signal. It is also known as the tracking or holding range. The lock range is directly proportional to the value of the loop gain. The *capture range* is the band of frequencies in the vicinity of f_o where the PLL can establish or catch or *acquire lock* with an input signal. It is also known as the acquisition range, and it is always smaller than the lock range. It is directly related to the bandwidth of the low-pass filter.

The PLL responds only to those input signals that are close to the VCO frequency f_o. The input frequency must fall within the lock or capture ranges of the system. The PLL performance characteristics offer a high degree of frequency selectivity, with the selectivity characteristics centered about f_o.

FM Demodulator

The 565 is a ten-pin (K package), general-purpose phase-locked loop (PLL) that can be directly used as an *FM demodulator*. The circuit for this application is shown in Figure 11-14. The input signal is a high-frequency carrier that is modulated with a low-frequency intelligence signal. This circuit removes the low-frequency signal from the high-frequency carrier.

The input signal $V_{in}(t)$ is applied to pin 2 with the reference for the input (pin 3) grounded. The output of the VCO (pin 4) is externally connected to the VCO input of the phase detector (pin 5). The VCO is identical to the VCO (566) previously discussed. Resistor R_1 and capacitor C_1 are used to establish f_o or the free-running frequency. The free-running frequency of the VCO is determined by

$$f_o = \frac{1.2}{4R_1C_1}$$

R_1 is restricted to the range of

$$2 \text{ k}\Omega \leqslant R_1 \leqslant 20 \text{ k}\Omega$$

but C_1 can be of any value.

The internal 3.6 kΩ resistor is used with an external capacitor C_2 to form a first-order low-pass filter. The 0.001 μF capacitor between pins 7 and 8 is used to suppress unwanted oscillation in the control current source of the VCO. The demodulated output is at pin 7. Pin 6 provides a dc voltage that is close to the dc potential of pin 7 and is used to decrease the gain of the amplifier and the lock range of the device. Capacitor C_2 establishes the cutoff frequency of the filter and the capture range of the PLL. The lock and capture ranges are determined by the following formulas.

$$f_L = \pm \frac{8f_o}{V_{CC}} \text{ Hz} \quad \text{(lock range)}$$

$$f_C = \pm \frac{1}{2\pi} \sqrt{\frac{2\pi f_L}{\tau}} \quad \text{(capture range)}$$

where $\tau = (3.6 \times 10^3) \times C_2$. The device is dc biased with a pair of supplies connected to pin 10 (V^+) and pin 1 (V^-).

Motor-Speed Control System

Many electromechanical systems, such as magnetic tape drives and disk or drum head drivers, require precise speed control. The block diagram of a PLL in a motor-speed control system is shown in Figure 11-15. The VCO section of the PLL is used to generate a voltage-controlled reference frequency (f_R) that is related to the desired *motor speed*. The tachometer on the motor shaft provides a voltage with a frequency (f_M) related to the actual motor speed. The two frequencies f_R and f_M are compared by the phase detector, whose output will cause the system to servo such that the two frequencies are equal. The motor controller is a power amplifier that drives the speed control windings of the motor.

11.4 LOCAL OSCILLATOR OF AN AM RADIO

The actual schematic of a popular, low-cost AM radio is shown in Figure 11-16. The block

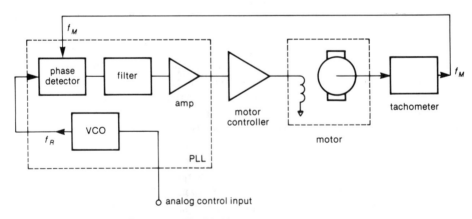

Figure 11-15. Motor-speed control system.

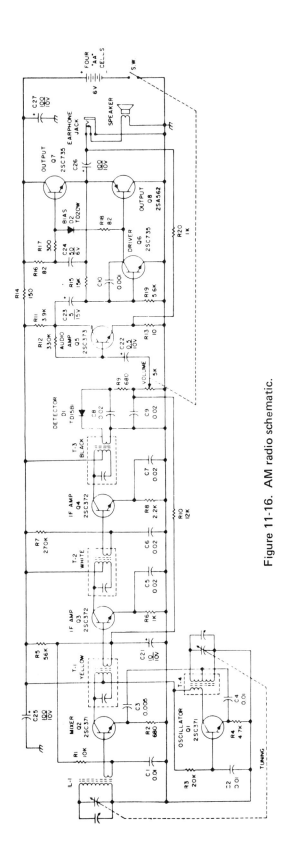

Figure 11-16. AM radio schematic.

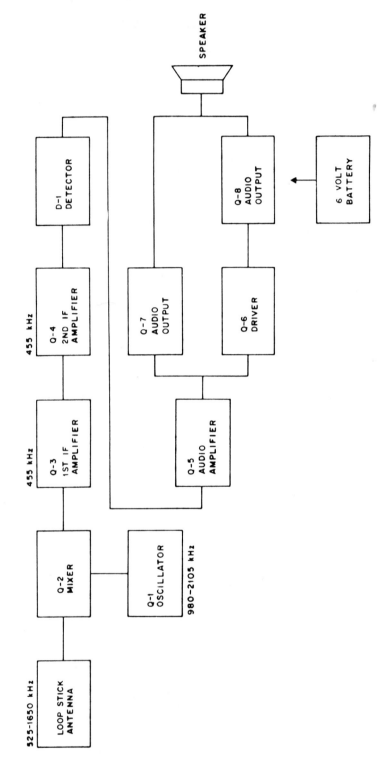

Figure 11-17. Block diagram of an AM radio.

diagram (Figure 11-17) is typical of most AM receivers. The high-frequency carrier, which is amplitude modulated with the intelligence, is received at the input transducer or antenna. The signal is processed or broken down until the intelligence or audio signal is recovered, which then is used to drive the output transducer or speaker.

The incoming RF (radio frequency) signal is converted to a lower or IF (intermediate frequency) signal by mixing the incoming RF with the output of an oscillator called the LO or local oscillator. The base of the mixer Q_2 contains the incoming RF, and the emitter of Q_2 contains the local oscillator signal. Capacitor C_3 couples the local oscillator (LO) output to the mixer. The collector of the mixer contains the sum, difference, and original input frequencies. The IF transformer T_1 selects the difference frequency, which is always 455 kHz. The LO and input antenna circuits are gang tuned to always produce a 455 kHz difference signal. The frequency of the *local oscillator* varies from 980 to 2105 kHz, whereas the incoming RF varies from 525 to 1650 kHz.

The local oscillator is a discrete component, Hartley LC circuit. The frequency of oscillation is determined by the LC tank in the secondary of T_4. The collector signal is transformer coupled to the tank circuit, inductive tapped, and ac coupled by C_4 to the emitter to complete the feedback path. The oscillator is a common-base configuration, and it is collector feedback dc biased.

Problems

1 What is the minimum number of cascaded RC phase shift networks in Figure 11-1 required to cause the circuit to oscillate? Is there theoretically a maximum?

2 What is the frequency of oscillation in the circuit of Figure 11-1 if $R = 5$ kΩ and $C = 0.001$ μF? Will the circuit oscillate if $R_2 = 3.1$ MΩ and $R_1 = 100$ kΩ?

3 Calculate the exact value (3 significant figures) of the frequency of oscillation of the circuit in Figure 11-3.

4 The gain of the individual circuits in Figure 11-3 can be computed by calculating the impedance of the capacitors at 1 kHz. A_1 can be thought of as a noninverting amplifier, A_2 an inverting amplifier, and the R_1-C_1 network as an ac voltage divider. Find the gain of each circuit at 1 kHz.

5 Calculate the frequency of oscillation of the circuit in Figure 11-4 if $R_1 = R_2 = R_3 = 5$ kΩ and $C_1 = 0.047$ μF. What is the slope of the triangular output ($V_Z = 10$ V)?

6 Specify the value of R_A and R_B in the circuit of Figure 11-6 if $C = 0.1$ μF, $f = 1$ kHz, and the duty cycle is 33.3%.

7 What must be the relationship between R_A and R_B if the duty cycle is to be 50%?

8 In Figure 11-6, $R_A = 5$ kΩ, $R_B = 10$ kΩ, and $C = 0.01$ μF. What is the duty cycle and frequency of oscillation?

9 What transistors in Figure 11-5 implement the threshold and trigger comparators in the block diagram of Figure 11-6? What resistors in Figure 11-5 correspond to the three resistors labelled R in the block diagram of Figure 11-6?

10 Resistor R_3 in Figure 11-11 is changed to 9.1 kΩ. What is the frequency of oscillation of the circuit if the external signal $V_{in} = 0$ V? What is the peak to peak output voltage of the square wave?

11 What will be the output frequency if the modulating input signal in Figure 11-11 causes V_C to drop to 9 V?

12 What is the free running frequency (f_o), the lock range (f_L), and the capture range (f_C) of the circuit in Figure 11-14 if $R_1 = 2$ kΩ, $C_1 = 5$ pF, $C_2 = 10$ pF, and a 12 V supply is used?

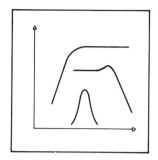

12
FILTER CONCEPTS AND PRINCIPLES

Filtering is the concept of selective removal. Filters are circuits that selectively attenuate a certain range of signal frequencies and pass the others. This chapter discusses the concepts and principles of filtering and basic filter circuits.

The primary characteristics of the filter are response type, characteristic frequency, passband gain, quality factor Q, and attenuation slope. The four common filter responses are the low pass, high pass, band pass, and band reject. The input and output relationship of a filter is mathematically described by the transfer function. The characteristic frequency is the frequency where the circuit gain begins to decrease and is no longer constant. The frequency region where the gain is constant is the frequency passband, and the value of gain that is constant is called the passband gain. The rate of attenuation of the decreasing gain outside the passband is measured by the attenuation slope. The peaking effect in band-pass, band-reject, and second-order filters is described by the quality factor or Q. The filter order identifies the complexity of the filter.

The basic implementation of filters uses discrete resistors, inductors, and capacitors. All the standard responses are possible.

The particular presentation of filter theory in this text is unique or rare in a technology level book; however, it is a most sound approach. The addition of the element of time (or frequency) presents instructional problems in describing the total behavior of a filter. The best way to develop a firm foundation in filter theory is to use mathematics as the vehicle in explaining what the filter does over the frequency spectrum and how to quantitatively assess the filter's characteristics. A filter's behavior is indirectly assessed by observing its mathematical model. The mathematical approach has far-reaching implications because almost any filter of any complexity (order) can be understood if the basis is the same. If simple or first-order filters are the only subject, this approach may not be the best; however, most current filter applications use complex, active RC networks whose analysis is simplified by looking at the circuit's mathematical model and interpreting its form.

12.1 THE CONCEPT OF FILTERING

Filtering is the process of selectively removing a part of a unit or whole. The process can be applied to gases, liquids, solids, and electrical signals. To filter means to strain, select, separate, remove, or screen.

A filter is a device or mechanism that filters. A porous cloth, paper, or material through which a liquid is passed to remove suspended impurities or to recover solids is a filter. A fine wire mesh screen is used as a gasoline filter in automobiles to remove any solids in the car's fuel system. The sieve or strainer is used as a kitchen utensil to separate a liquid and a solid.

Electronically, filters are circuits that selectively remove certain signal frequencies. Any complex-shaped signal can be thought of as the composite of a number of sinusoidal signals of various frequencies. The electronic filter selectively removes or attenuates certain of these signals while passing others.

The performance, or response, of a filter is frequency dependent. Hence, it must contain components whose response is also frequency dependent. All filters must contain one or more inductors and/or capacitors whose impedances are a function of frequency.

Filter integrated circuits are few in number. Most filters are implemented with discrete components or are a combination of ICs and discrete components. The IC op amp is frequently employed to implement the more advanced filters.

12.2 FILTER TYPES

A filter is a functional block or circuit that exhibits a preference for signals of a certain range of frequencies over others that are outside this range.

The filter input signal, a voltage or a current, is time varying. The analysis of filters is eased by considering only *sinusoidal input voltages* whose magnitude is maintained constant but whose frequency varies. The magnitude and phase of the output voltage is then examined and compared with the input. The overall response of the filter is established by examining the output over the entire frequency spectrum. This analytical approach is in line with most filter applications.

A filter's response refers to what the magnitude of the output does relative to the input over the frequency spectrum. This is shown on a graph of voltage gain (in decibels) versus frequency (f or ω). Gain is expressed on a linear scale and frequency on a logarithmic scale. There are five types of filter responses:

1. Low pass
2. Band pass
3. High pass
4. Band reject
5. All pass

The response title or name refers to what the filter does. A low-pass filter *passes* input signals of low frequencies to the output but rejects or attenuates all others. The band-reject filter rejects input signals of a certain band of frequencies and passes all others. The all-pass filter passes input signals of all frequencies, but the phase between the input and output is different. This filter type or response is the least encountered of the group. Figure 12-1 graphically illustrates the dominant four filter responses.

12.3 FILTER CHARACTERISTICS

The primary characteristics of filters are

1. Response type
2. Passband gain, A_o
3. Characteristic frequency, ω_o or f_o
4. Quality factor, Q
5. Attenuation slope or the order of the filter

FILTER CHARACTERISTICS

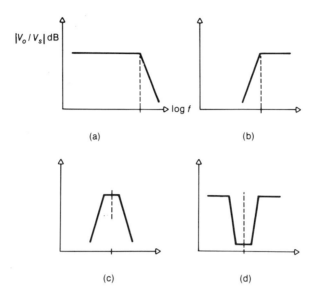

Figure 12-1. Straight-line approximations of filter responses: (a) low pass; (b) high pass; (c) band pass; (d) band reject.

Secondary characteristics include the delay, phase, and sensitivity functions, and those characteristics unique to the components and circuit implementing the filter. A graphical meaning of the primary characteristics is shown in Figure 12-2.

Response Types

The response of a filter refers to what the magnitude of the output does relative to the input over the frequency spectrum. Mathematically, this is done by an equation relating

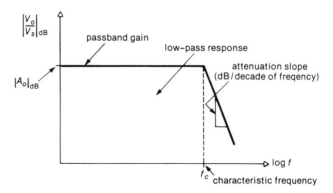

Figure 12-2. Primary filter characteristics.

voltage gain, as a function of frequency, in terms of the circuit components. Voltage gain is the ratio of the output and input voltages and is frequency dependent. It is a dimensionless number and is usually expressed as the ratio of two impedance expressions. The frequency dependence of the voltage gain comes from the frequency dependence of the impedance of the reactive components (L and C). In filter theory, the gain function is usually the ratio of two impedance expressions and is called the *transfer* function. The input signal is transferred to the output by multiplying it with the transfer function. Mathematically,

$$\frac{V_o}{V_s} = T \quad \text{or} \quad V_s \cdot T = V_o$$

where T designates the transfer function.

The *magnitude* and *phase* of the output voltage relative to the input voltage can be obtained from the transfer function. The phase information for the filter comes from the phase information contained in the impedances of the reactive components. For sinusoidal signals, the impedance of a capacitor is

$$Z_C = \frac{V}{I} = \frac{1}{j2\pi fC}$$

and for an inductor,

$$Z_L = \frac{V}{I} = j2\pi fL$$

The j in these expressions is the mathematical operator representing a phase difference of 90° between the voltage and the current. Where phase information is not important, the j is omitted and we talk about capacitive and inductive reactance; that is,

$$X_C = \frac{1}{2\pi fC}$$

and $X_L = 2\pi fL$

The frequency f is called cyclic frequency. Its unit of measurement is Hertz or cycles per second. $2\pi f$ is also a frequency and is called the radian frequency, whose symbol is the Greek letter omega, ω. There are 2π radians in a cycle and 360° in a cycle. Hence, 2π radians equals 360° and 1 radian is approximately 57.3°.

$$\omega = 2\pi f$$
$$360° = 2\pi \text{ radians}$$
$$57.3° \cong 1 \text{ radian}$$

Filter gain expressions or transfer functions can be simplified by representing C and L impedances using $2\pi f$ abbreviated as ω; thus,

$$Z_C = \frac{1}{j\omega C} = -jX_C$$

and $Z_L = j\omega L = jX_L$

A further simplification can be made by combining the j and ω and defining a new frequency s.

$$s \triangleq j\omega$$

s is called the complex frequency (because of the complex operator j) and basically is used here for convenience. However, a significant amount of filter theory has been developed utilizing the complex frequency s and the abstract s plane. The use of s simplifies (1) the derivation of the transfer function, (2) the examination of the form of the transfer function, and (3) the identification of the filter characteristics.

Using s, the capacitive and inductive impedances are written as

$$Z_C = \frac{1}{sC}$$

and $Z_L = sL$

If the impedances are expressed as a function of s, the transfer function or gain expression will be a function of s.

$$T(s) = \frac{V_o}{V_s} = \frac{Z_N(s)}{Z_D(s)}$$

The symbol $T(s)$ is interpreted as "T is a function of s." Similarly, $Z(s)$ means that the impedance Z is a function of s.

If the numerator and denominator impedance expressions (Z_N and Z_D) are a function of s, their ratio can be ultimately reduced to a complex number (dimensionless) whose magnitude and phase can be determined from complex number theory. For a complex number c, then

$$c = a + jb$$

and $|c| = (a^2 + b^2)^{1/2}$ (magnitude)

$$\phi = \tan^{-1}\left(\frac{b}{a}\right) \quad \text{(phase)}$$

The quantity $a + jb$ represents the Cartesian or rectangular coordinate form of representing the complex number c. The quantities $|c|$ and ϕ represent the polar coordinate form of representing the complex number c. The real part, a, of the complex number, $a + jb$, arises from the filter resistances. The imaginary part of the number, b, represents the inductive and capacitive reactances.

The response of a filter is found by plotting the magnitude of the gain or transfer function versus frequency. For a given frequency, the gain magnitude is found by finding the magnitude of a complex number.

The filter's gain-versus-frequency relationship can also be expressed graphically on a plot of gain magnitude (in decibels) versus frequency (log ω or log f). The graph is a pictorial version of the gain-versus-frequency equation. Similarly, the phase relationship between the output and input can be shown graphically.

The response of simple filters, that is, first-order filters, can be quickly sketched using the straight-line approximation technique of the Bode plot. This technique cannot be used for the more sophisticated second- and higher-order filters.

Filter Order

The order of the filter is identified by the largest exponent of s in the denominator of the transfer function. First-order filters have transfer functions with only s and non-s terms. They usually contain only one reactive element. Second-order filters have transfer functions containing an s^2 term in the denominator, and they must have a minimum of two reactive elements. Similarly, third-order filters have an s^3 term in the denominator. The filter analysis techniques employed in the various orders of filters are somewhat different and become difficult for the higher-order filters.

Standard Forms of Filter Transfer Functions

It is very difficult to analyze filter circuits because of the added dimension of time (or frequency) and the somewhat bulky equations. The analysis is eased by using the extensive experience of the people who work in this topic area. The transfer functions of all the standard filter circuits have been previously derived, and their terms have been arranged in a standard form so that it is easy to recognize the filter's characteristics. In this treatment of filters, we will derive the transfer function of a circuit and match it with one of these standard forms. By comparing our circuit's transfer function with its matching standard form, we will obtain the key characteristics of the filter of interest. Figure 12-3 lists the standard forms for the first-order low- and high-pass cases, and the second-order low-pass, high-pass, band-pass, and band-reject responses. These standard forms are *not* all inclusive.

First order low pass

$$T(s) = \frac{A_o \omega_c}{s + \omega_c}$$

First order high pass

$$T(s) = \frac{A_o s}{s + \omega_c}$$

Second order low pass

$$T(s) = \frac{A_o \omega_o^2}{s^2 + (\omega_o/Q)s + \omega_o^2} \quad (Q > \tfrac{1}{2})$$

Second order high pass

$$T(s) = \frac{A_o s^2}{s^2 + (\omega_o/Q)s + \omega_o^2} \quad (Q > \tfrac{1}{2})$$

Second order band pass

$$T(s) = \frac{A_o s(\omega_o/Q)}{s^2 + (\omega_o/Q)s + \omega_o^2} \quad (Q > \tfrac{1}{2})$$

Second order band reject

$$T(s) = \frac{A_o(s^2 + \omega_o^2)}{s^2 + (\omega_o/Q)s + \omega_o^2} \quad (Q > \tfrac{1}{2})$$

Figure 12-3. Standard forms of filter transfer functions.

Characteristic Frequency

A low-pass filter passes input voltages of low frequency and attenuates voltages of higher frequencies. Low and high are relative terms and are quantitatively defined by the filter's characteristic frequency, usually abbreviated as ω_o or ω_c. For the low-pass filter, all frequencies below ω_c are low, and all frequencies above ω_c are high. ω_c has an analogous meaning for the high pass. For the band-pass and band-reject cases, ω_o is the *center frequency of the pass band* or the center frequency of the reject band of frequencies. The value of ω_c and ω_o is a function of the filter circuit's compo-

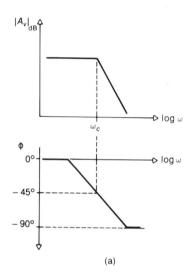

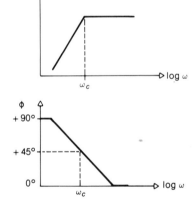

Figure 12-4. Gain and phase bode plots: (a) low-pass filter; (b) high-pass filter.

nents, and they are identified for the basic filter responses in Figure 12-1. The meaning of ω_o for the higher-ordered low- and high-pass filters is somewhat obscured because of the more complex filter response shapes.

For first-order filters, the characteristic frequency ω_c is the frequency where the magnitude of the gain is 0.707, or -3 dB, of its value in the passband region. This is also the frequency where the two straight lines intersect on the Bode plot. The actual response is -3 dB from the straight-line approximation and represents the region of greatest deviation between the approximate and actual responses.

For the first-order low- and high-pass filters, the characteristic frequency is the frequency where the phase angle between the output and input is 45°. For the low-pass case, the phase is -45° at ω_c and +45° for the high-pass circuit. The low-pass filter is equivalent to a single lag network, and its phase goes from 0° to -90°. The high-pass filter is equivalent to a single lead network, and its phase goes from +90° to 0°. Bode plots of gain and phase for the low- and high-pass filters are shown in Figure 12-4.

Passband Gain

Passband gain refers to the *magnitude of the voltage gain* in the band or region of frequencies that the filter is passing. It is usually expressed in decibels and is constant for some frequency range.

Filters that are implemented with discrete resistors, capacitors, and/or inductors have a typical gain of 1 or 0 dB. They are referred to as passive filters. Filters that are implemented with resistors, capacitors, and/or inductors *and* at least one active or gain-producing device can have a passband gain of greater than 1. *An active filter is a filter implemented with resistors, capacitors, and/or inductors, with at least one gain-producing device, that is, a transistor or integrated circuit.*

Active and passive filters can be of any order. Most low-frequency active filters are implemented using the operational amplifier and discrete resistors and capacitors. They are also referred to as active *RC* networks.

Quality Factor or Q

Q is an abbreviation for *quality factor*. In band-pass and band-reject filters it *defines* the *sharpness of the response curve*, and in second-order low- and high-pass filters it *defines the degree of peaking* at ω_o. It is formally defined as

$$Q = \frac{2\pi \cdot \text{maximum energy stored}}{\text{total energy lost per cycle}}$$

It is a dimensionless constant and is a function of the circuit elements. The meaning of *Q* for a second-order band-pass filter is illustrated in Figure 12-5. For this type of response, *Q* can be interpreted as

$$Q = \frac{\omega_o}{\omega_2 - \omega_1}$$

where ω_o is the center frequency and ω_2 and ω_1 are the -3 dB frequencies of the passband.

Q has a somewhat different graphic interpretation for the low- and high-pass second-order filters. Figure 13-6 shows the various responses for different values of *Q*.

Attenuation Slope

Attenuation slope is the *rate of change of gain versus frequency*. This characteristic describes how the voltage gain (in decibels) rolls

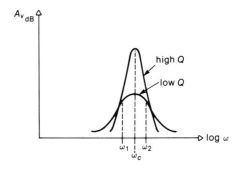

Figure 12-5. Band-pass filter response and *Q*.

off (per decade of frequency) for frequencies past ω_c. The slope is primarily a function of the order of the filter. First-order low- and high-pass filters have a rolloff of -20 dB/decade. This means that the gain is reduced by a factor of 10 (that is, $\frac{1}{10}$) for every decade change in frequency. For frequencies beyond ω_c, the attenuation slope of second-order filters is -40 dB (that is, $\frac{1}{100}$) per decade of frequency. Many applications require the rapid attenuation of signals above ω_c (low pass) and below ω_c (high pass), and hence require the use of the higher-ordered filter types.

12.4 BASIC FILTERS

Low Pass: RC

The circuit in Figure 12-6 is that of a low-pass filter and is usually implemented with discrete components. This circuit can be intuitively analyzed by considering the impedance (or reactance) of the capacitor at very low frequencies and at very high frequencies. At low frequencies, Z_C is high and approaches an open circuit as a limit. R and Z_C form an ac voltage divider. At low frequencies, Z_C is high, and the voltage drop across the capacitor is large, approaching V_s in magnitude. The passband gain approaches 1 ($A_o = 1$) for this case. At high frequencies, that is, frequencies

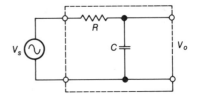

Figure 12-6. *RC* low-pass filter.

above the filter's characteristic frequency ω_c, the impedance of the capacitor decreases, and the output voltage decreases in magnitude. For this circuit,

$$\omega_c = \frac{1}{RC} \quad \text{or} \quad f_c = \frac{1}{2\pi RC}$$

The Bode plot can be sketched by drawing a straight horizontal line (0 dB) from $\omega = 0$ to $\omega = \omega_c$, and then a sloped line at $\omega = \omega_c$ with a tilt of -20 dB/decade. The straight-line approximation and the actual response are shown in Figure 12-7. For this filter,

$$A_o = 1 = 0 \text{ dB}$$

$$\omega_c = \frac{1}{RC}$$

attenuation slope = -20 dB/decade

Low-Pass Transfer Function

The circuit resistance and capacitance form an ac voltage divider. If the impedance of the capacitor is represented by $1/sC$, then the out-

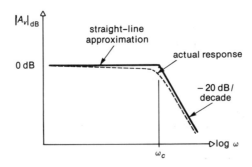

Figure 12-7. Bode plot and actual response of a low-pass filter.

BASIC FILTERS

put voltage is determined by impedance divider action; that is,

$$V_o = \frac{Z_C}{R+Z_C} V_s = \frac{1/sC}{R+(1/sC)} V_s$$

$$= \frac{1/RC}{s+(1/RC)} V_s$$

$$\frac{V_o}{V_s} = \frac{1/RC}{s+(1/RC)} = \frac{1/RC}{j\omega+(1/RC)}$$

The above equation states that the magnitude of the gain of the circuit (and also the phase) is a function of frequency. If various values are assigned to ω and the gain magnitude is determined, a plot of gain (in decibels) versus frequency (log of f) can be made. The graph or plot will reveal a low-pass characteristic. The gain (1) will remain constant at low frequencies until ω_c is reached. At the characteristic frequency ω_c, the magnitude of the gain will be down 3 dB or 0.707 of its passband value. At ω_c, the phase angle ϕ of the low-pass filter is $-45°$. For frequencies above ω_c, the gain will decrease by 20 dB for every decade increase in frequency.

The transfer function or gain expression for first-order low-pass filters has the form

$$T(s) = \frac{V_o}{V_s} = \frac{A_o \omega_c}{s + \omega_c}$$

Any filter circuit whose transfer function has the above form will be a first-order low-pass filter. If the terms are arranged properly, the filter characteristics can be immediately identified. The highest exponent of s is 1, which indicates a first-order filter whose attenuation slope is 20 dB/decade. If the coefficient of s is set to 1, the term summed with s will be the characteristic frequency ω_c expressed in terms of the circuit components. The numerator is the product of the passband gain A_o and the characteristic frequency ω_c.

For this low-pass filter, the passband gain is 1 or 0 dB. The denominator expression in the transfer function is a complex number with real ($1/RC$) and imaginary (ω) parts. When they are equal, that is,

$$\omega = \frac{1}{RC}$$

then

$$\left|\frac{V_o}{V_s}\right| = \frac{A_o \omega_c}{(\omega_c^2 + \omega_c^2)^{1/2}}$$

$$= \frac{1}{\sqrt{2}} = 0.707$$

and $\phi = -\tan^{-1}(\omega/\omega_c)$
$= -\tan^{-1}(1) = -45°$

At ω_c, the magnitude of the voltage gain is 0.707 of its passband value and the phase is $-45°$.

At a decade above ω_c, that is, $\omega = 10\omega_c$,

$$\left|\frac{V_o}{V_s}\right| = \frac{A_o \omega_c}{(100\omega_c^2 + \omega_c^2)^{1/2}} \cong \frac{1}{10}$$

A gain reduction of 10 equals -20 dB; that is,

$$|A_v| \text{ dB} = 20 \log \left|\frac{V_o}{V_s}\right| = 20 \log \frac{1}{10} = -20 \text{ dB}$$

Example

If $R = 1$ kΩ and $C = 0.1$ μF, then

$\omega_c = 1/RC = 10$ k radians/second

or $f_c = 1.59$ kHz

$A_o = 1 = 0$ dB

slope $= -20$ dB/decade

High Pass: RL

The RL circuit in Figure 12-8 is a high-pass filter. The filter will ideally pass all signal frequencies above the characteristic or corner frequency and attenuate or amplitude reduce those below it. This circuit can be intuitively

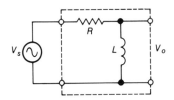

Figure 12-8. *RL* high-pass filter.

analyzed in a manner similar to that employed in the *RC* low-pass circuit.

The circuit resistance and inductance form an ac voltage divider. If the impedance of the inductor is represented by sL, then

$$V_o = \frac{Z_L}{R + Z_L} V_s = \frac{sL}{R + sL} V_s$$

$$= \frac{s}{s + (R/L)} V_s$$

or $T(s) = \dfrac{V_o}{V_s} = \dfrac{s}{s + (R/L)}$

The above transfer function, or gain expression, is the ratio of two complex numbers. The complex number in the numerator is expressed in rectangular coordinate form as $0 + j\omega$. The equivalent complex number in polar coordinate form is

$$0 + j\omega = \omega, \measuredangle\, 90°$$

where ω is the magnitude and $90°$ is the phase angle. Similarly, for the denominator,

$$j\omega + \frac{R}{L} = \sqrt{\omega^2 + \left(\frac{R}{L}\right)^2},\; \measuredangle\, \tan^{-1} \frac{\omega}{R/L}$$

Carrying out the division of the complex numbers in polar coordinate form results in the gain magnitude and phase expressions; thus,

$$\left|\frac{V_o}{V_s}\right| = \frac{\omega}{\sqrt{\omega^2 + \omega_c^2}}$$

$$= 90° - \tan^{-1}(\omega/\omega_c)$$

where $\omega_c = R/L$.

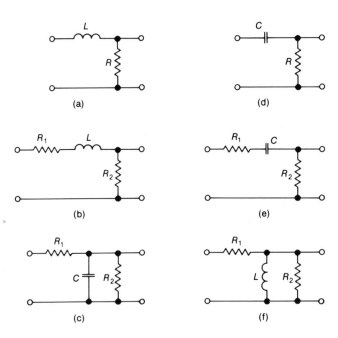

Figure 12-9. Other *RC* and *RL* filters: (a), (b), (c) low pass; (d), (e), (f) high-pass.

BASIC FILTERS

The transfer function of the first-order high-pass filter has the form

$$T(s) = \frac{A_o s}{s + \omega_c}$$

The value of the filter's characteristics can be determined by comparing the standard form and the transfer function of the specific circuit. For this circuit, the passband gain is 1 or 0 dB, the characteristic frequency is $\omega_c = R/L$, and its attenuation slope is a constant -20 dB/decade. Other RC and RL filter circuits are shown in Figure 12-9.

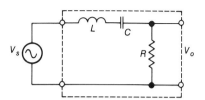

Figure 12-10. *RLC* band-pass filter.

Example

If $R = 1$ kΩ and $L = 1$ mH, then

$$\omega_c = \frac{L}{R} = 1M \text{ radians/second}$$

or $\quad f_c = \frac{\omega_c}{2\pi} = 159$ kHz

$A_o = 1 = 0$ dB

slope $= -20$ dB/decade

Band Pass: *RLC*

A band-pass filter can be made by cascading a low- and a high-pass filter. The transfer functions of each will multiply when they are cascaded, and the overall transfer function will have the form

$$T(s) = \frac{a_1 s}{s^2 + b_1 s + b_o}$$

Any circuit that exhibits the above transfer function will be a band-pass filter. It is a second-order filter that requires a minimum of two reactive elements.

The circuit of Figure 12-10 is that of a band-pass filter. It requires one less resistor than the cascaded low- and high-pass version. The $R, L,$ and C form an ac voltage divider with the filter output taken from across the resistor. It is a series resonant circuit. The output voltage is

$$V_o = \frac{R}{R + Z_c + Z_L} V_s = \frac{R}{R + sL + (1/sC)} V_s$$

$$T(s) = \frac{s(R/L)}{s^2 + s(R/L) + (1/LC)}$$

If $Q > \frac{1}{2}$, band-pass filter transfer functions have the form

$$T(s) = \frac{A_o(\omega_o/Q)s}{s^2 + s(\omega_o/Q) + \omega_o^2}$$

From the above,

$$\omega_o = \sqrt{\frac{1}{LC}}$$

$$\frac{\omega_o}{Q} = \frac{R}{L}$$

$$Q = \frac{1}{R}\sqrt{\frac{L}{C}}$$

$A_o = 1$

For the band-pass filter, Q is the ratio of the center frequency to the -3 dB bandwidth.

$$Q = \frac{\omega_o}{BW} = \frac{\omega_0}{\omega_2 - \omega_1}$$

The gain at any frequency may be determined from

$$A(\omega) = \frac{A_o}{\left\{1 + Q^2\left[\dfrac{\omega}{\omega_o} - \dfrac{\omega_o}{\omega}\right]^2\right\}^{1/2}}$$

Band Reject: RLC

The band-reject, or band-stop, filter passes signals of all frequencies, except for a select region where it attenuates them. The center of the reject band of frequencies is the characteristic frequency ω_o.

The RLC circuit in Figure 12-11 is a band-reject filter. The output voltage is taken from across the resistor, which forms an ac voltage divider with the parallel combination of L and C.

$$V_o = \frac{R}{R + Z_L \| Z_C} V_s$$

$$= \frac{R}{R + \frac{(sL/sC)}{sL + (1/sC)}} V_s$$

Simplifying the expression yields

$$T(s) = \frac{V_o}{V_s} = \frac{s^2 + (1/LC)}{s^2 + s(1/RC) + (1/LC)}$$

Band-reject filter transfer functions have the form

$$T(s) = \frac{A_o(s^2 + \omega_o^2)}{s^2 + s(\omega_o/Q) + \omega_o^2}$$

From the above,

$$\omega_o = \sqrt{\frac{1}{LC}}$$

$$\frac{\omega_o}{Q} = \frac{1}{RC}$$

$$Q = R\sqrt{\frac{C}{L}}$$

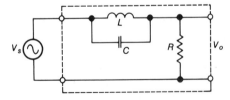

Figure 12-11. RLC band-reject filter.

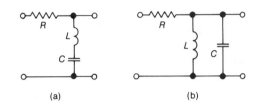

Figure 12-12. Other RLC circuits: (a) band reject; (b) band pass.

Other versions of the RLC band-pass and band-reject filters are shown in Figure 12-12.

Problems

1. Determine the total impedance of a series RLC circuit if the frequency of the applied source is 10 kHz, $R = 100 \, \Omega$, $C = 0.1 \, \mu F$, and $L = 1 \, mH$.

2. For the conditions specified in problem 1:
   (a) Determine the magnitude and phase of the impedance.
   (b) At what frequency will the circuit appear resistive?

3. What is the magnitude and phase of the gain expression for the RC low-pass filter in Figure 12-6 at:
   (a) $\omega = 0.1 \, \omega_c$ and
   (b) $\omega = 10 \, \omega_c$?

4. What is ω_c, $0.5\omega_c$, $5\omega_c$, f_c, A_o, and the attenuation slope of the circuit in Figure 12-8 if $R = 100 \, \Omega$ and $L = 100 \, \mu H$? What is the magnitude and phase of the gain expression or transfer function at $\omega = 0.5\omega_c$ and $\omega = 5\omega_c$?

5. Sketch the Bode plot of the filter circuit in Figure 12-9c. On the graph, identify ω_c and A_o in terms of the circuit's components R_1, R_2, and C. Find the value of A_o and ω_c for $R_1 = R_2 = 2 \, k\Omega$ and $C = 0.047 \, \mu F$.

6. Derive the transfer function $T(s)$ for the filter circuit in Figure 12-9d.

7. Sketch the Bode plot of the filter circuit in

Figure 12-9e. On the graph, identify ω_c and A_o in terms of the circuit's components R_1, R_2, and C. Find the value of A_o and ω_c for $R_1 = R_2 = 2$ kΩ and $C = 0.047$ μF.

8 For the band-pass filter of Figure 12-10, $L = 2.2$ mH. Find the value of R and C for $f_o = 10$ kHz and $Q = 10$.

9 Find f_o and the -3 dB bandwidth for the band-reject filter in Figure 12-11 if $R = 910$ Ω, $C = 0.047$ μF, and $L = 10$ μH.

10 Derive the transfer function for the RLC circuit in Figure 12-12a.

11 For the band-pass circuit in Figure 12-12b, find ω_o and Q in terms of the circuit's components.

12 The inductance in the circuit in Figure 12-12a is a 1 H simulated inductor. Find the value of R and C for a f_o of 60 Hz and a Q of 10.

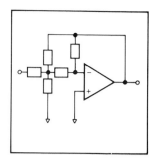

13
ACTIVE FILTERS AND APPLICATIONS

Filters that are implemented with at least one active or gain-producing device are called active filters. This chapter looks at several active-filter implementations and illustrates their applications.

Discrete RC and RL first-order filters can be replaced with basic op amp filter configurations and with circuits using simulated inductors. The active-filter versions are usually capable of pass-band gains greater than 1. Second-order active filters are typically implemented with operational amplifiers and RC networks using various configurations. The infinite-gain, multiple-feedback circuit is an example of a second-order active filter, and audio scratch and rumble filters are applications of this particular technique. All practical components are frequency limited and display filterlike characteristics.

13.1 ACTIVE FILTER CHARACTERISTICS

Active filters represent a large applications area for the operational amplifier. The operational amplifier, especially the IC version, proves to be an extremely useful active device in implementing active *RC* networks.

Several common circuit techniques are used in making *active filters:*

1. Simulated inductor
2. Infinite gain, multiple feedback
3. Infinite gain, single feedback
4. Controlled source
5. State variable
6. Negative immitance convertor

This chapter looks at the *simulated inductor*, *multiple-feedback*, and *basic op amp filter circuits*. The state variable active filter is available in IC form; however, most techniques only require a small number of external, passive components (and the op amp) to implement them.

Active filters purposely avoid the use of discrete inductors. Discrete inductors, that is, those made by winding wire on a bobbin, are lossy, expensive, heavy, and large in size when used in the audio range. These disadvantages are overcome through circuit techniques and the use of capacitors and simulated inductors. Capacitors, when compared to inductors, have a greater range of values, are less expensive, smaller, weigh less, and are nearer to the ideal.

The operational amplifier and most active-filter circuits that use it have a moderate-to-high input impedance and a low output impedance. This allows filter circuits to be cascaded with minimal interaction between them. The behavior of one filter circuit does not affect the previous or subsequent filter's behavior. The circuits, in essence, are isolated from each other.

There are a few disadvantages in using active filters. Active filters usually have single-ended inputs and outputs and thus do not float with respect to the system ground or common as a passive RLC network. The performance of the filters that use the op amp is affected by the data sheet parameters of the amplifier. As an example, the temperature dependence of the amplifier's characteristics is reflected to the filter.

13.2 FIRST-ORDER ACTIVE FILTERS

High Pass: Simulated Inductor

The behavior and performance of the active filter in Figure 13-1 is similar to that of the discrete RL high-pass filter in Figure 12-8. This *active filter uses a simulated inductor circuit* called a gyrator (see text of Figure 7-23). The simulated inductor is implemented by a multi-ple amplifier circuit that generates the same current and voltage relationship at its terminals as the discrete inductor. The transfer function for this filter,

$$T(s) = \frac{s}{s + (R_1/L_{eq})} = \frac{s}{s + (R_1/R^2C)} = \frac{A_o s}{s + \omega_c}$$

is the same as that of the discrete component version, noting that the equivalent inductance of the gyrator is

$$L_{eq} = R^2 C$$

For reasonable values of R (1 to 100 kΩ) and C (100 pF to 0.1 μF), L_{eq} will vary from 0.1 mH to 100 H. Low-frequency *discrete inductors* of high values are not practical.

Other filter responses are possible using the simulated inductor. However, this particular simulated inductor must always be used such that one terminal is grounded.

Low Pass: Operational Amplifier

An active circuit version (Figure 13-2) of the first-order low-pass RC filter consists of an operational amplifier with an input resistor R and a parallel R_1-C_1 feedback circuit.

Assuming the amplifier has ideal characteristics, the input and output voltage of the inverting amplifier circuit are related by

$$V_o = -\frac{Z_F}{Z_N} V_s = -\frac{R_1 \| (1/sC_1)}{R} V_s$$

The transfer function is

$$T(s) = \frac{V_o}{V_s} = -\frac{(R_1/R)[1/(R_1 C_1)]}{s + (1/R_1 C_1)} = \frac{A_o \omega_c}{s + \omega_c}$$

The transfer function and the filter characteristics for the passive and active low-pass filters are the same, except the passband gain for the active version is

$$A_o = -\frac{R_1}{R}$$

which may be greater than 1 or 0 dB.

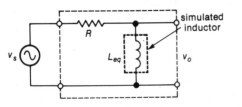

Figure 13-1. High-pass filter using a simulated inductor.

FIRST-ORDER ACTIVE FILTERS

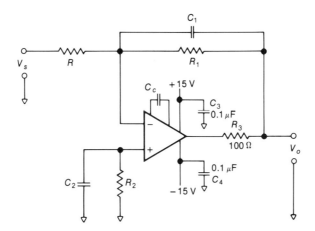

Figure 13-2. Low-pass filter using an operational amplifier.

The positions and values of R, R_1, and C_1 determine the filter characteristics of the circuit. R_2 is equal to the parallel combination of R and R_1, and it is used to reduce the amplifier dc bias current error. The capacitor C_2 shunts R_2 and keeps the noninverting amplifier input at ac ground. C_3 and C_4 are supply bypass capacitors. They should be located physically near the amplifier. Resistor R_3 isolates the amplifier output from capacitive loads. The compensation capacitor C_c should be chosen such that the open-loop bandwidth of the amplifier is significantly greater than the highest frequency of the signal that the circuit must pass.

High Pass: Operational Amplifier

An active circuit version (Figure 13-3) of the first-order high-pass RC filter consists of an operational amplifier with a feedback resistor R and a series R_1-C_1 input circuit.

The input and output voltage of the inverting amplifier circuit are related by

$$V_o = -\frac{Z_F}{Z_N} V_s = -\frac{R}{R_1 + (1/sC_1)} V_s$$

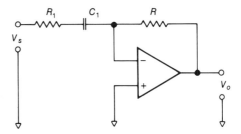

Figure 13-3. High-pass filter using an operational amplifier.

where Z_F and Z_N are the impedances of the feedback and input circuits.

The transfer function or gain relationship is

$$T(s) = \frac{V_o}{V_s} = -\frac{(R/R_1)(s)}{s + (1/R_1 C_1)} = \frac{A_o s}{s + \omega_c}$$

The transfer function and the filter characteristics for the passive and active high-pass filters are the same, except the passband gain for the active version is

$$A_o = -\frac{R}{R_1}$$

whose magnitude may be greater than 1 or 0 dB.

Example

If $R = 10$ kΩ, $R_1 = 2$ kΩ, and $C_1 = 0.1$ μF, then

$$\omega_c = \frac{1}{R_1 C_1} = 5\text{k rad/s} = 796 \text{ Hz}$$

$$|A_o| = \frac{R}{R_1} = 5 = 14 \text{ dB}$$

slope $= -20$ dB/decade   (first order)

13.3 SECOND-ORDER ACTIVE FILTERS

Low Pass: Infinite Gain, Multiple Feedback

A second-order low-pass active filter is shown in Figure 13-4. It is called an *infinite-gain* (high amplifier open-loop gain), *multiple-feedback* (C_5 path and R_4-R_3 path) *circuit*. It is a commonly used circuit technique of implementing *second-order active filters*, and the low-pass, band-pass, and high-pass responses are possible depending on the location of the capacitors and resistors.

The components of the *model* for this filter (Figure 13-5) are represented by their admittances. If the component is a resistor,

$$Y_R = G$$

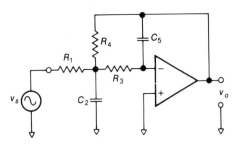

Figure 13-4. Second-order low-pass filter: infinite gain, multiple-feedback type.

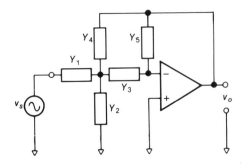

Figure 13-5. Infinite-gain, multiple-feedback filter model.

and if the component is a capacitor, its admittance is

$$Y_C = \frac{1}{Z_C} = sC$$

It can be shown that the transfer function for this model is

$$T(s) = \frac{-Y_1 Y_3}{Y_5(Y_1 + Y_2 + Y_3 + Y_4) + Y_3 Y_4}$$

The admittances Y_1 through Y_5 represent the admittances of the five components.

Several low-pass, high-pass, and band-pass responses are possible depending on the location of the resistors and capacitors. At least two capacitors are necessary to achieve a second-order filter.

For the circuit in Figure 13-4,

$$Y_1 = G_1$$
$$Y_2 = sC_2$$
$$Y_3 = G_3$$
$$Y_4 = G_4$$
$$Y_5 = sC_5$$

If the admittances of the components are substituted in the model's transfer function, then

$$T(s) = \frac{(-G_1/G_4)(G_4 G_3/C_2 C_5)}{s^2 + s(1/C_2)(G_1 + G_3 + G_4) + (G_3 G_4/C_2 C_5)}$$

A standard form for a second-order low-pass filter is

$$T(s) = \frac{A_o \omega_o^2}{s^2 + (\omega_o/Q)s + \omega_o^2}$$

Comparing the circuit's transfer function and its model,

$$A_o = -\frac{R_4}{R_1} = -\frac{G_1}{G_4}$$

$$\omega_o = \sqrt{\frac{G_3 G_4}{C_2 C_5}}$$

and $\quad Q = \dfrac{1}{G_1 + G_3 + G_4} \sqrt{\dfrac{G_3 G_4 C_2}{C_5}}$

This active filter is capable of a passband gain of greater than 1.

The gain magnitude versus frequency response curve is shown in Figure 13-6. The *effect of Q* on the response shape is illustrated for several values of Q. For frequencies beyond ω_o, the slope of the response is -40 dB/decade. For frequencies near ω_o, the slope and the degree of peaking is a function of the circuit's Q.

The low-pass circuit may be converted to a *high-pass filter* by changing resistors R_1, R_3, and R_4 to capacitors and capacitors C_2 and C_5 to resistors. The expression in the denominator of the high-pass filter's transfer function will be similar to the low pass. However, the capacitors in the Y_1, Y_3, and Y_4 positions will cause the expression in the numerator to have an s^2 term. The s^2 term in the numerator is characteristic of a high-pass filter.

Physical Interpretation

Second-order low- and high-pass filter responses can be identified by *replacing the reactive elements with their equivalent circuit* at the frequency spectrum extremes. At dc or 0 Hz, capacitors are equivalent to opens and induc-

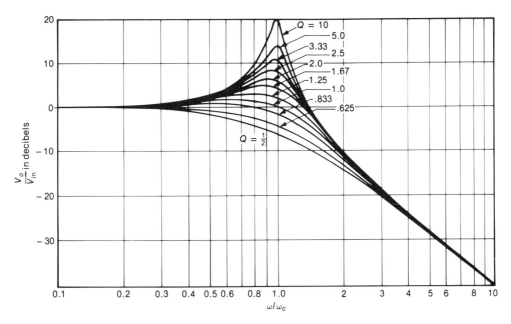

Figure 13-6. Magnitude response of second-order low-pass filters for several values of $Q \geqslant \frac{1}{2}$.

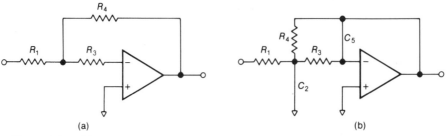

Figure 13-7. Physical interpretation of the filter at the frequency extremes: (a) low-frequency equivalent circuit; (b) high-frequency equivalent circuit.

tors are equivalent to shorts. At extremely high frequencies, capacitors are equivalent to shorts and inductors are equivalent to opens. The low- and high-frequency equivalent circuits of the multiple-feedback low-pass filter are shown in Figure 13-7. At low frequencies, the filter behaves as an inverting amplifier with a gain of $-R_4/R_1$. R_3 is in series with the amplifier's high input resistance and does not affect the closed-loop performance. At high frequencies, C_2 and C_5 are replaced with short circuits. C_2, in essence, grounds the input of the circuit, and C_2 shorts the output back to the inverting (−) input. Since the amplifier's noninverting (+) is at ground and the differential input voltage is extremely small, the output will be very near zero volts. The circuit passes signals of low frequencies and attenuates signals of high frequencies. It is a filter with a low-pass response.

Band Pass: Infinite Gain, Multiple Feedback

There are four versions of the *band-pass multiple-feedback circuit*. The band-pass five-element combinations are shown in Figure 13-8.

The R_1-R_2-C_3-C_4-R_5 band-pass configuration is shown in Figure 13-9. If the admittances of the circuit components are substituted for the admittances in the transfer function of the model, one obtains

Y_1	Y_2	Y_3	Y_4	Y_5
R	R	C	C	R
R	C	C	C	R
C	R	R	R	C
C	C	R	R	C

Figure 13-8. Multiple-feedback band-pass configurations.

More on Characteristic Frequency

The characteristic frequency of a filter has three slightly different interpretations depending on the filter it is associated with.

For *first-order low- and high-pass filters*, the characteristic frequency, ω_c, is the −3 dB frequency. It is the frequency where the gain is 0.707 of its passband value, and the phase between the input and output is 45°.

For *band-pass and band-reject filters*, the characteristic frequency, ω_o, is the center frequency of the passband or the reject band.

For *second-order low- and high-pass filters*, the meaning of ω_o, and also Q, is obscure. For moderate values of Q and greater (>5), ω_o is the frequency of maximum peaking. At this frequency, the amplitude is equal to Q times the passband gain. For most applications, the

$$\frac{V_o}{V_s} = \frac{-(1/R_1 C_4)s}{s^2 + \left(\frac{(C_3+C_4)}{R_5 C_3 C_4}\right)s + \left(\frac{1}{R_5}\right)\left(\frac{1}{R_1}+\frac{1}{R_2}\right)\frac{1}{C_3 C_4}}$$

FILTER APPLICATIONS

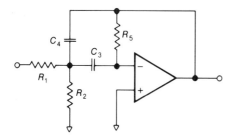

Figure 13-9. Multiple-feedback band-pass filter.

Q	ω_c (LP)	ω_c (HP)
0.707	1.000 ω_o	1.000 ω_o
1	1.272 ω_o	0.786 ω_o
3	1.523 ω_o	0.657 ω_o
5	1.543 ω_o	0.648 ω_o
10	1.551 ω_o	0.645 ω_o
100	1.554 ω_o	0.644 ω_o

Figure 13-10. Q and ω_c relationship in second-order filters.

frequency where peaking occurs is *not* the frequency of interest. The most meaningful frequency is the −3 dB frequency, which is ω_c for the first-order low- and high-pass circuits. There is an ω_c for the second-order filters, but it is not strictly a function of ω_o, but is also a function of Q. This can be seen in Figure 13-6 where the higher Qs cause a greater degree of peaking *and* an extended −3 dB frequency.

For second-order low- and high-pass filters, ω_o and ω_c (−3 dB frequency) are related as follows:

High pass: $\omega_c = \dfrac{\omega_o}{\beta}$

Low pass: $\omega_c = \beta \omega_o$

where β is a function of Q; that is,

$$\beta = \sqrt{\left(1 - \frac{1}{2Q^2}\right) + \sqrt{\left(1 - \frac{1}{2Q^2}\right)^2 + 1}}$$

A table relating ω_c and ω_o for several different values of Q is provided for convenience (Figure 13-10). The frequencies ω_o and ω_c are equal only for the special case where $Q = 0.707$. The filter for $Q = 0.707$ is called a Butterworth filter and has a maximally flat response, that is, no peaking. Butterworth filters are frequently used in audio systems.

The characteristic curves of the second-order low- and high-pass filters are not unique in electronics. The −3 dB bandwidth of amplifiers can be extended through circuit techniques that cause a peaking effect near the cutoff frequency similar to that of the filters. This circuit technique is referred to as shunt peaking.

Specialized Filters

Theoretically, a filter can be made of any order by increasing the number of reactive components. These components are arranged in the circuit such that they add s terms with higher exponents in the denominator of the transfer function. However, most of the higher-order filters are designed to maximize a certain filter characteristic.

The *Butterworth filter* is a filter that is designed to provide a maximally flat magnitude (MFM) response. The higher the order of the Butterworth filter, the flatter the response in the filter's passband region.

The *Chebyshev or equal ripple filter* provides an equal error evenly throughout the passband in an oscillating manner. In this type of filter, the largest peak of the error is minimized, and the magnitude response varies between equal maximum and minimum values in the passband.

The *Thomson or maximally flat delay (MFD) filter* provides a linear phase relationship between the input and output voltages.

13.4 FILTER APPLICATIONS

Nonideal or Practical Components and Circuits

Circuits that are designed for a specific frequency characteristic are called filters. However, *all real components and circuits display filter-like characteristics*. The inductance and capacitance in a filter places frequency limitations on the signals it will pass. Since all real components and circuits have some associated inductance and capacitance, they too must also be frequency limited. The L and C may be discrete, distributed, or device.

The circuit of Figure 13-11 models a *real resistor*. The L represents the lead inductance (≈ 10 nH), and the C represents the end-to-end or body capacitance (≈ 1 pF). For low frequencies and moderate values of R, the circuit behavior is that of an ideal resistor, because the inductive reactance is much smaller than R and the capacitive reactance is much greater than R. The physical component is then modeled by the resistance symbol. At some frequency, the reactances will become a significant factor, and the physical resistor must then be modeled with its associated L and C. This model can then be analyzed to predict the frequency performance of the real resistor.

In a way, all amplifiers may be viewed as band-pass filters. They will have some upper cutoff frequency that is determined by circuit design, component limitations, or distributed inductance and capacitance. For most cases, amplifier circuits are limited by all three factors. The objective of a good design is to have the gain–frequency behavior of an amplifier much lower in frequency than that determined by the other factors. It is easier to control and predict the amplifier's performance if it is established by discrete components rather than distributed ones. The same can be said for the lower cutoff frequency.

The real transformer embodies many of the filter concepts and principles discussed. While the transformer is not used as a filter, the inductance and capacitance associated with the physical device limit the upper and lower cutoff frequencies and the response shape of its application.

Practical Transformer

The physical or *practical transformer* and its equivalent circuit are shown in Figure 13-12. The ideal transformer relationships only serve as an introduction to describe the basic, *ideal* capabilities of the device. In an actual application, one must contend with the equivalent circuit to ensure a guaranteed outcome.

In the equivalent circuit, L_p and L_s represent the leakage inductance of the primary and secondary windings, and R_p and R_s represent the winding resistances. An ideal (no loss) transformer is shown with R_s and L_s connected in series with the secondary, so an output (load) current causes a voltage drop across R_s and L_s. Similarly, R_p and L_p are connected in series with the primary windings, and voltage drops are produced across them when a primary current flows.

Leakage inductance is an effect due to leakage flux. Leakage flux is those magnetic lines of force that do not pass from the primary to the secondary via the core. It produces the same effect as an unwanted inductance in series with each winding. In a practical transformer, it is not possible to get perfect flux-linkage. The magnetic flux produced by a coil current does not link uniformly with each of that coil's turns, nor with each of the turns of a neighbor-

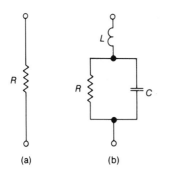

Figure 13-11. Resistor: (a) ideal; (b) nonideal or practical.

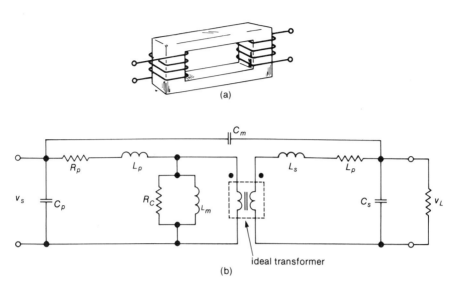

Figure 13-12. Practical transformer: (a) pictorial representation; (b) equivalent circuit.

ing winding. This imperfect linkage is equivalent to inductance that fails to contribute usefully to transformer action.

The lumped capacitances C_p and C_s represent the distributed primary and secondary interwinding capacitance. These capacitances arise from the separation of conducting wires by an insulating material. C_m, or the mutual capacitance, models the stray capacitance between the primary and secondary windings.

The resistance R_c represents the transformer energy losses due to hysteresis and eddy currents. In the real transformer, current is required to magnetize the core. A magnetic field is concentrated in the core and, since the core is a conducting material, a voltage is induced. This induced voltage produces currents in the core called eddy currents. Currents within the core, due to the induced emf, constitute energy loss.

As transformer currents change, the magnetic flux changes. The crystal structure of an iron core also changes with the changing magnetic flux; but there is a form of inertia that inhibits the crystal structure. This inertia results in energy loss and is called hysteresis.

The inductance L_m, called the magnetizing or incremental inductance, is the inductance associated with the magnetization of the core, that is, the establishing of the core flux.

Typical values for the components in the equivalent circuit of a signal transformer are as follows:

R_p, R_s: 100 Ω

C_p, C_s: 100 pF

C_m: 10 pF

R_c: 100 kΩ

L_m: 5 H

L_s, L_p: 5 mH

Frequency Considerations

Lower Cutoff Frequency

The practical transformer is frequency limited. This is due to the reactive components shown

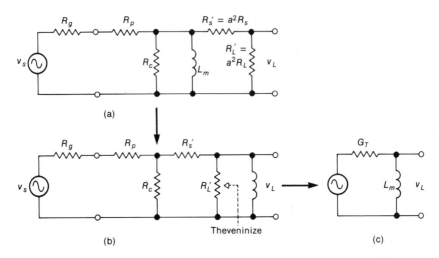

Figure 13-13. Low-frequency equivalent circuit of the transformer: (a) equivalent circuit; (b) rearranging the resistances; (c) reducing the circuit to an RL filter.

in the equivalent circuit of Figure 13-12b. The transformer will have a lower cutoff frequency, an upper cutoff frequency, and there will be some frequency range where its gain will be constant. In addition, the transformer gain-frequency response curve will display a peaking effect for a certain range of values of load resistance.

The magnetizing inductance L_m dominates in establishing the *lower cutoff frequency*. For the lower frequencies, the low-value capacitances appear as open circuits, and the low-value leakage inductances appear as short circuits. The low-frequency equivalent circuit of the transformer is shown in Figure 13-13. The load and secondary-winding resistances are reflected to the primary, and together with L_m and the source and primary resistances establish an RL filter circuit. The RL circuit is a high-pass filter and can be used to determine the transformer's lower cutoff frequency f_{cl}. The resistive portion of this circuit (Figure 13-13b) is Theveninized and results in the simple RL circuit of Figure 13-13c. For the RL circuit,

$$A_V = T(s) = \frac{V_L}{V_s} = \frac{A_o}{s + (2\pi f_c)}$$

where $f_c = \dfrac{1}{2\pi L_m G_T}$

and $G_T = \dfrac{1}{R_p + R_g} + \dfrac{1}{R_c} + \dfrac{1}{R'_s + R'_L}$

The generator resistance R_g, or the Thevenin resistance of the circuit driving the transformer, can significantly influence the value of the lower cutoff frequency. The lower the value of R_g, the lower the value of f_{cl}.

Mid-Frequency

In the *mid-frequency range of the transformer*, the capacitances and L_m appear as open circuits, and the leakage inductances are modeled as short circuits. The mid-frequency equivalent circuit (Figure 13-14) is a resistor-divider network. For this circuit the gain is constant; that is,

$$A_v = \frac{a^2 R_c R_L}{R_p R_c + a^2 R_p (R_g + R_L) + a^2 R_c (R_g + R_L)}$$

FILTER APPLICATIONS

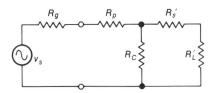

Figure 13-14. Mid-frequency equivalent circuit of the transformer.

Upper Cutoff frequency

The *high-frequency equivalent circuit* (Figure 13-15a) is an RLC circuit. The components are lumped and the circuit is simplified to the form shown in Figure 13-15b. This circuit can display a resonant effect (peaking) depending on the value of the reflected load resistance R'_L. The capacitance C' is a combination of the reflected secondary and mutual capacitances. It is given as

$$C' = \frac{C_s}{a^2} + C_m \left(1 \pm \frac{1}{a}\right)^2$$

The plus and minus signs (±) in the above equation are a function of the dot or phase orientation between the primary and secondary windings. The minus sign is used when the phase dots are opposite each other.

The key to finding the conditions for resonance is to convert or transform the parallel combination of C' and R'_L to an equivalent $R''_L C''$ series circuit. The end result of this transform is a series RLC circuit. For a series RLC circuit, resonance occurs when

$$X_L = X_{C''}$$

The above condition will be true when

$$R'_L > \sqrt{\frac{L_p + L'_s}{C'}}$$

If the reflected load resistance R'_L is low, resonance or peaking will *not* occur, and the upper cutoff frequency is a function of the leakage inductances.

$$f_{cu} \cong \frac{1}{[2\pi(L_p + L'_s)]/R_T}$$

where $R_T = R_g + R_p + R'_s + R'_L$

If R'_L is high, that is, the condition for resonance is met, the frequency where peaking occurs is a function of the leakage inductances *and* the reflected capacitance C'.

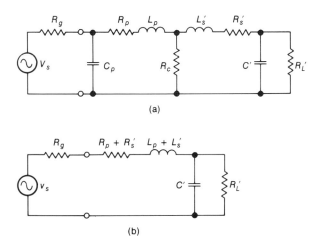

Figure 13-15. High-frequency equivalent circuit: (a) exact; (b) approximate.

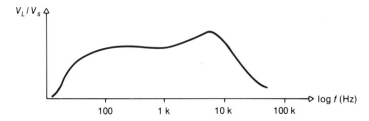

Figure 13-16. Typical transformer frequency response curve.

$$f_o \cong \frac{1}{2\pi C'} \sqrt{\frac{C'}{L_p + L_s'} - \left(\frac{1}{R_L'}\right)^2}$$

Peaking is usually a positive circuit characteristic because it extends the circuit's -3 dB bandwidth. A typical gain–frequency response curve is shown in Figure 13-16.

Example

Given: $N_p = 2.5$, $N_s = 1$, $C_p = C_s = 100$ pF, $C_m = 10$ pF, $L_s = L_p = 5$ mH, $L_m = 5$ H, $R_s = R_p = 100 \ \Omega$, $R_g = 500 \ \Omega$, and $R_c = 100$ kΩ. The square of the turns ratio must be found to reflect the components in the secondary to the primary.

$$a^2 = \left(\frac{N_p}{N_s}\right)^2 = (2.5)^2 = 6.25$$

The secondary winding resistance is reflected to the primary by multiplying it by the square of the turns ratio.

$$R_s' = a^2 R_s = 625 \ \Omega$$

The load resistance is reflected in a similar manner. For $R_L = 10$ kΩ,

$$R_L' = a^2 R_L = 62.5 \text{ k}\Omega$$

The reflected secondary leakage inductance is

$$L_s' = a^2 L_s = 32.5 \text{ mH}$$

The secondary and mutual capacitances are reflected to the primary capacitance C' by

$$C' = \frac{C_s}{a^2} + C_m \left(1 - \frac{1}{a}\right)^2$$
$$= 16 \text{ pF} + 10 \text{ pF } (1 - 0.4)^2 = 19.6 \text{ pF}$$

The low-frequency equivalent circuit is shown in Figure 13-13b. In this circuit, R_s' and R_L' are in parallel with R_c, which, upon Theveninizing the circuit, is in parallel with R_g and R_p. The total conductance is

$$G_T = \frac{1}{R_g + R_p} + \frac{1}{R_c} + \frac{1}{R_s' + R_L'}$$
$$= 1.667 \text{ mS} + 0.010 \text{ mS} + 0.0158 \text{ mS}$$
$$= 1.693 \text{ mS}$$

The lower cutoff frequency is a function of L_m and G_T.

$$f_{c1} = \frac{1}{2\pi L_m G_T} = 18.81 \text{ Hz}$$

The response will exhibit peaking if

$$R_L' > \sqrt{\frac{L_p + L_s'}{C'}}$$

$$\sqrt{\frac{L_p + L_s'}{C'}} = \sqrt{\frac{36.25 \text{ mH}}{19.6 \text{ pF}}} = 43.01 \text{ k}\Omega$$

The reflected load resistance, 62.5 kΩ, is greater than the required minimum value for resonance. The frequency where peaking occurs is a function of the leakage inductances and the reflected secondary and mutual capacitances. Thus,

$$f_o \cong \frac{1}{2\pi C'} \sqrt{\frac{C'}{L_p + L_s'} - \left(\frac{1}{R_L'}\right)^2} = 137.1 \text{ kHz}$$

Audio Filters

Two of the most popular filters found in audio equipment are the scratch and rumble filters. The scratch filter has a low-pass characteristic and is used to roll off excess high-frequency noise appearing as hiss, ticks, and pops from worn records. The rumble filter has a high-pass characteristic and is used to roll off low-frequency noise associated with worn turntable and tape transport mechanisms.

By combining the low- and high-pass filters, a broadband band-pass filter can be created to limit the audio bandwidth to include only the speech frequencies. The bandwidth filter used for this purpose would have a passband from 300 Hz to 3 kHz.

Scratch and Rumble Filters

A typical *scratch filter* is shown in Figure 13-17a. For this filter,

$$f_c = 9.72 \text{ kHz} \cong 10 \text{ kHz}$$
$$Q = 0.68$$
$$A_o = -1$$

attenuation slope = −40 dB/decade

The scratch filter will attenuate (at a rate of −40 dB/decade) signals of a frequency of 10 kHz or greater to inhibit high-frequency disturbance

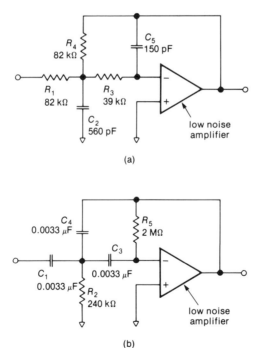

Figure 13-17. Audio filters: (a) scratch filter; (b) rumble filter.

voltages. The Q of the circuit is within 4% of 0.707. It is a Butterworth filter with a maximally flat response in the passband. The passband gain is -1, and the gain–frequency response characteristic will not exhibit any peaking. If a low-noise amplifier is used, the total harmonic distortion (THD) for the filter will be less than 0.1%.

A *multiple-feedback rumble filter* is shown in Figure 13-17b. The infinite-gain, multiple-feedback circuit is a simple, low-cost, and effective second-order audio filter implementation.

Problems

1. Calculate the value of L_{eq} in the high-pass-filter circuit of Figure 13-1 if R is 680 Ω and the cutoff frequency is 50 Hz. Specify a reasonable set of values for R and C in the simulated inductor circuit of Figure 7-23 to achieve the above determined value of L_{eq}.

2. In Figure 13-2, $R = 1$ kΩ. Specify values for all components in the filter circuit for a cutoff frequency of 3 kHz and a passband gain of 2.

3. Specify the value of R_1 and C_1 in Figure 13-3 if $R = 8$ kΩ, $f_c = 8$ kHz, and $A_o = -2$.

4. The low-pass and high-pass operational amplifier filters in Figures 13-2 and 13-3 can be combined to form a single-amplifier band-pass filter.
   (a) Specify the circuit.
   (b) Find the transfer function.
   (c) Identify the filter's characteristics.

5. For the circuit in Figure 13-4, $R_1 = 1125$ Ω, $R_3 = 1021$ Ω, $R_4 = 11.25$ kΩ, $C_2 = 2.2$ μF, and $C_5 = 0.1$ μF. Calculate f_o, A_o, and Q. Does the filter have an MFM or Butterworth response?

6. In the multiple-feedback filter model of Figure 13-5, $Y_1 = G_1$, $Y_2 = G_2$, $Y_3 = G_3$, $Y_4 = sC_4$, and $Y_5 = G_5$. For this configuration,
   (a) Find the transfer function.
   (b) Identify the filter response, the filter order, the characteristic frequency, Q (if applicable), passband gain, and the attenuation slope.

7. Derive the transfer function of the infinite-gain, multiple-feedback filter model in Figure 13-5 if the circuit components are C_1, R_2, R_3, R_4, and C_5.

8. In the multiple-feedback filter model of Figure 13-5, $Y_1 = G_1$, $Y_2 = G_2$, $Y_3 = sC_3$, $Y_4 = sC_4$, and $Y_5 = sC_5$. For this configuration,
   (a) Find the transfer function.
   (b) Identify the filter response, the filter order, the characteristic frequency, Q (if applicable), passband gain, and the attenuation slope.

9. For the circuit in Figure 13-9, $R_1 = 10$ kΩ, $R_2 = 528$ Ω, $R_5 = 200$ kΩ, $C_3 = C_4 = 10$ μF. Calculate f_o, A_o, and Q.

10. In Figure 13-12, $N_p = N_s$, $R_p = R_s = 75$ Ω, $C_p = C_s = 120$ pF, $C_m = 8$ pF, $R_c = 120$ kΩ, $L_m = 4.7$ H, $L_p = L_s = 6$ mH, $R_g = 50$ Ω, and $R_L = 10$ kΩ. For the transformer, find
    (a) the lower cutoff frequency;
    (b) the mid-frequency gain;
    (c) the reflected capacitance C' (the phase dots are opposite of each other);
    (d) the upper cutoff frequency.
    (e) Will the transformer's response have a peaking effect?

11. Derive the transfer function for the multiple-feedback filter in Figure 13-5 if the circuit components are C_1, R_2, C_3, C_4, and C_5. Identify ω_o^2, A_o, and ω_o/Q.

12. Calculate f_c, A_o, and Q for the low-pass rumble filter circuit in Figure 13-17.

SEMICONDUCTOR, TRANSISTOR, AND INTEGRATED CIRCUIT THEORY

The purpose of this section is to provide the necessary coverage of semiconductors, *pn* junctions, transistors, and integrated circuit theory to support the concepts, principles, and circuits presented in Section I. Many students will have previously taken a course in solid-state theory and devices. For these students, this material represents a review and/or additional reference. For those who have had no previous experience or education in solid-state device theory, Section II provides the essentials necessary to understand the concepts and theory behind semiconductor devices. A list of references at the end of the section provides further information for each of the individual topics.

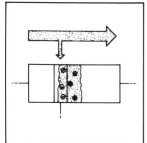

SEMICONDUCTOR, pn JUNCTION, AND TRANSISTOR THEORY

The most important electronic device is the transistor. It is comprised of two pn junctions that are made of semiconductor materials with added impurities. This chapter discusses the theory of semiconductors, pn junctions, and transistors.

Two conceptual pictures or models commonly used to describe the electrical conduction mechanism at the atomic level in materials are the valence-bond model and the energy-band model. The valence-bond model consists of a nucleus and orbiting electrons with current flow described in terms of charge particles moving from one orbit to another. The current flow in the energy-band model is described in terms of charge particles moving between energy levels. The two models are complementary.

Impurities added to a semiconductor material create additional mobile charge carriers. External sources can cause a net movement of these mobile carriers or current to flow. The two types of doped semiconductor materials are n and p type, and the two types of charge carriers found in these materials are holes and electrons, or positive and negative charge carriers. The joining of n- and p-type semiconductor material forms a pn junction or a diode, which, in a circuit, can be forward or reverse biased. The current and voltage in a diode are logarithmically related.

The transistor operates in the forward-reverse bias mode in amplifier applications. The forward-biased B–E pn junction causes the emitter charge carriers to diffuse into the base. The base region is very narrow, and although a few carriers recombine, most are swept into the collector region. The ratio of the collector current to the small base current defines the transistor's base-to-collector gain β.

14-1 SEMICONDUCTOR THEORY

Two pictures of a Semiconductor

The most sophisticated electronic devices used are called integrated circuits. Within the integrated circuits are a large number of components. The transistor is one of the more common components and it is also the most important. The transistor is physically comprised of two *pn* junctions, with each *pn* junction made of two different types of semiconductor materials. Semiconductor materials possess certain special properties, and in order to relate to these properties, one must talk about the atomic structure of materials. Hence, our coverage of semiconductors begins with the atomic structure of materials and ends with integrated circuits.

In electronics, and in semiconductors, we are concerned with electrical conduction. *Electrical conduction refers to the ability of a device or material to transfer charge or conduct a current.* Conduction refers to the motion of electrons, or charge particles.

There are several views, models, conceptual pictures, or sets of words that are used to explain the causes, effects, and mechanisms of *electrical conduction at the atomic level.* Two views or models will be presented. The first view, called the *valence-bond model*, emphasizes the events in space and time, whereas the second model, called the *energy-band model*, emphasizes them in terms of energy.

The two views are *complementary* rather than duplicative. The valence-bond model presents a visual picture of the conduction mechanism. It relies on the notion of a valence bond and an atom with a nucleus and orbiting electrons. The energy-band model describes the conduction mechanism in terms of charge particles moving between energy levels.

It should be remembered that these pictures or models are explanations that give us a mental picture and help us to understand what is happening at the atomic level. These pictures are only valid to explain the basic electrical characteristics of semiconductors. The models have to be modified or completely replaced by more sophisticated models to explain the advanced features of semiconductor behavior. The more sophisticated models require a knowledge of advanced physics and mathematics.

Valence-Bond Model

Atoms and Matter

The materials we see with our eyes (macroscopic) are comprised of an extremely large number of particles called atoms (microscopic). The *atom is the smallest unit of a material* that retains the identity of that material. As an example, the smallest unit of a piece of copper wire is an atom of copper. A copper atom is the smallest particle that possesses the characteristics of copper as we know it.

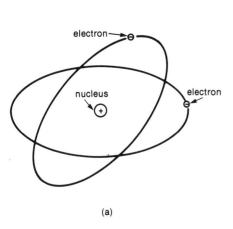 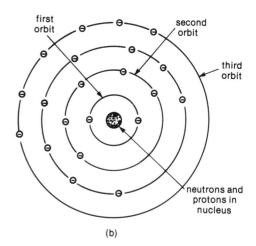

Figure 14-1. The atom: (a) three-dimensional representation (two electrons); (b) two-dimensional representation.

The *atom is comprised of a core called a nucleus* (Figure 14-1) *and electrons.* The electrons are arranged in orbits, encircling the nucleus as planets encircle the sun. There are a number of orbits, and the maximum number of electrons in each orbit is fixed. Each unique type of material is made up of a unique atom. Atoms are distinct from each other by, primarily, the number of orbiting electrons and the contents of the nucleus.

The nucleus is primarily comprised of subatomic particles called protons and neutrons. The protons possess a property called charge, which, for the protons, is given the positive polarity. The orbiting electrons are also charged particles, but possess the opposite, or negative, polarity. In the natural or stable state, the number of negatively charged electrons is the same as the positively charged protons. There is a force of attraction between these two charged particles. This force of attraction is balanced by the centrifugal force associated with the motion of the electrons in their orbits around the nucleus. The electrons thus maintain a fixed distance from the nucleus, with the force of attraction of the charged particles balanced with the force of the motion of the electrons in the orbit.

Whether a material is a gas, liquid, or solid depends upon the separation and distance between atoms. If atoms are far apart, compared to the diameters of their outer electronic orbits, they cannot exert strong forces upon one another and therefore form a gas. When atoms form a solid, they arrange themselves in an orderly three-dimensional array called a crystal structure. The atoms in this type of structure are close to each other and, in fact, share electrons in the outermost orbit. The *electrons in the outermost orbit are called valence electrons.* Figure 14-2 illustrates examples of atoms arranged in a common crystal structure.

Conductors, Insulators, and Semiconductors

Copper, aluminum, and silver are examples of metals or solids that are excellent conductors. The crystal structure of these metals is such that the outer electrons are shared by all atoms. In these metals, each atom supplies one such free electron so that the number of electrons or charged particles per unit volume is very large (10^{23} electrons/cm^3). A potential or voltage applied to a material will induce an electric field and cause electrical conduction. In metals, conduction will occur for very small electric fields, and we say the electrical conductivity is very high.

Insulators are materials in which electrical conductivity is very low. Relatively few charged particles are available to move through the material, even with moderate electric fields applied. The structure of solid insulators is such that over a wide range of temperature almost all electrons are bound to their constituent atoms. Examples of insulators include quartz, mylar, mica, air, and silicon dioxide.

Semiconductors are materials whose electrical conductive abilities lie between the insulator and the conductive metals. Germanium and silicon dominate in the production of semiconductor devices. They are brittle, metallic-appearing crystalline solids. Our attention will center on silicon. Many transistors and almost all integrated circuits are made from silicon.

A two-dimensional representation of the *silicon atom* is shown in Figure 14-3. A neutral

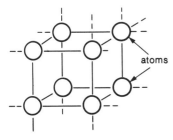

Figure 14-2. Common crystal structure (cubic).

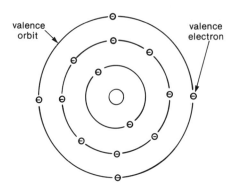

Figure 14-3. Silicon atom.

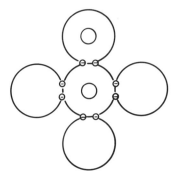

Figure 14-4. Covalent bonds in a crystal structure.

silicon atom contains 14 orbiting negatively charged electrons and 14 positively charged protons in the nucleus. The maximum number of electrons that can exist in the first orbit is 2. This orbit is filled. The maximum number of electrons that can exist in the second orbit is 8. The second orbit is also filled. The maximum number of electrons that can exist in the third orbit is 18. This orbit is *not* filled and contains 4 electrons. This last orbit for silicon is called the valence orbit, and its electrons are called valence electrons.

Electrical Conduction and Charge Carriers

Germanium and silicon have diamondlike crystalline structures. Each of their atoms is surrounded in the structure by four neighboring atoms. Each semiconductor atom has four valence or outermost electrons and shares one of its own electrons with each of its neighboring atoms. This arrangement of shared electron pairs is called a *covalent bond* or simply a *valence bond*. A two-dimensional representation of these bonds is shown in Figure 14-4. These bonds are the cement that hold together the atoms in the structure, and they will establish the electrical behavior of the crystal.

In extremely pure semiconductor material, the structure is stable and electrical conductivity is low. Pure semiconductor material is called *intrinsic*, and it has less than 1 part in 10^{10} impurities. At low temperatures, the behavior of the material approaches that of an insulator. At high temperatures, the thermal vibration of the atoms frees some of the electrons and electrical conductivity increases. The energy required to free the electron from the symmetric stable crystalline structure is called the ionization energy (for silicon, 0.7 eV).

When the energy level of an atom is increased such that an electron is liberated, a vacancy in the atom's structure is left behind (Figure 14-5). The positively charged nucleus is greater than the negative charge of the electrons, and the net charge of the atom is now positive. The vacancy attracts a nearby bound electron and creates a new vacancy at the site the electron left. This procedure is repeated, and the vacancy behaves as if it were a new free particle with a positive charge. This apparent particle is quite logically called a *hole*.

When a valence electron is liberated, a hole is created. The positively charged hole attracts a neighboring charged electron. When it does attract an electron, a new hole or vacancy is created from whence the electron came. The result of this process is the effect of the movement of a hole or positively charged particle. The original electron set free by thermal vibration moves independently of the hole and is called a conduction electron.

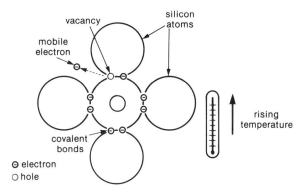

Figure 14-5. Thermal generation of electron–hole pairs.

The valence-bond model provides a set of words and a model that helps us to mentally visualize and understand the behavior of electrical conduction at the atomic level. This understanding gives more meaning to what happens at the macroscopic level, that is, in the laboratory. This model is only valid for explaining certain basic concepts, characteristics, and properties. For more advanced concepts, a much more sophisticated model will have to be developed.

The movement of charge in the valence-bond model can be described three ways:

1. The movement or apparent movement of *conduction band electrons* and *valence band electrons*.

2. The movement or apparent movement of *electrons* (conductions) and *holes*.

3. The movement or apparent movement of *negatively and positively charged particles or carriers*.

Impurities in Semiconductors: Increasing the Number of Free Charge Carriers

Electrical conductivity is relatively low in pure semiconductor materials. Free or mobile charge particles are only created as a result of temperature. The number, however, of mobile or free charge carriers can be significantly increased if appropriate foreign substances (impurities) are added to the otherwise pure material. By adding controlled amounts of these impurities (called doping), the semiconductor material will have either more holes than electrons, or vice versa. The impurities (about 1 part per million) that are added must be very close in structure to silicon so that the only significant change in the material is the number of free charge particles. In fact, the impurities (called dopants) are almost identical to silicon or germanium. Each dopant, however, has either one *more* valence electron than silicon or one *less*.

Donors and Acceptors:
Making *n*- and *p*-type Semiconductors

A *donor* is an impurity atom with *one more* electron in its outermost orbit than silicon. Except for the additional valence electron, both materials are similar in structure. When a donor is added to the crystalline structure of silicon, it fits into the valence bond structure without much difficulty. The difference, however, is the fifth electron. Four electrons of the impurity atom form covalent bonds (sharing) with the neighboring atoms of silicon. The fifth electron (Figure 14-6) becomes a conduction electron

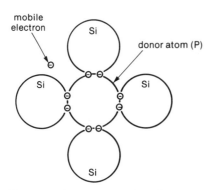

Figure 14-6. *n*-type semiconductor.

or mobile charge carrier. Hence, mobile or free carriers are available in the doped semiconductor material that are not thermally generated.

Examples of *donor impurity materials* are phosphorus (P), arsenic (As), and antimony (Sb). Semiconductor material that has been doped with donor impurities is called *n-type semiconductor* material. The majority carriers in *n*-type material are the conduction electrons. There are some minority carriers, holes, that are thermally generated.

An *acceptor* is an impurity atom with one *less* electron in its outermost orbit compared to silicon. Acceptor atoms have a similar structure to silicon atoms and fit into the valence bond structure (Figure 14-7) without difficulty. The

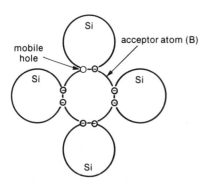

Figure 14-7. *p*-type semiconductor.

difference, however, between the impurity and silicon is the lack of the fourth valence electron in the impurity atom. This creates a vacancy in the crystalline structure, which accepts nearby bound electrons and behaves as a mobile, positive charge carrier. Hence, mobile carriers are available in the doped semiconductor material that are not thermally generated.

Examples of *acceptor impurity materials* are boron (B), aluminum (Al), and gallium (Ga). Semiconductor material that has been doped with acceptor impurities is called *p-type material*. The majority carriers in the *p*-type material are the holes (or positive charge carriers). There are some minority carriers or electrons that are thermally generated. Doped semiconductor materials are referred to as *extrinsic*.

Energy-Band Model of a Semiconductor

The valence-bond model discussed semiconductors using a structural atom with its components occupying a given space and location and relationship to each other. The energy-band model discusses semiconductors in terms of energy. The valence-bond model already introduced two energies. The ionization energy was the energy required to break a valence bond (and free a charge particle), and the thermal energy associated with a crystalline structure caused the generation of electron-hole pairs.

In the valence-bond model, raising the temperature (or increasing the energy level) of an atom caused a valence electron to be freed. Actually, in this model the electron was raised to a higher orbit. In the energy-band model, the orbits are represented by energy levels (Figure 14-8). Hence, raising the temperature causes the electron to move from one energy level (valence) to a higher energy level (conduction). The energy levels are actually energy bands; that is, each level is a range of energy

SEMICONDUCTOR THEORY

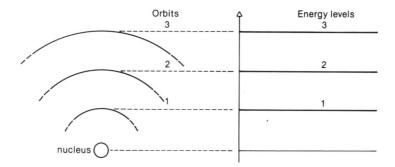

Figure 14-8. Orbits and energy levels.

values. Since no two electrons can have the same state in the same system, each orbit is represented by a *band of energy values*. The distance between the energy bands is the energy required to raise an electron from one level to the next higher level.

Interpretation of the Bands

In intrinsic (pure) silicon, at low temperature, few charge carriers are free. Most electrons are bound in the crystal structure. In terms of the energy-band picture (Figure 14-9), most electrons are within the valence energy band of values. Few electrons are at the next orbit level or the conduction band of energies, which is where we say the electrons are free or mobile As temperature is increased, electrons are excited to the higher energy level, leaving holes behind. *Thermal energy, thus, generates electron-hole pairs*. The electron becomes a free negative-charge particle, and the hole behaves as a free positive-charge particle. The hole in the valence band is quickly filled by a nearby valence band electron, which leaves behind a hole. The process is continually repeated, and motion in the valence band can either be described as a moving hole or a mov-

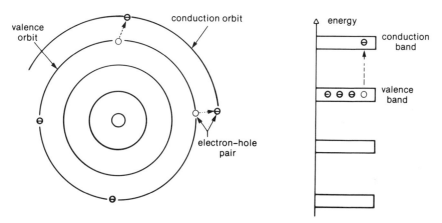

Figure 14-9. Creating mobile charge carriers in the silicon atom: (a) valence-band model; (b) energy-band model.

ing valence band electron that fills it. Current can then be described in terms of the net movement of conduction and valence band electrons *or* the net movement of electrons (conduction band) and holes.

Current is defined as the net movement of positive charge carriers. It is the rate of change of charge per unit time.

Above 0°K, there is always movement of charge carriers in materials; however, there is a *net movement of these carriers only when an electric field is induced* as the result of an applied voltage or potential.

Insulators, Conductors, and Semiconductors

The energy-band representations of an insulator, conductor, and an intrinsic semiconductor are shown in Figure 14-10.

For the insulator, there is a significant energy difference between the valence and conduction bands. All levels or positions within the valence band are occupied by electrons, while all levels in the conduction band are empty. Small electric fields cannot raise electrons in the valence band to the conduction band level. In terms of the physical structure, the electrons form strong bonds with neighboring atoms. These bonds are difficult to break; thus, there will be no free or mobile electrons. Hence, electrical conduction in insulators is minimal.

In the energy-band picture of metals, the valence and conduction bands overlap. The application of very small electric fields will cause the movement of charge carriers. It is relatively easy to impart kinetic energy to the electrons and, thus, electrical conduction in metals is high.

The energy difference between the valence and conduction bands in semiconductors is moderate. Electron–hole pairs are thermally generated, and a few electrons appear in the conduction band with a corresponding number of holes in the valence band. The application of an electric field will cause the electrons and holes to gain kinetic energy. Conduction is limited in the intrinsic semiconductor, but it can be significantly increased by adding impurities that increase the number of free electrons or free holes.

n- and *p*-type Semiconductors

Semiconductor material that has been doped with donor impurities (for example, boron) is called *n*-type semiconductor. Similarly, material

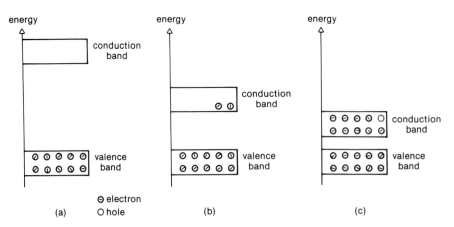

Figure 14-10. Energy-band pictures: (a) insulator; (b) semiconductor (intrinsic at room temperature); (c) conductor.

doped with acceptor impurities (for example, phosphorus) is called *p*-type semiconductor.

The majority carriers in *n*-type semiconductor material are electrons (conduction band), and the minority carriers are holes (or valence band electrons). In *p*-type semiconductors the majority carriers are holes (or positive charge carriers), and the minority carriers are electrons (or negative charge carriers).

The *n*- and *p*-type materials by themselves are of little value. The value of semiconductors occurs when the two types are joined together and form a *pn* junction.

Current Flow in a Semiconductor

Conduction in a semiconductor material may occur by either of two mechanisms: (1) drift and (2) diffusion.

Drift is the mechanism by which current is conducted in a copper wire. The mobile charge particles in the wire are moved by the force *produced by an electric field*. Drift current in a semiconductor is governed by

$$J = \sigma E$$

where J is the current density in amperes per square centimeter (A/cm^2), σ is conductivity in siemens per centimeter (S/cm), and E is the applied electric field in volts/centimeter (V/cm).

The second mechanism of current flow in semiconductor materials is referred to as current by diffusion. *Diffusion* occurs in a semiconductor whenever there is a *difference in the concentration of carriers* in any adjacent regions in the crystal. The current that flows by diffusion is a function of the concentration gradient, the cross-sectional area of the material, the charge, and a constant. Diffusion current (electron) in a semiconductor is governed by

$$I_n = -D_n A q \frac{dn}{dx}$$

D_n is a proportionality constant and is called the diffusion constant. A is the cross-sectional area of the semiconductor block, and q is the electric charge. The concentration gradient, dn/dx, defines the rate of change of the number of electrons (n) per distance (x) from the interface of the two semiconductor materials that have the different concentration of carriers.

The intent of presenting the equations for the two current mechanisms is to make the reader aware of the mechanisms themselves and the factors that establish their values.

14.2 *pn* JUNCTION THEORY

Making *pn* Junctions

The dominant technology employed in making single and multiple *pn* junction devices is called the planar process or planar technology. The *pn* junction or interface is formed by the physical process called diffusion, and the *pn* junctions are called diffused junctions. *Diffusion* is the intermingling of the atomic structures of two materials. This method has evolved from two early procedures called the grown junction and alloy junction methods of fabricating *pn* junctions.

Diffused junctions are formed by exposing one type of semiconductor material (in solid form) to a gas with a high concentration of the opposite-type impurity. The impurities in the gas, precisely controlled by temperature and time, diffuse into the semiconductor solid and form a *pn* interface or *pn* junction. The making of a *pn* junction by diffusion is illustrated in Figure 14-11.

Depletion Layer

The *n*-type material at the *pn* interface has a surplus of negative charge carriers. The *p* material (Figure 14-12) has a surplus of positive charge carriers. There is a force of attraction and the carriers cross the junction (by diffusing). The conduction electrons in the *n* material cross the interface and become minority carriers in the *p* material. Here the electron

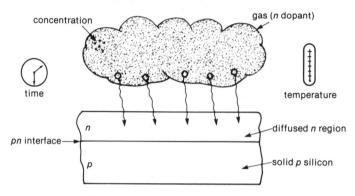

Figure 14-11. Making a *pn* junction.

surrounded by holes quickly falls into a hole. This process is called *recombination;* the positive and negative charge carriers *combine*. The hole disappears and the conduction electron becomes a valence electron. In other words, an electron falls from the conduction energy band back to the valence band where a hole is filled.

Each electron from the *n* material that recombines leaves behind an atom with one less electron than it has in its natural state. The nucleus then has one more positively charged proton than electrons, and the atom becomes positively charged. *An atom with a net positive charge is called a positive ion.* Ions are charged atoms but they are not mobile. A similar situation occurs in the *p* material. The electron filling the hole causes the atom in the *p* material to have one more electron than proton. The atom has a net negative charge and is called a *negative ion*. It is not a mobile charge particle.

The result of the diffusion of the carriers across the *pn* junction is recombination and the formation of a depletion layer or region. The *depletion region* (Figure 14-13) *is an area near the pn interface that is void of any mobile or free charge carriers.* This region behaves like an insulator.

Barrier Potential

The diffusion of carriers across the *pn* junction and recombining is not a continuous process. As recombination takes place, the negative ions in the *p* material repel further diffusion of the negative charge particles from the *n* region. Similarly, the positive ions in the

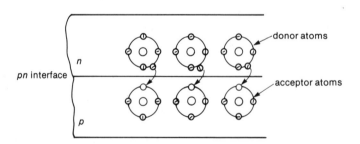

Figure 14-12. *pn* junction before diffusion.

pn JUNCTION THEORY

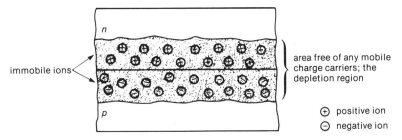

Figure 14-13. *pn* junction after diffusion; the depletion region.

n material repel further diffusion of the mobile positive charge particles from the *p* region. Ultimately, an equilibrium condition (Figure 14-14) is reached where the attraction of opposite-charged particles is balanced by the repulsion of like-charged particles and ions.

Associated with the ions in the *p* and *n* regions is an *electric field*, and it is the field that stops the diffusion of the carriers across the junction. Having an electric field between the regions of ions is *equivalent to a difference in potential* between them. This difference in potential is called the *barrier potential*. At 25°C, the barrier potential of silicon is approximately 0.65 V and is 0.3 V for germanium. At 25°C, the thermal generation of electron–hole pairs still occurs. A few carriers do move across the junction in both directions, but the net effect does not change the equilibrium condition at the junction.

A very *important characteristic* of the barrier potential is its *dependence on temperature*. As temperature increases, more electron–hole pairs are created. The drift of minority carriers across the junction increases and forces the equilibrium to occur at a slightly lower barrier potential. The characteristic that describes the temperature dependence of the barrier potential is the temperature coefficient $\Delta V/\Delta T$. For silicon, the temperature coefficient is approximately *-2 mV per degree Celsius*. In other words, the barrier potential will decrease by 2 mV for every increase in temperature of 1°C.

Bias Modes

A diode is a two-terminal component. The junction diode is an encapsulated *pn* junction with leads attached to the *n* and *p* materials. The physical diode and its symbol is shown in Figure 14-15. The diode has two modes of operation: (1) forward bias and (2) reverse bias.

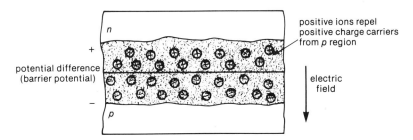

Figure 14-14. Barrier potential; *pn* junction after equilibrium is reached.

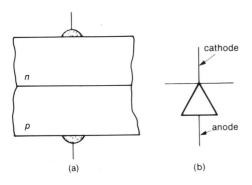

Figure 14-15. Diode: (a) physical structure; (b) schematic symbol.

Forward Bias

Figure 14-16 shows a forward-biased diode. The positive terminal of the dc potential source is connected to the anode (*p* material), and the negative terminal is connected to the cathode (*n* material). The potential source sets up an electric field that opposes the field of the depletion region. The potential source forces majority carriers from the source terminals into the *p* and *n* regions and deionizes the edges of the depletion region. The current through the diode is very small for low values of applied voltage (compared to 0.65 V). The greater the applied potential, the narrower the depletion region becomes. As the source potential reaches a value near that of the barrier potential, the diode will begin to conduct a heavy current. After that, very slight increases in potential across the diode will cause the junction to conduct very large currents. *The voltage and current are logarithmically related.*

Reverse Bias

Figure 14-17 shows a dc potential source across the diode. The diode is reverse biased. The applied voltage of the dc source establishes an electric field. This electric field reinforces or adds to the electric field of the depletion region. The applied voltage draws a greater number of carriers away from the *p* and *n* regions, creating more ions, and widening the depletion region further. The depletion region stops growing when its difference in potential (due to the ions) equals the applied reverse voltage.

To a first approximation, a reverse-biased diode is equivalent to an open circuit. However, thermal energy does create electron–hole pairs in the reverse-bias condition. This causes a small amount of reverse current to flow. This current is caused by the minority carriers and is called the *reverse saturation current*, I_s.

Current and Voltage Relationships

A graph of diode current versus diode voltage is shown in Figure 14-18. For low values of

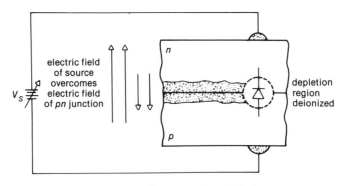

Figure 14-16. Forward-biased diode.

pn JUNCTION THEORY

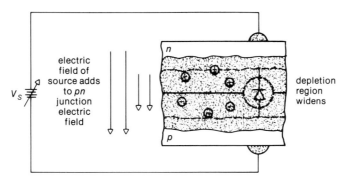

Figure 14-17. Reverse-biased diode.

forward voltage ($\ll 0.65$ V), the diode current is small. As the forward voltage approaches the barrier potential, the current begins to increase dramatically. At or slightly beyond the nominal barrier potential, small increases in voltage cause significant increases in diode current.

The I-V characteristic (forward bias) is called the diode curve, and the part of the characteristic that turns upward is called the knee. For high values of current, the characteristic becomes linear, indicating a resistive effect.

For reverse voltage conditions, the diode

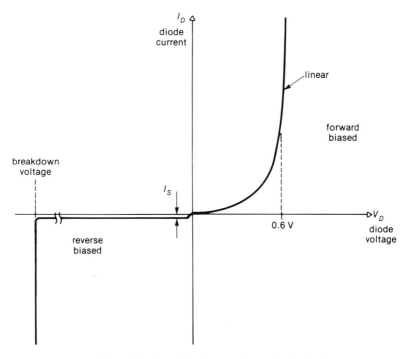

Figure 14-18. I-V characteristic of the diode.

current is extremely small (I_s). This current remains small until the breakdown voltage is reached, where the diode will conduct very large values of current and maintain a near-constant breakdown voltage.

The graph is a pictorial way of showing the relationship between a diode's current, I_D, and voltage, V_D. This relationship can also be expressed mathematically. The equation or mathematical model that best fits the diode's actual I-V characteristic is as follows:

$$I_D = I_S(e^{V_D/0.026\ V} - 1)$$

I_S is the saturation current. For forward diode currents that are much greater in value than I_S, that is, $I_D \gg I_S$, then

$$I_D \cong I_S(e^{V_D/0.026\ V})$$

This equation states that the diode voltage is proportional to the log of the diode current.

Diode Resistances

Two resistances are associated with the forward-biased diode: (1) the ac (small-signal) resistance, r_d, and (2) the dc diode resistance, R_D.

For high values of diode current, the diode characteristic is linear, indicating a resistive effect. This resistive effect is due to the bulk resistance of the n and p semiconductor materials in the device. The value of the bulk resistance may be found by finding the reciprocal of the slope at the higher currents, or it may be estimated from

$$R_D \cong \frac{0.3\ V}{I_D}$$

Typical values of bulk resistance vary from 0.5 to 50 Ω.

Since diodes are generally used with ac or time-varying potentials, an important quantity is their ac (dynamic, small signal, or incremental) resistance, r_d. The value of r_d is obtained from the ratio of a small change in diode voltage to the resulting change in diode current, or

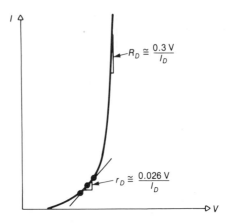

Figure 14-19. Diode resistances.

$$r_d = \frac{\Delta V_D}{\Delta I_D} = \frac{V_D}{I_D}$$

The symbol Δ (delta) is used to denote a small change.

Because of the curvature of the diode characteristic, the ac resistance will vary depending on the operating (dc) point. The dynamic resistance can be approximated from

$$r_d \cong \frac{0.026\ V}{I_D}$$

where I_D is the dc forward biased diode current. Typical values of dynamic resistance vary from 5 to 50 Ω. The graphical interpretation of these resistances is shown in Figure 14-19.

Diode Capacitances

Two capacitances are associated with the diode. They are (1) the depletion-layer capacitance of a reverse-biased diode, and (2) the diffusion capacitance of a forward-biased diode.

When a pn junction is reverse biased, a depletion region is formed. This region is similar to an insulator; that is, it is a material with no free charge carriers. The depletion region and the outer areas of the device form the equivalent

pn JUNCTION THEORY

of a parallel plate capacitor, that is, two conducting surfaces separated by a dielectric. However, unlike the parallel plate capacitor, the value of the diode capacitance is voltage dependent. The greater is the applied reverse potential, the wider the depletion region and the smaller the value of capacitance. A change in the voltage will cause a change in the charge in the device.

The value of the depletion capacitance is in the low picofarad (5 to 100 pF) range, and it is determined by the following expression.

$$C = k(V_d - V)^{-1/n}$$

In the above expression, k is a proportionality constant, V_d is the junction barrier potential, and n is a number between 2 and 3 and is a function of the physical characteristics of the device.

Capacitance is defined as the ratio of a change in charge (ΔQ) to a change in voltage (ΔV).

$$C = \frac{\Delta Q}{\Delta V}$$

A capacitive effect exists if, in a device, a change in voltage will cause a change in charge (or vice versa).

The value of capacitance is equal to the ratio of these changes. Diffusion capacitance is due to the change in minority carrier distribution (number of charge carriers) as a result of a change in applied voltage. Figure 14-20 is a graph of the number of minority charge carriers (for a given value of applied voltage) versus distance (from the pn junction). The number of minority carriers (for a given value of forward voltage) decays exponentially as the distance increases from the pn junction. Because minority carriers are surrounded by majority carriers, the chance for recombination is high, and few minority carriers exist a great distance from the junction. The greater is the applied voltage, the greater the number of carriers (charge) at the various positions. A change in the voltage causes a change in the amount of charge stored in the semiconductor region. This effect in the forward-biased diode is modeled by a capacitance called the diffusion capacitance. The diffusion capacitance, at room temperature, is determined by

$$C_d \cong C_1 \frac{I_F}{0.026 \text{ V}}$$

where I_F is the forward current and C_1 is a constant related to the physical characteristics

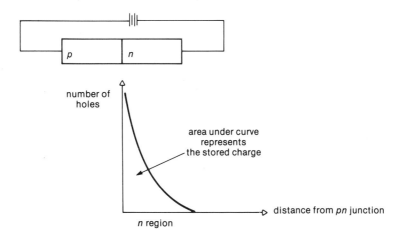

Figure 14-20. Stored charge in a forward-biased diode.

of the device. A typical value of diffusion capacitance is 250 pF.

14.3 TRANSISTOR THEORY: BIPOLAR TRANSISTORS

A junction diode is a two-terminal device consisting of a single *pn* junction. A junction transistor is a three-terminal device consisting of two *pn* junctions. The transistor is the most important electronic device, and its discovery has had an enormous impact on the electronics industry.

Junction transistors are called bipolar transistors. *Bipolar* is an abbreviation for "two polarities" and refers to the two types of charge carriers, that is, the majority carriers or electrons in the *n* region and the majority carriers or holes in the *p* region.

The bipolar transistor can have one of two crystal forms: *npn* or *pnp*. For either case, each of the three materials is connected to a device lead or terminal. The terminals of the bipolar transistor are called emitter, base, and collector. The schematic symbols for the *npn* and *pnp* transistors are shown in Figure 14-21.

Physical Structure

A cross-sectional view of the physical structure of the *npn* and *pnp* transistors found in integrated circuits is shown in Figure 14-22. To facilitate explaining transistor action, a sandwichlike representation of the structure is used (Figure 14-23). Physcially, the emitter region is heavily doped with majority carriers. The emitter "emits or injects" electrons into the base in a *npn* transistor, and emits or injects holes into the base in a *pnp* device. The base region is lightly doped and very thin or narrow. It passes most of the emitter's injected charge carriers to the collector. The collector is moderately doped and is the largest of the three regions. It is so named because it collects or gathers charge carriers from the base.

The base and emitter form one *pn* junction and the base and collector form the other. The base and emitter (*B-E*) junction is called the *B-E* diode. Similarly, the *B-C* junction is called the *B-C* diode.

Bias Modes of Operation

The transistor has two *pn* junctions, and each *pn* junction can either be forward or reverse biased. Thus, the transistor has four possible bias modes. They are called forward (*B-E*)-forward (*B-C*) bias, forward-reverse, reverse-

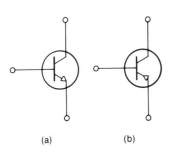

Figure 14-21. Schematic symbols: (a) *npn* transistor; (b) *pnp* transistor.

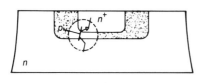

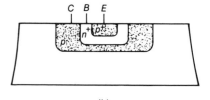

Figure 14-22. Physical structures (IC): (a) *npn*; (b) *pnp*.

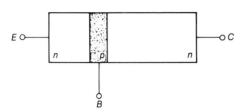

Figure 14-23. Sandwich-like representation of the *npn* transistor.

forward, and reverse-reverse bias. Transistors used in amplifier and linear circuits generally operate in the forward (*B-E*)-reverse (*B-C*) mode. Transistors used in digital circuits generally operate in the forward–forward or reverse–reverse modes.

Transistor Operation

Current amplification in a transistor occurs when the device is forward-reverse biased. In this mode, the base-to-emitter *pn* junction is forward biased and the base-to-collector *pn* junction is reverse biased. A *npn* transistor is shown in the forward-reverse bias mode in Figure 14-24. The V_{EE} and V_{CC} supplies establish the forward-reverse bias condition.

With the *B-E* junction forward biased, large amounts of majority carriers (conduction band electrons) from the *n* emitter region diffuse into the *p* region of the base. In the *p* base, the electrons become minority carriers and recombination occurs. The result of recombination is a base current. However, the base region is very thin and the number of diffusing electrons so large that most of the emitter electrons reach the collector region. At the other end of the collector region, electrons leave and enter the external collector lead, where they return to the positive terminal of the collector supply. Approximately 95% of the carriers leaving the emitter are collected by the collector.

The base-emitter junction is voltage forward biased. The larger V_{BE} is made, the greater the number of injected electrons. The reverse bias of the base-collector junction has little influence on the number of electrons that enter the collector. A large current flows in the reverse-biased *C-B* junction owing to the existence of a forward-biased junction in the vicinity.

The thin base region is an important physical characteristic of the transistor. The region must be narrow so that the minority carriers in the base (the electrons from the emitter) have a long enough lifetime such that most of them diffuse into the collector region.

The electrons from the emitter that cross into the base region form the emitter electron current. The electrons leaving the base region are the base current, and the electrons leaving the collector are the collector electron current. We say the transistor is a current amplifier when we consider the input to be the base current and the output to be the collector current. The ratio of I_C to i_B can be very large.

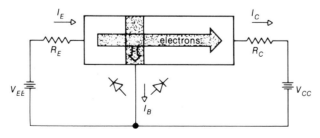

Figure 14-24. Electron currents in a forward-reverse biased transistor.

Transistor Currents and Voltages

Three currents and three voltages, or differences in potential, are associated with the transistor. The three transistor currents are I_B, I_C, and I_E. The uppercase letters indicate *dc currents*. The currents i_b, i_c, and i_e are the transistor *ac currents*. The potential differences (dc) associated with the transistor are V_{CB}, V_{BE}, and V_{CE}, where the second letter in the subscript is the reference. The transistor ac voltages are v_{cb}, v_{be}, and v_{ce}.

Kirchhoff's current and voltage laws can be applied to the transistor as a device. The current entering the transistor must equal the current leaving it. Applying KCL yields

$$I_E = I_B + I_C$$

Applying KVL, we arrive at

$$V_{CE} = V_{CB} + V_{BE}$$

Gain Characteristics

Base-to-Collector Gain β

Gain is the ratio of the output over the input. What is considered the output and input of a device may vary depending on how the device is used in a circuit. If the input current is I_B and the output current is I_C, the ratio I_C/I_B defines the base-to-collector current gain, which is called beta (β).

$$\beta = \frac{I_C}{I_B}$$

The current gain β of a transistor is typically large. Its nominal range of values is from 35 to 300, with 100 an average number.

If, in a circuit, we can make the signal current the same as the base current, the collector current will be a magnified (in amplitude or magnitude) version of the base current. A transistor can be viewed as a current-amplifying device if the base current is the input and the collector current is the output. This case is true in many applications. It must be remembered that we are not getting something (extra current) for free. Kirchhoff's current law for the device still holds true; that is, remember I_E. In other words, the conservation-of-energy principle is being maintained for the system (transistor and circuit). However, from a particular input of the system (I_B) to another particular output (I_C), we have current amplification.

In a typical circuit, the transistor's currents will be comprised of a dc and an ac component. Each component of I_C will be a multiple (β) of I_B; however, the ac and dc values of β are slightly different. Circuit techniques minimize the dependence of the circuit's characteristics on the exact value of β. The current gain β is frequency dependent and decreases, as one might expect, when frequency increases. Other aspects of β and additional transistor characteristics are covered as they relate to the circuits in which the transistors are used.

Emitter-to-Collector Current Gain α

In describing transistor action, we said that most of the charge carriers emitted by the emitter were collected by the collector. The ratio of the current I_C to the current I_E defines the emitter-to-collector current gain alpha (α).

$$\alpha_{dc} = \frac{I_C}{I_E}$$

The theoretical maximum value of α is 1. In practice, this can never be achieved because there will always be some emitter carriers that recombine in the base region. There will be some small value of base current, and thus I_C will always be less than I_E. The thinner and more lightly doped the base region is, the higher the value of α. The typical value of α is between 0.95 and 0.99. Alpha can be approximated as equal to 1 in most preliminary analysis.

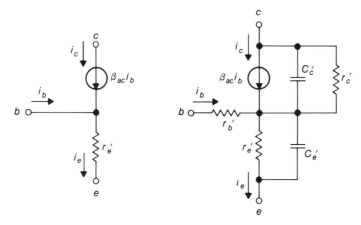

Figure 14-25. Transistor ac equivalent circuits: (a) first-order or ideal model; (b) higher-order model.

14.4 TRANSISTOR AC EQUIVALENT CIRCUITS

The first and higher order ac equivalent circuits of the transistor are shown in Figure 14-25. These device circuit models are valid for small-signal conditions. A brief explanation of each component in the models follows; however, additional information is provided in the text material where these components affect circuit characteristics.

r'_e: Emitter-base junction ac resistance, emitter diffusion resistance, or incremental ac resistance. This resistance represents the constant of proportionality or ratio of the base-emitter voltage (ac) and emitter current (ac). Nominal value: 25 Ω.

r'_b: Base spreading resistance. The ohmic resistance between the base lead and the center of the base region is modeled by r'_b. Its value depends on the geometry and current flow pattern in the base region. Nominal value: 150 Ω.

r'_c: Even though the collector base junction is reverse biased, some current conduction may be expected. This conduction is represented by the ohmic resistance r'_c between the collector and base, and is due to the minority carriers in the collector region. Nominal value: 100 kΩ.

C'_e: Diffusion capacitance of the forward-biased B-E junction. A change in the base-emitter voltage will cause a change in the stored charge in the base region. The ratio of these changes ($\Delta Q/\Delta V$) defines a capacitive effect identified as C'_e. Nominal value: 250 pF.

C'_c: Depletion layer capacitance of the reverse-biased C-B junction. The reverse-biased C-B region is analogous to a parallel plate capacitor. This junction contains a depletion region separated by two conducting surfaces. This voltage-dependent capacitance is typically called C'_c, C_c, C_{cb}, or C_{obo}. Nominal value: 8 pF.

$\beta_{ac} i_b$: Dependent current source. This current source establishes the value of i_c and reflects the current gain capability of the transistor. It can also be described by $\alpha_{ac} i_e$.

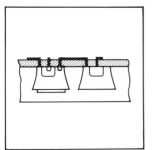

15 INTEGRATED CIRCUIT THEORY

The monolithic integrated circuit is an organized configuration of components manufactured on a continuous or single piece of semiconductor material. This chapter discusses the characteristics, fabrication, and components of the integrated circuit.

The underlying technology of the electronics industry is "solid state." Knowledge of the technology and its prime product, integrated circuits, is now required not only by IC manufacturing people but is essential for all electronics personnel. The basic manufacturing process used in making solid-state devices is the planar process, which embodies four key principles: (1) diffusion, (2) oxidation, (3) selective oxide removal, and (4) epitaxy.

The IC is the result of a rather complicated multistep physical and chemical process. The process begins with a highly purified silicon slice or wafer and ends with a packaged component that performs a high-level function. Silicon is the dominant semiconductor material used in the fabrication of ICs. Linear integrated circuits are predominantly bipolar devices. The bipolar process or procedure used to produce these types of devices includes the following steps: diffusion, epitaxy, oxidation, oxide removal, masking, metallization, probing, dicing, assembly, and testing.

The bipolar components in ICs are, in general, restricted to diodes, transistors, resistors, and low-value capacitors. While the types of components are restricted, the number of the components are not. The physical characteristics of the components illustrate the small size and uniqueness of the devices made with this IC technology.

15.1 DEFINITIONS

Every science, and even its branches, has a vocabulary that is unique in describing its parts and principles. This section defines those key words that are necessary for the understanding of integrated circuits.

Integrated Circuit. A *circuit* is a collection of components connected together in a unique and complete configuration to perform some useful electronic function. An *integrated circuit*, or IC, is a circuit whose components are manufactured on a continuous or single piece of semiconductor material. If all the components are on a single piece of material (or chip), the integrated circuit is called *monolithic*, which literally means *one-stone*. If two or more interconnected chips or components comprise the circuit, the IC is said to be *hybrid*.

Solid State. The control of current by electronic devices that are continuous in structure and thus contain no moving parts, filaments, or vacuum gaps. Examples include crystals, transistors, and integrated circuits.

Solid-State Technology. The science of producing solid-state devices.

Planar Process. The basic manufacturing process of solid-state technology. It derives its name from the fact that all integrated circuit components are interconnected on the same surface or *plane* of the semiconductor material. This process embodies four key principles: (1) diffusion, (2) oxidation, (3) selective oxide removal, and (4) epitaxy.

Diffusion. The intermingling of the atomic structures of two materials. The penetration or permeation of a gas or liquid in a solid material. Examples include water (or steam) and wood, ink and paper, and a phosphorous gas and silicon.

Oxidation. The conversion of an element to a new element through its combination with oxygen. Examples include ferric oxide (rust) and silicon dioxide.

Selective Oxide Removal. The ability to remove an element containing oxygen, for example, silicon dioxide, from a selected area.

Deposition. The act of precipitating or laying down by a natural process. The procedure of evaporating a metal, usually aluminum, and precipitating or depositing it on a surface, usually silicon, to form metallic interconnections and pads for IC components.

Wafer. A slice or circular disc of purified silicon. It is usually 20 mils thick and 3 to 4 inches in diameter.

Die. A portion of a wafer containing an integrated circuit.

Dopant. An element, or impurity, added to silicon to modify the silicon's electrical properties. Examples include boron and phosphorus.

Epitaxy. The growing of one crystalline material on another crystalline material. The two materials have similar physical structures but are chemically different.

15.2 CLASSIFICATIONS

The two general fabrication categories of ICs, shown in Figure 15-1, include the monolithic and hybrid types. The number of specific manufacturing processes within these two categories is large, and each one is individually unique and complex. The greatest number of devices sold today are *bipolar* and *unipolar* monolithic ICs. To this category we exclusively direct our attention.

Bipolar ICs are those devices whose components are *current controlled* and require both positive (+) and negative (-) polarity currents for operation in their active elements. Active elements refer to those leads of a device that directly affect the input-output characteristics for that device. An example of a bipolar component is the *npn* bipolar junction transistor. This is the original, common transistor.

Unipolar monolithic ICs pertain to those

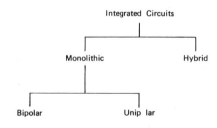

Figure 15-1. Classification of integrated circuits by fabrication.

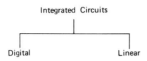

Figure 15-2. Classification of integrated circuits by function.

15.3 FEATURES

Integration refers to the manufacturing process that combines separate electronic components into an integral whole or single chip. This process possesses the advantage of being able to produce numerous devices at one time and is called *batch processing*. The control and uniformity within the process allow devices to be produced at a low unit cost.

devices whose components are primarily *voltage controlled* and have a single polarity operational current in its active elements. An example of a unipolar component is the MOS (*m*etal-*o*xide-*s*emiconductor) transistor.

The two general functional categories of ICs shown in Figure 15-2 are of the digital and linear types. Although the present trend in IC technology is to build increasingly more complex circuits, some of which do both, the great majority of devices perform either digital *or* linear functions. Neither category is the more important, since each type does a unique job.

Digital ICs are those devices whose inputs and outputs have two discrete states; *linear* ICs are those devices whose inputs and outputs are proportionally or mathematically related.

In addition to cost, the IC has small size and weight, is reliable, and consumes a small amount of power. These features have led to countless applications in the military, aerospace, consumer, business, and industrial fields. The success of recent aerospace endeavors would not be possible without the high reliability and small size of electronic components in communications and control equipment. Battery-operated consumer products, such as calculators and watches, depend on low power consumption to maximize the operating life of the product. The reduction in size and the increased reliability of digital computers have led to their application in nonelectronic indus-

Figure 15-3. Semiconductor devices and products.

tries and business. Of course, low cost has pushed the usage of electronics in all fields.

The IC is a development of solid-state technology. However, this technology has also spawned transducers, sensors, and optoelectronic devices. Figure 15-3 illustrates the many sizes and shapes of semiconductor devices and products.

15.4 IC FABRICATION

The integrated circuit is the result of a multistep physical and chemical process. This chapter discusses the procedural steps in converting raw material to a complex electronic device.

The manufacture of integrated circuits is an extremely complex process. This process is radically different from the discrete-circuit fabrication procedure. The discrete circuit starts with individual components such as resistors, transistors and diodes. These components are inserted into printed circuit (pc) boards and soldered. The components are interconnected by the etched copper strips on the board. For complex circuits involving several pc boards, a mechanical assembly or rack is used to hold the pc boards. The boards are then interconnected with discrete wires.

The IC is the result of a *multistep photochemical manufacturing process.* Although the process is complex and as a whole costly, the net result is miniature devices that are inexpensive and reliable. The cost per IC is small because from one hundred to several thousand integrated circuits are made at one time. This batch-processing type of production and the device's miniaturization are the result of intricate photography and masking. The creation of the device is physical and chemical in nature, and the procedure has evolved to a preciseness whereby device parameters are repeatable and highly acceptable by today's standards.

The study of IC manufacturing processes represents a significant departure from customary electronic topics. Instead of components, circuits, voltages, and currents, the primary emphasis shifts to physical and chemical *processes.* The conceptual understanding of these processes, without a reasonably comprehensive background in science, may be a problem for the electronics student. However, the information is so fundamental to the understanding of integrated circuits that it cannot be avoided. The information is also essential to an appreciation of the impact of electronic products in the marketplace and their *full* potential.

Synopsis

Materials are divided, electrically, into three classes: conductors, insulators, and semiconductors. The electrical properties of the semiconductor class are in between those of the conductor and the insulator, and members of this class have atomic structures that are amenable to alteration. Integrated circuits are made from semiconductor materials.

If the atomic structure of a semiconductor material is modified to produce an excess of negatively charged mobile particles (conduction band electrons), the material is identified as n-type semiconductor. In n material, we talk about the flow of negatively charged carriers. If the atomic structure of a semiconductor material is modified to produce a deficiency of negatively charged mobile particles (valence band electrons), the material is identified as p-type semiconductor. The movement of these deficiencies, or holes, is conceptually represented as the flow of positively charged carriers that we refer to in p-type semiconductors. The negatively charged carriers in n-type material are called *majority carriers.* Positively charged carriers also exist in the n-type material, but they are fewer in number and are referred to as *minority carriers.* Similarly, the positively charged carriers in *p*-type material are called majority carriers, and the negatively charged carriers are the minority.

Silicon is the basic element of the semiconductor class. It is rarely used in its natural or purified state, but is converted to either *n* or *p* type. The atomic structure of silicon is altered to form either of the two types by mixing into the silicon another element whose atomic structure in conjunction with silicon will produce the desired type. The added elements represent impurities in the silicon and are called *dopants*. An extremely small amount of dopant is required to change the conduction properties of silicon. A dopant used as an impurity in silicon to produce *n*-type material is phosphorus. A dopant used as an impurity in silicon to produce *p*-type material is boron.

N- and *p*-type materials, by themselves, are of little value. They can be used to fabricate resistors but little else. Their true value lies in the joining of the two to form a *pn* junction. The operation of most solid-state devices is dependent on the properties of one or more *pn* junctions incorporated into their structure.

Several methods have been developed to form *pn* junctions, but the technique employed in the manufacturing of ICs is the most popular. It is called *diffusion*. In this technique, the dopants exist as gases and the silicon as a solid. In diffusion, the dopant atoms of the gas penetrate, or permeate, and intermingle in the atomic structure of the solid silicon. This procedure takes place in an environment where temperature and time are of the utmost importance. The principle of diffusion is not uncommon to us. A piece of dry wood, left outside overnight, is saturated with water in the morning because of the diffusion of the water molecules from the water vapor of the dew into the wood. Ink dropped on a piece of paper diffuses through the paper structure, with the blot increasing in size as the two elements intermingle.

Diffusions can be made into other diffusions. A solid-state *pn* junction is the diffusion of a *p*-type dopant into a previously diffused *n*-type region, or vice versa. A solid-state *npn* transistor is *n* diffused into *p* diffused into *n*. (In practice, the procedure is more complicated, but the principle is correct.)

The IC fabrication process begins with raw silicon. Raw silicon is purified and then converted to either *p* or *n* type. At this stage it exists as a cylinder, called an *ingot*. From the ingot, thin circular discs called *wafers* are cut. The wafer is the foundation for hundreds of identical ICs where multiple diffusions are made. Diffusions made into other diffusions must be made with precision. There must be no overlap of the second diffusion over the first, and the second must be shallower. In fact, all dimensions of one with respect to the other must be accurately controlled. This precision must carry over to every device, since all are made simultaneously.

Diffusions are made into selected areas or locations of the wafer. This selectivity depends upon oxidation, oxide removal, and masking. The three combined simulate the use of a template in which holes are made where the diffusions can occur and elsewhere are prevented. Oxidation is the chemical process by which oxygen combines with another element to form a new compound called an *oxide*. When water vapor, which contains oxygen, is passed over silicon in a high-temperature environment, a layer of silicon dioxide (SiO_2) is formed at the silicon's surface. Silicon dioxide is a high-grade insulator and is impervious to the diffusion process. It is the material that masks out the area of the wafer that is *not* to be diffused. Normally, an oxide layer is grown over the entire wafer. Holes are then cut or etched in the silicon dioxide for those areas that are to be diffused.

The selective removal of the silicon dioxide is carried out by a photolithographic process using photoresistant material. In this process, a photoresistant lacquer is applied to the wafer surface. Then the photoresistant lacquer is exposed to an ultraviolet light, with a photomask being used as a template for the diffused and nondiffused areas. The ultraviolet light polymerizes the exposed photoresistant lacquer. The unexposed photoresist along with the selected oxide layer area is then removed with a solvent. The result is an opening through the oxide layer in which diffusion can take place. The wafer, covered with an oxide layer with holes in it, is placed in a high-temperature oven. A gas containing the dopant is passed over the wafer, and the impurity atoms of the gas diffuse into the unprotected areas of the silicon.

For subsequent diffusions, the process is basically repeated. A new oxide layer is grown over the entire wafer, and photoresistant lacquer is applied to the surface. The lacquer is exposed to ultraviolet light, but through a different photomask. The exposed photoresist is polymerized or hardened, and the unexposed photoresist and silicon dioxide are removed with a solvent. A new area, or window, is now opened for diffusion. This procedure is repeated about six times to make an integrated circuit.

The photomasks are glass templates made to fit over the wafer. Each wafer contains hundreds of ICs, and each mask contains hundreds of patterns. Each pattern has the same dimensions as the IC. They are not made directly. The masks are initially laid out many times larger than the IC to obtain the required accuracy. They are then photographically reduced, in several steps, to the IC dimensions.

All the integrated components are made by using diffusion, oxidation, oxide removal, and microphotographic masks. Once made, the components must then be interconnected. In printed circuit boards, interconnections are made with etched copper strips. An analogous procedure is used in ICs. It is called *metallization through deposition*. In this process, aluminum is evaporated onto the entire surface of the wafer, and a photoresist sequence is repeated with a "reverse" contact photomask. The aluminum is removed from all areas except for the interconnecting traces and component contact pads.

The batch fabrication of the ICs is now complete. The wafer is cut up into individual devices, and each device is mounted and bonded to the frame of its own package. The IC device (now called a *die*) is wired by means of hairlike gold wires to the package's terminals, sealed, and then encapsulated. The device is then electrically tested, marked, packaged, and stored for distribution.

15.5 PLANAR PROCESS

The basis of solid-state technology is the planar process. It derives its name from the fact that all leads are brought out to the top surface or *plane* of the wafer, where they are interconnected or bonded. Four key concepts form the basis for this process: (a) oxidation, (b) selective dioxide removal, (c) diffusion, and (d) epitaxy. Oxidation, the formation of the silicon dioxide layer, is essential to diffusion masking, for sealing junctions, for making dielectrics for capacitors, and for acting as insulating layers for the metal interconnections. Photographic masks, an organic substance called *photoresist*, ultraviolet light, and an acidic etchant enable predetermined areas of the silicon dioxide to be etched to form the windows for diffusion. During diffusion, dopant in vapor form is deposited through the windows in the dioxide to form on the silicon's surface. Time and temperature cause these dopants to move into the selected region. Epitaxy is the *growing* of one semiconductor upon another. Diffusion or epitaxial techniques are used to fabricate *pn* junctions.

The IC structures formed by multiple diffusions are interconnected by the condensation of aluminum vapor on the wafer surface in a process called *deposition*.

15.6 SEMICONDUCTOR MATERIAL

The foundation of the IC is a slice of semiconductor material. It is the basement of the device to which layers of other related materials will be chemically added. The electrical performance of semiconductors is between that of the conductors, such as copper and aluminum, and insulators, of which glass and quartz are examples. The most important feature of the semiconductor class of materials is that its structure and electrical properties can be modified through chemical techniques.

Although there are several members in the semiconductor class of materials, two have dominated in the production of semiconductor-related components. Germanium is widely used in the manufacturing of transistors. However, today, all integrated circuits and many transistors are made with silicon. It is chemically simpler than most other semiconductors and has very good high-temperature electrical properties. This is important in the military and aerospace applications of ICs, where operating environments reach temperature highs in excess of 100°C and temperature lows approaching -50°C. However, the prime attribute of silicon is its ability to grow a stable oxide. This means that the silicon can combine, at its surface, with gaseous oxygen to form a layer of insulating material called silicon dioxide (SiO_2). The process is called *oxidation*. Silicon dioxide, the product of oxidation, plays a key role in IC fabrication. It is one of the layers that is chemically built on the slice of silicon. The time and extent to which silicon has been used in the production of ICs have made it a highly developed industrial technology.

There are two main requirements in preparing the silicon for solid-state devices. First, the silicon must be extremely pure. Unwanted impurities must be down to a level of 1 part in 10^{10}, that is, one impurity atom for 10,000,000,000 atoms of silicon. Second, the silicon must have a continuous and regular crystal structure. Silicon atoms have an irregular structure. The orientation of the atoms must be aligned into what is called *single-crystal form*.

The purification of silicon is performed through the chemical processes of distillation and reduction. The result of these processes is the depositing of silicon onto the surface of a high-purity silicon rod, building it up to a diameter between 1 and 4 inches. The silicon deposits on the rod are in polycrystalline form; that is, their atomic structures are joined together in random orientation.

The procedure generally used to produce single-crystal silicon for ICs is called *crystal pulling*. Purified, solid silicon is placed in a crucible within a furnace and heated to a temperature just above its melting point. A seed crystal, a small piece of single-crystal silicon, is lowered until it barely enters the melt or molten silicon. Because the seed crystal is at a lower temperature than the melt, heat flows from the melt to the seed. The temperature of the melt in contact with the seed falls, and some of the melt solidifies onto the seed, with the atoms arranging themselves to have the same orientation as the atoms in the seed crystal. The seed crystal is rotated and slowly raised, growing larger as more silicon solidifies onto it. Typically, pulled silicon crystals are cylindrical in shape, between 2 and 4 inches in diameter, and 8 to 40 inches long; they are called *ingots*.

An impurity dopant, p or n, is added in the initial melting process so that the pulled crystal has the required conduction properties.

In the fabrication of transistors and ICs, the silicon is used in the form of thin circular

300 INTEGRATED CIRCUIT THEORY

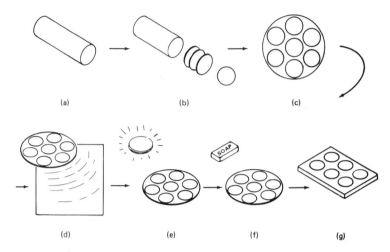

Figure 15-4. Preparation of the IC wafer: (a) the basic ingot; (b) slicing the wafer; (c) mounting the wafer in wax on a plate; (d) lapping; (e) polishing; (f) cleansing; (g) storage and handling.

slices. The p or n ingots are sawed into slices about 20 mils thick by a thin diamond-impregnated wheel rotating at high speed. The slices are mechanically lapped to smooth the surface. Finer and finer polishing compounds, or chemical etchants, are used until the surface is flat and mirrorlike. The slices of silicon, called *wafers*, pure and single crystal in structure, are then ready for subsequent processing. Ultimately, from each wafer will come one hundred to several thousand ICs. Each IC on the wafer is referred to as a die (the plural is dice). Depending on the complexity of the IC, die sizes vary from 0.03 to 0.250 inch square. Figure 15-4 highlights the steps in the preparation of silicon for IC manufacturing, and Figure 15-5 illustrates the resultant product.

15.7 BIPOLAR PROCESSING

Diffusion

Solid-state diffusion is the process of implanting impurity or dopant atoms in a single-crystal structure. The doping atoms, depending on their number, are capable of modifying the electrical characteristics of the n- or p-type *wafer or substrate*. In diffusion, the wafers are placed in carriers or boats and loaded into a furnace through glass tubes. This is shown in Figure 15-6. In the furnace, which is about $1100°C$,

Figure 15-5. A silicon ingot and wafers.

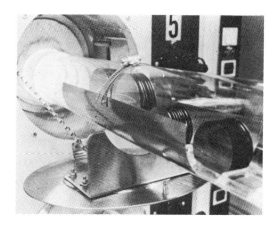

Figure 15-6. Wafers loaded into a diffusion furnace.

dopant gases are passed over the wafers. Dopant atoms from the gas diffuse into the wafers. The results of this process are dependent on (1) concentration, type, and distribution of dopant, (2) time, (3) temperature, (4) pressure, and (5) environmental conditions. Two of the common dopant sources are phosphorus oxychloride ($POCl_3$) and boron tribromide (BBr_3). The phosphorus in $POCl_3$ converts silicon to n-type, and the boron in BBr_3 converts silicon to p-type.

Several diffusions are made in fabricating ICs. Each one is labeled with a name related to the function or component lead with which the diffusion is associated. The isolation diffusion creates separate regions in the epitaxial n layer (on a p wafer) for each IC component. The epitaxial n layer is a nondiffused layer and forms the collectors of *npn* transistors and the cathodes of circuit and isolation diodes. The isolation diffusion electrically separates the various areas of the wafer containing circuit elements. The base diffusion forms the base of *npn* transistors and the anodes of circuit diodes and p-type resistors. It is a shallow diffusion requiring closer process control. The emitter diffusion forms the transistors' emitters, low-value n resistors, MOS capacitor plates, and low-resistance terminals for n collector regions. A fourth diffusion, called a *buried layer*, is sometimes used and is discussed under IC components. Its function is to form a low-resistance region under the collectors of *npn* transistors. The four key diffusions are illustrated in Figure 15-7.

Epitaxy

A second, often used technique of fabricating a *pn* junction is called *epitaxy*. Epitaxy is the growing or deposition of one semiconductor upon another. It is a growing process in which gaseous molecules collect in regularly oriented patterns on the outside surface of a solid material. The process is electrically and thermally induced. Epitaxy differs from diffusion in that the epitaxial interface is grown, whereas the diffused interface is the result of a molecular formation. An epitaxial grown layer is more easily process controlled and possesses certain advantageous electrical properties, but is more expensive.

An epitaxial layer is usually grown over a p

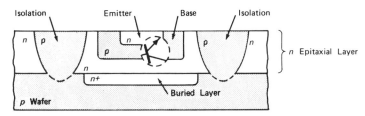

Figure 15-7. Key diffusions in the IC fabrication process.

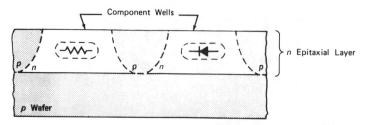

Figure 15-8. Epitaxial layer split into component wells.

wafer as one of the initial steps in the formaton of integrated *pn* structures. This layer takes on the same crystal orientation as the single-crystal wafer. *P* isolation diffusions split the epi layer (Figure 15-8) into sections or wells for the individual components. The epitaxial *n* layer is the collector of *npn* transistors. The *pn* junction formed by the epitaxial grown layer of *n*-type material upon a *p*-type substrate is also used extensively as part of the component isolation necessary for ICs.

Oxidation

Silicon chemically reacts with the oxygen of water vapor to form an oxide called silicon dioxide (SiO_2). Basically, time and temperature determine the thickness of the oxide on the wafer surface. The properties of silicon dioxide make it one of the most useful of microcircuit materials. The formation of the oxide layer is easily controlled, the layer is uniform and continuous, and it closely adheres to the silicon surface. It is also an effective barrier for the more important diffusants but is easily removed by using a chemical solvent. These two factors make SiO_2 ideal as a diffusion template or screen. The oxide layer is a good dielectric, finding further value in the fabrication of capacitors and passivating or protecting surfaces.

Oxide Removal

The photochemical procedure of removing selected areas in the oxide layer is similar to that used in the manufacturing of printed circuit boards. The wafer with the oxide layer is uniformly coated with a layer of a chemical called *photoresist*. The photoresist is dried and covered with a glass photomask that contains the desired patterns. The photoresist is then exposed to ultraviolet light through the mask. The nonopaque areas are polymerized, that is, converted to a new compound, which resists attack by acids and solvents. The nonpolymerized photoresist and oxide are etched away with an acidic solvent. The wafer is now ready for diffusion, which will occur only in the openings, called *windows*, in the oxide. The complete photoresist process is repeated each time the silicon dioxide is selectively removed. The selective oxide removal by the photoresist process is illustrated in the step-by-step procedure of Figure 15-9.

Masking

The templates or screens used to define the regions for diffusions and metallization are called *masks*. The fabrication of an IC requires from five to as many as twenty different masks, with six a typical number. A six-mask set is shown in Figure 15-10. The masks are used one at a time, during each of the processing steps. Each one consists of a unique pattern established by a photographic emulsion on a flat glass plate. Areas of the photomask are either transparent or opaque to light. The masks are the same size as the wafer and must be highly accurate to define the proper regions for

BIPOLAR PROCESSING

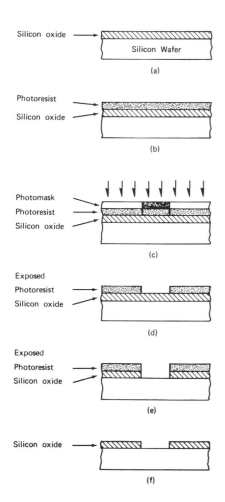

Figure 15-9. Selective oxide removal by the photoresist process: (a) silicon wafer with oxide formed on the surface; (b) photoresist lacquer applied to the surface; (c) photoresist exposed to ultraviolet light through a photomask; (d) unexposed photoresist removed with solvent; (e) silicon oxide removed by etching; (f) photoresist removed to leave window in silicon oxide.

processing. This high degree of accuracy and resolution is achieved through a photographic process. The patterns are initially drawn, a Mylar base and tape being used, 200 to 1000 times the actual size; they are called, at this stage, the artwork. The artwork is then reduced, in several steps, to the IC dimensions by using microphotographic techniques. The images are transferred to a glass disc, which is now called a *photomask* (Figure 15-11). Since a wafer contains many ICs, the pattern is photographically duplicated and repeated many times to complete the wafer mask.

Metallization

Integrated-circuit components are interconnected with metallic, usually aluminum, traces during the metallization phase of the fabrication process. Also during this phase, component and circuit contact pads are formed. The component contacts must be ohmic or non-rectifying, and the circuit pads are provided for the bonding of small wires from the die to the package's leads. Metallization occurs through a process called *deposition*. During deposition, aluminum is evaporated in a high-vacuum system, and enough aluminum is permitted to deposit over the entire wafer surface. The photoresist process is again employed to protect the metal in the contact and interconnection areas, while the remainder of the metallization is exposed for removal by etching. The mask and process are inverted from those used in the diffusion steps.

Probing

The last step in wafer processing is to test the die. At this stage, a machine called a wafer prober makes contact to the die through minute, pointed probes. These probes touch the aluminum pads of the IC and provide the electrical connection from a test instrument or system to the IC. Each die is electrically tested against predetermined specifications. Those dice that are determined to be faulty are inked and rejected. Figures 15-12 and 15-13 show two perspectives of the wafer probe procedure. Figure 15-12 is a photomicrograph of the

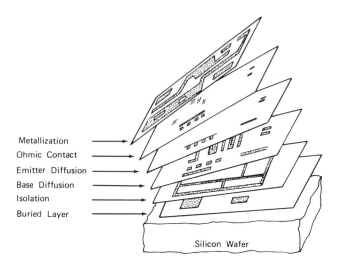

Figure 15-10. Photomask set used to fabricate ICs. Buried layer mask—used to form a low-resistance N+ region under the active devices (optional). Isolation diffusion mask—used to create separate N regions or component wells for the IC components. Base diffusion mask—used to form the base region of transistors, one side of a diode, and resistor areas. Emitter diffusion mask—used to form the transistor's emitter, low-value resistors, and one side of a capacitor plate. Ohmic contact mask—used to make holes through the oxide layer to permit the metallization to make contact to the terminal pads of the circuit elements. Metallization mask—used to form the metal interconnects between the circuit elements and to provide external bonding pads.

prober needles on the circuit pads of an IC. Figure 15-13 is a macroscopic view of the probe heads with their needles making contact with a wafer die. The prober automatically indexes from die to die, resting only to perform the electrical tests and inking, if necessary.

Dicing

After the dice have been tested, they are ready to be separated by using a procedure called *dicing*. The most common method is scribing and breaking. Fine, diamond-cut lines are scribed vertically and horizontally across the wafer but between each circuit or die. The wafer is then placed on a rubber pad, and pressure from a roller is exerted on the wafer, breaking it up into individual chips.

Assembly

Assembly is the most expensive part of the fabrication process. In wafer processing, large

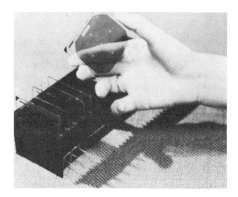

Figure 15-11. Photomask.

BIPOLAR PROCESSING

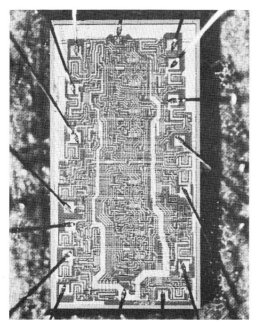

Figure 15-12. Prober needles on an IC die.

numbers of circuits are processed simultaneously, but during assembly the dice are handled individually, and connections to each die are made separately. Assembly operations are required to protect the die, to facilitate handling, and to provide a package and contacts for equipment application.

Die Bonding

Die bonding is the soldering, brazing, or glassing of the die to a package frame. This attachment provides a mechanical and, occasionally, an electrical contact from die to package. This bond also acts as a thermal path for internally generated heat to flow to the surroundings.

Wire Bonding

The electrical connections from the chip to the package's terminals are made by one of several wire bonding techniques. One-mil (0.001 inch) wire, either gold or aluminum, is connected to an IC aluminum pad and then

Figure 15-13. Prober with needles testing a die on a wafer.

extended to the package terminal, where a similar connection is made. Ultrasonic, pressure, or temperature techniques are used to form the wire to pad and terminal welds. Great emphasis is being placed on eliminating this costly and labor-oriented method and replacing it with an automated procedure. Presently, most IC manufacturers have shifted to an automated bonding technique where connections are made from die to the package by using metal tabs.

Packaging

The individual IC die is too small and delicate to handle. To facilitate handling and protect it from damage, the die is welded to a frame, its leads are connected, and the unit is sealed, passivated, and encapsulated. Glass, ceramic, and plastic are common materials that encase the die. There are many types of IC packages, but the most popular are the axial lead (TO5), flat pack, and dual-inline (DIP) packages. They are illustrated in Figure 15-14.

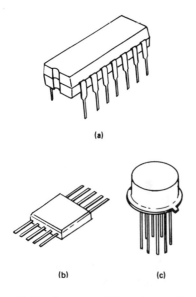

Figure 15-14. Common IC packages: (a) dual inline (DIP); (b) flat pack; (c) TO-5.

Testing

The last step in the manufacturing process is the electrical testing of the IC. Dc, ac, and functional measurements are made to ensure that the IC will perform to established standards. Large-volume products are automatically transported by mechanical autohandlers. A computer-controlled IC test system, in conjunction with the autohandler, systematically and quickly screens each IC to verify its performance. Bad ICs are rejected, and those accepted are marked and stored for distribution. A summary of the IC processing steps is illustrated in Figure 15-15.

15.8 IC COMPONENTS: BIPOLAR

Before one can understand the whole, he must know the parts. So it is in integrated circuits. This section discusses the parts or components that make up the integrated circuit.

The designer of discrete circuits has a large number and wide variety of components that he can use. He can select from resistors, inductors, and capacitors to transformers, switches, and indicators. He can specify accuracy, voltage, current, and power ratings, size, and even shape. The designer of monolithic circuits is not so fortunate. He has a restricted number of components that he can use, and their performance and ratings are process limited. However, he is not totally at a disadvantage. He is less restricted by space and can use a greater number of components. Most important, he can produce his circuit in large quantity for significantly less cost.

The components that may be integrated are resistors, capacitors, diodes, and transistors. Capacitors are costly to integrate and are used sparingly. Inductors cannot be integrated, but their effect can be simulated through circuit techniques. In fact, the circuit design techniques of solid-state technology have allowed designers to duplicate most electronic functions.

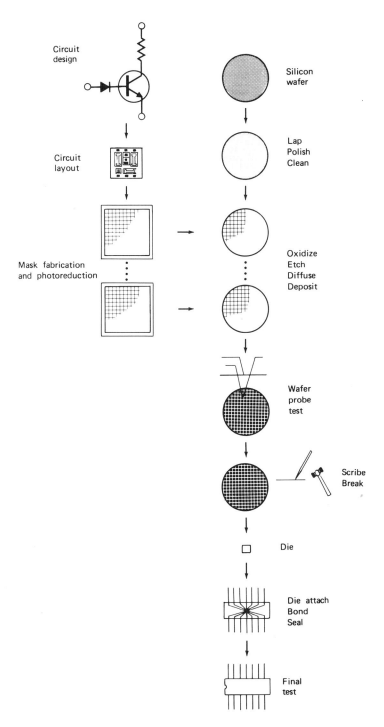

Figure 15-15. Monolithic IC processing steps.

Diodes

A solid-state diode is formed by making an *n* diffusion into a *p* wafer (or vice versa). The *n* area is the diode's cathode, and the *p* area is the anode. Diode action occurs at the *pn* junction. When the positive terminal of a voltage supply is connected to the anode and the negative terminal is connected to the cathode (Figure 15-16), the diode is forward biased. The positive voltage attracts the negatively charged carriers of the *n* region and they flow through the *pn* junction. The carriers flowing from the *n* region are replenished from the source's negative terminal. Similarly, the negative voltage attracts the positively charged carriers of the *p* region. They flow through the *pn* region to the source's negative terminal, resulting in the flow of current. As the voltage is increased, the current rapidly increases.

When the positive and negative terminals of the supply are connected to the cathode and anode, respectively, the diode is reverse biased. The negative voltage on the anode attracts the positive carriers of the *p* region but repels the negative carriers of the *n* region. A similar action occurs at the cathode. The repelling of the positive and negative carriers in the diode leaves the *pn* junction void of any carriers. No current flows, and the diode appears as an open circuit.

Circuit and Isolation Diodes

A diode is a *pn* junction. The *pn* junctions that function as diodes in intergrated circuits are used for two purposes: (a) circuit action and (b) isolation.

Figure 15-17 shows two integrated diodes. They are formed as the result of two diffusions.

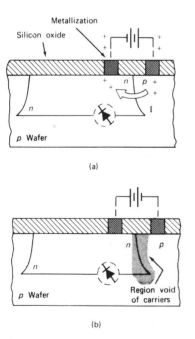

Figure 15-16. A *p-n* junction: (a) forward-biased; (b) reverse-biased.

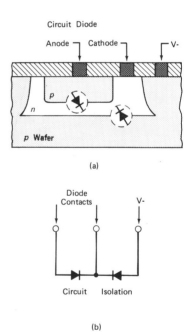

Figure 15-17. IC diodes, circuit and isolation: (a) integrated structure; (b) equivalent circuit.

The *pn* junction formed by the second diffusion is the *circuit diode*. The first *pn* function, formed from the *p* wafer and *n* diffusion, is an *isolation diode*. (For simplicity in this seciton, this *n* region is generally implied to be diffused. In most ICs, it is epitaxially grown.) The *n* region may be diffused or may be an epitaxially grown layer. Isolation diodes are used to isolate integrated-circuit components from each other. To provide isolation, a contact is made to the *p* wafer and is externally attached to the most negative voltage supply used by the IC. Since this negative voltage is on the anodes, the isolation diodes are reverse biased and appear as open circuits. Each circuit component sees an open or reverse-biased diode to the wafer, and hence to each other. Thus IC components can operate independently and with minimal interaction.

Transistors

npn Transistor

The basic structure of an *npn* transistor is shown in Figure 15-18. It is similar to the diode structure of Figure 15-17, but contains a third, *n*-type diffusion made into the second diffusion's *p* region. The *pn* junction formed by the wafer and the first diffusion is used to provide electrical isolation for the components in the circuit. The first deep *n* diffusion, or epitaxially grown layer, is the transistor's collector, and the subsequent *p* and *n* diffusions are the base and emitter, respectively. This simplified structure is usually modified to improve the performance of the transistor.

Minimizing Resistance of Diffused Regions and Component Contacts in IC Transistors

The resistance of the collector region of an *npn* transistor adversely affects the performance of the device in amplification and switching circuits. This resistance can be reduced by

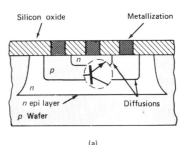

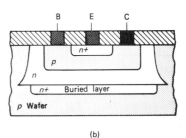

Figure 15-18. Integrated structure of a bipolar (*npn*) transistor: (a) simplified; (b) with n^+ diffusions.

shunting it with a smaller value. In an IC, this is accomplished through an additional diffusion of *n* material that has an extremely high concentration of dopant. This diffusion, called n^+, is located below the collector region and is referred to as a buried layer. The resistance of the n^+ layer is much smaller than the collector because of the greater amount of carriers in the dopant. Some n^+ and p^+ diffusions are also used to minimize the contact resistance between the die and the interconnection metallization.

Figure 15-18 shows an IC structure for an *npn* transistor including the n^+ buried layer and n^+ emitter diffusions. The shallow emitter diffusion is usually entirely n^+.

Windows are cut in the oxide layer for deposited metal to interconnect the leads of the transistor to the circuit. The mask associated with this step is called the *ohmic contact mask*.

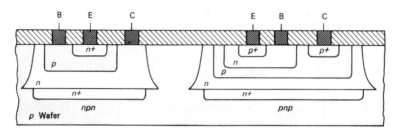

Figure 15-19. Integrated structures for the *npn* and complementary *pnp* bipolar transistors.

pnp Transistor

The bipolar *npn* transistor is the most widely used three-terminal device. Its overall performance is superior to that of its complement, the *pnp* transistor, and is less difficult to manufacture. However, the *pnp* transistor is important in many applications.

The structures of the *npn* and *pnp* are shown in Figure 15-19. The *pnp* requires an additional p^+ diffusion step for the device's emitter and is referred to as a *complementary pnp*, because its structure is analogous to the *npn*. Two additional methods of fabricating *pnp* transistors are used: (1) the lateral *pnp* and (2) the vertical *pnp*. They are shown in Figure 15-20.

The lateral *pnp* starts with two *p* regions diffused into an isolated *n*-type island and is made at the same time as the base diffusion (*p*) for the *npn* transistor. Two *p* diffusions, both circular, are made into the isolated *n* region for the device's collector and emitter. The current flows laterally from the emitter through the two base regions to the collector. This device is used extensively in linear ICs.

The vertical *pnp* uses the *p* substrate for its collector. Since the substrate is connected to the most negative IC voltage, it is restricted in the IC design to those *pnp* applications that tie the collector to V^-. The emitter is a *p* diffusion into the *n* base. Current for this device flows from emitter to base and vertically to the collector.

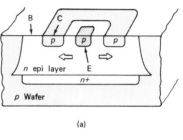

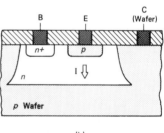

Figure 15-20. Integrated structures for the lateral and vertical *pnp*.

Whether the bipolar transistor is *npn* or *pnp*, it fundamentally depends on current for its operation. When positively charged carriers (*npn*) or negatively charged carriers (*pnp*) are injected into the transistor's base lead, a resultant gain in current is present in the device's emitter and collector leads. The mode of operation of the current-controlled bipolar transistor is significantly different from the voltage-controlled field-effect (MOS) transistor.

Resistors

Silicon is a resistive material. In fact, all materials have resistance. For insulators such as glass, mica, or silicon dioxide, the resistance is very high but not infinite. For conductors such as copper, gold, or silver, the resistance is relatively low but not zero. We approximate a short length of copper wire as a short circuit or 0 Ω, but our approximation would not hold true for a mile of wire. In fact, the resistance of a mile of wire is several hundred ohms (24 AWG), and increases as we increase the length of the wire. The value of resistance of a piece of wire not only depends on its length but also is a function of the diameter and the type of material. The characteristic that relates length, diameter, and material to resistance is called *resistivity*. The concept of resistivity applies to all materials, insulators, conductors, and semiconductors.

Figure 15-21 shows the cross section and top view of a *p*-type monolithic resistor. The *pn* junction of the *n* epi layer and *p* wafer forms the isolation diode. A *p* diffusion is then made into the *n* area. The monolithic *p* resistor that exists is continuous; that is, its resistance increases as we leave one terminal and move toward the other. Its final or maximum value is a function of the *p* diffusion's thickness, its length, width, the type of material, or, collectively, the material's resistivity. Unlike the pure copper wire, the diffused *p*-type material can be altered to change the resistance. The alteration is in the form of controlling the concentration and type of dopant or impurity of the *p* material. Thus, dopant concentration and type control the resistivity of the semiconductor material, and this, with length, width, and thickness of the diffused material, controls the resistor's value.

There are several types of IC resistors. The previously mentioned resistor is called a *diffused-base type*, because it is formed simultaneously with the transistor base. Similarly, *diffused emitter* resistors are formed simultaneously with the transistor emitters. These resistors are useful where low values are required. To achieve high resistance values, *base pinch resistors* are employed.

Large values of resistance may be obtained with minimum die area by squeezing or pinching the base region. The *p* base region is pinched by the emitter diffusion (see Figure 15-22), which results in a greater resistance per length. The smaller the *p* region is made, the fewer carriers it will have, and hence the greater its resistance. No contact is made to the emitter diffused area, and it serves only to increase the *p* region resistance. The nominal pinched base resistance is 10 kΩ per square mil (0.001 in. × 0.001 in.).

Diffused base resistors have a fairly good temperature coefficient, but, as is true with any IC resistor, they have poor absolute accuracy. This is somewhat compensated for by the excellent relative accuracy and tracking ability they have in conjunction with other device resistors. A typical resistor requires at least 1 square mil (0.001 in. × 0.001 in.) per 100 Ω.

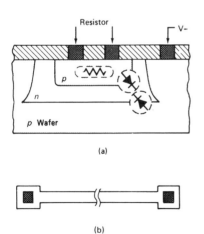

Figure 15-21. IC resistor, diffused base type: (a) integrated structure; (b) top view.

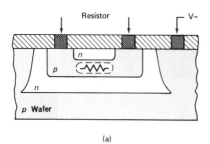

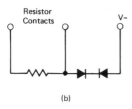

Figure 15-22. IC resistor, pinched base type: (a) integrated structure; (b) equivalent circuit.

For a width of 1 mil, the resistor would have to be 20 mils long for a value of 2 kΩ. Large-valued base diffused resistors can increase the die size and the cost through a loss in yield and materials.

Capacitors

Diodes, resistors, and transistors are the fundamental bipolar integrated components; however, a limited range of low-value capacitors is fabricated and used. A reverse-biased *pn* junction, while theoretically an open, is in actuality a small capacitor. Capacitor action occurs when a change in voltage causes a corresponding change in stored charge or carriers. To understand how a diode behaves as a capacitor, we examine its structure under the reverse-biased condition as shown in Figure 15-23. As is true in any IC, the first diffusion is used to isolate the IC components.

The second *p* and *n* diffused regions shown in the figure have an excess of positive and negative carriers, respectively. This is the result of the introduction of the impurity dopants used

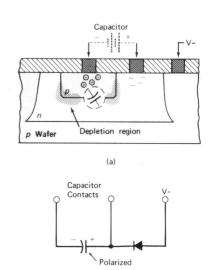

Figure 15-23. IC capacitor, junction type: (a) integrated structure; (b) equivalent circuit.

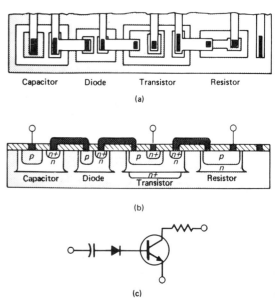

Figure 15-24. Component profiles and equivalent circuit: (a) top view; (b) side view including metallization; (c) equivalent circuit.

in the diffusion process. When a positive voltage is applied to the diode's cathode (n) and a negative voltage is applied to the anode (p), these excess carriers are drawn to the voltage source. The positive carriers are drawn to the −voltage, and the negative carriers are drawn to the +voltage, depleting the region of carriers at the pn junction. This depletion region is like an insulator. That is, it has no carriers, no current flow, and an extermely high resistance. The magnitude of voltage influences the amount of stored charge. As the voltage increases and decreases, a proportional amount of charge is increased and decreased. Therefore, the junction behavior exhibits the same characteristics as a parallel plate capacitor. Unilke the discrete capacitor, the value of the bipolar junction capacitor is voltage dependent, since the applied voltage also affects the depletion region width. Since the more positive voltage must always be applied to the cathode, the junction capacitor is polarized. The value of capacitance per diffused area is low, and economic considerations restrict the size and usage. A typical value of an integrated bipolar capacitor is 30pF, with a maximum of about 100 pF.

Physical Characteristics

Component Profiles

Figure 15-24 shows a side section and top view of the four bipolar components, including the interconnection metallization. It should

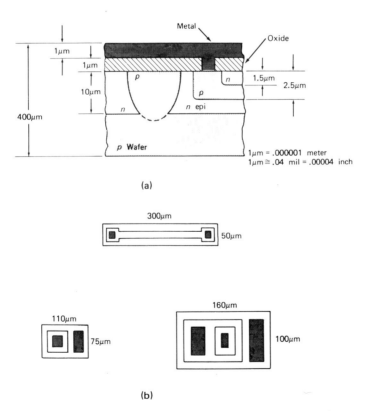

Figure 15-25. (a) Typical vertical dimensions of an integrated circuit; (b) typical horizontal dimensions of IC components.

be noted that the *n* and *p* diffusions for *all* components of *all* dice are done simultaneously and that their integrated structures are similar. The geometries of the diffusions will vary, depending on the desired electrical performance.

Component Dimensions

The approximate vertical dimensions of a component on a wafer are shown in Figure 15-25a. This figure graphically illustrates the preciseness and miniaturization associated with the production of the IC. These dimensions are brought into proper perspective when one considers that the diameter of a thick hair is about 10 mils.

Figure 15-25 (b, c, and d) shows the top view of the three fundamental IC components and the typical horizontal dimensions of each. Great emphasis is placed on minimizing each component's surface area to ensure the smallest die size possible. Yield, that is, the number of good dice per wafer, is directly related to the area of the integrated circuit. The greater the yield, the lower is the cost.

15.9 REFERENCES FOR TOPICS IN CHAPTERS 14 AND 15

Adler, R. B., Smith, A. C., and Longini, R. L. *Introduction to Semiconductor Physics*. New York: John Wiley and Sons, Inc., 1964.

Bell, D. A. *Fundamentals of Electronic Devices*. Reston, Va.: Reston Publishing Company, Inc., 1975.

Comer, D. T. *Large Signal Transistor Circuits*. Englewood Cliffs, N.J.: Prentice-Hall, Inc., 1967.

Cooper, W. D. *Solid-State Devices: Analysis and Application*. Reston, Va.: Reston Publishing Company, Inc., 1974.

Deboo, G. J., and Burrous, C. N. *Integrated Circuits and Semiconductor Devices*, 2nd ed. New York: McGraw-Hill Book Company, 1977.

Grove, A. S. *Physics and Technology of Semiconductor Devices*. New York: John Wiley and Sons, Inc., 1963.

Hibberd, R. G. *Integrated Circuits*. New York: McGraw-Hill Book Company, 1969.

Kiver, M. S. *Transistor and Integrated Electronics*, 4th ed. New York: McGraw-Hill Book Company, 1972.

Malvino, A. P. *Electronic Principles*, 2nd ed. New York: McGraw-Hill Book Company, 1979.

Wojslaw, C. F. *Integrated circuits: Theory and Applications*. Reston, Va.: Reston Publishing Company, Inc., 1978.

Appendix A
DATA SHEETS OF REPRESENTATIVE ICs AND TRANSISTORS USED IN TEXT

INTEGRATED CIRCUITS

Device Number	Manufacturer	Function
µA741C	Fairchild	Operational amplifier
LM301A	National	Operational amplifier
LM340	National	Voltage regulator
SE555	Signetics	Timer/oscillator
SE565	Signetics	Phase locked loop
MC1306	Motorola	Audio power amplifier

TRANSISTORS

Device Number	Type
2N2222A	Bipolar—*npn*
2N3905	Bipolar—*pnp*

μA741
FREQUENCY-COMPENSATED OPERATIONAL AMPLIFIER
FAIRCHILD LINEAR INTEGRATED CIRCUIT

GENERAL DESCRIPTION — The μA741 is a high performance monolithic Operational Amplifier constructed using the Fairchild Planar* epitaxial process. It is intended for a wide range of analog applications. High common mode voltage range and absence of latch-up tendencies make the μA741 ideal for use as a voltage follower. The high gain and wide range of operating voltage provides superior performance in integrator, summing amplifier, and general feedback applications. Electrical characteristics of the μA741A and E are identical to MIL-M-38510/10101.

- NO FREQUENCY COMPENSATION REQUIRED
- SHORT CIRCUIT PROTECTION
- OFFSET VOLTAGE NULL CAPABILITY
- LARGE COMMON MODE AND DIFFERENTIAL VOLTAGE RANGES
- LOW POWER CONSUMPTION
- NO LATCH-UP

ABSOLUTE MAXIMUM RATINGS

Supply Voltage
  μA741A, μA741, μA741E   ±22 V
  μA741C   ±18 V
Internal Power Dissipation (Note 1)
  Metal Can   500 mW
  Molded and Hermetic DIP   670 mW
  Mini DIP   310 mW
  Flatpak   570 mW
Differential Input Voltage   ±30 V
Input Voltage (Note 2)   ±15 V
Storage Temperature Range
  Metal Can, Hermetic DIP, and Flatpak   −65°C to +150°C
  Mini DIP, Molded DIP   −55°C to +125°C
Operating Temperature Range
  Military (μA741A, μA741)   −55°C to +125°C
  Commercial (μA741E, μA741C)   0°C to +70°C
Lead Temperature (Soldering)
  Metal Can, Hermetic DIPs, and Flatpak (60 s)   300°C
  Molded DIPs (10 s)   260°C
Output Short Circuit Duration (Note 3)   Indefinite

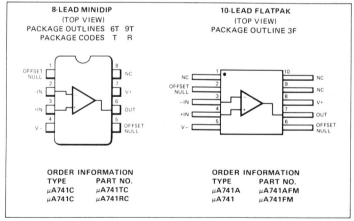

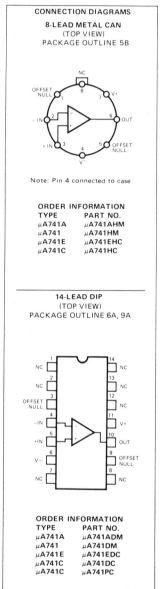

Notes on following pages.

*Planar is a patented Fairchild process.

FAIRCHILD LINEAR INTEGRATED CIRCUITS • μA741

μA741C

ELECTRICAL CHARACTERISTICS ($V_S = \pm 15$ V, $T_A = 25°C$ unless otherwise specified)

PARAMETERS (see definitions)	CONDITIONS	MIN	TYP	MAX	UNITS
Input Offset Voltage	$R_S \leq 10$ kΩ		2.0	6.0	mV
Input Offset Current			20	200	nA
Input Bias Current			80	500	nA
Input Resistance		0.3	2.0		MΩ
Input Capacitance			1.4		pF
Offset Voltage Adjustment Range			±15		mV
Input Voltage Range		±12	±13		V
Common Mode Rejection Ratio	$R_S \leq 10$ kΩ	70	90		dB
Supply Voltage Rejection Ratio	$R_S \leq 10$ kΩ		30	150	μV/V
Large Signal Voltage Gain	$R_L \geq 2$ kΩ, $V_{OUT} = \pm 10$ V	20,000	200,000		
Output Voltage Swing	$R_L \geq 10$ kΩ	±12	±14		V
	$R_L \geq 2$ kΩ	±10	±13		V
Output Resistance			75		Ω
Output Short Circuit Current			25		mA
Supply Current			1.7	2.8	mA
Power Consumption			50	85	mW
Transient Response (Unity Gain) Rise time	$V_{IN} = 20$ mV, $R_L = 2$ kΩ, $C_L \leq 100$ pF		0.3		μs
Overshoot			5.0		%
Slew Rate	$R_L \geq 2$ kΩ		0.5		V/μs

The following specifications apply for $0°C \leq T_A \leq +70°C$:

Input Offset Voltage				7.5	mV
Input Offset Current				300	nA
Input Bias Current				800	nA
Large Signal Voltage Gain	$R_L \geq 2$ kΩ, $V_{OUT} = \pm 10$ V	15,000			
Output Voltage Swing	$R_L \geq 2$ kΩ	±10	±13		V

EQUIVALENT CIRCUIT

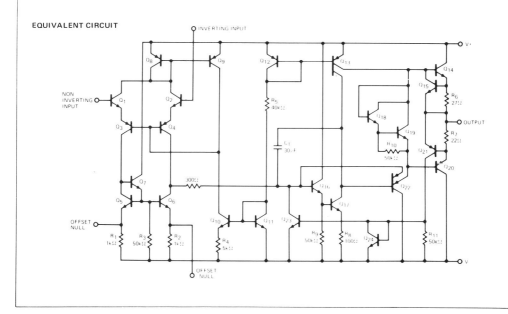

317

Operational Amplifiers

LM301A operational amplifier
general description

The LM301A is a general-purpose operational amplifier which features improved performance over the 709C and other popular amplifiers. Advanced processing techniques make possible an order of magnitude reduction in input currents, and a redesign of the biasing circuitry reduces the temperature drift of input current.

This amplifier offers many features which make its application nearly foolproof: overload protection on the input and output, no latch-up when the common mode range is exceeded, freedom from oscillations and compensation with a single 30 pF capacitor. It has advantages over internally compensated amplifiers in that the compensation can be tailored to the particular application. For example, as a summing amplifier, slew rates of 10 V/μs and bandwidths of 10 MHz can be realized. In addition, the circuit can be used as a comparator with differential inputs up to ±30V; and the output can be clamped at any desired level to make it compatible with logic circuits.

The LM301A provides better accuracy and lower noise than its predecessors in high impedance circuitry. The low input currents also make it particularly well suited for long interval integrators or timers, sample and hold circuits and low frequency waveform generators. Further, replacing circuits where matched transistor pairs buffer the inputs of conventional IC op amps, it can give lower offset voltage and drift at reduced cost.

schematic** and connection diagrams

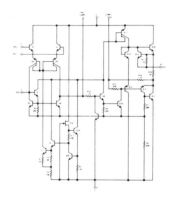

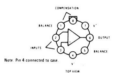

Note: Pin 4 connected to case.
TOP VIEW

Order Number LM301AH
See Package 11

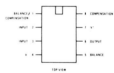

TOP VIEW

Order Number LM301AN
See Package 20

typical applications **

Integrator with Bias Current Compensation

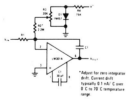

*Adjust for zero integrator drift. Current drift typically 0.1 nA/°C over 0°C to 70°C temperature range.

Low Frequency Square Wave Generator

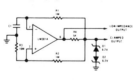

Voltage Comparator for Driving DTL or TTL Integrated Circuits

**Pin connections shown are for metal can.

absolute maximum ratings

Supply Voltage	±18V
Power Dissipation (Note 1)	500 mW
Differential Input Voltage	±30V
Input Voltage (Note 2)	±15V
Output Short-Circuit Duration (Note 3)	Indefinite
Operating Temperature Range	0°C to 70°C
Storage Temperature Range	−65°C to 150°C
Lead Temperature (Soldering, 10 sec)	300°C

electrical characteristics (Note 4)

PARAMETER	CONDITIONS	MIN	TYP	MAX	UNITS
Input Offset Voltage	$T_A = 25°C$, $R_S \leq 50\ k\Omega$		2.0	7.5	mV
Input Offset Current	$T_A = 25°C$		3	50	nA
Input Bias Current	$T_A = 25°C$		70	250	nA
Input Resistance	$T_A = 25°C$	0.5	2		MΩ
Supply Current	$T_A = 25°C$, $V_S = ±15V$		1.8	3.0	mA
Large Signal Voltage Gain	$T_A = 25°C$, $V_S = ±15V$, $V_{OUT} = ±10V$, $R_L \geq 2\ k\Omega$	25	160		V/mV
Input Offset Voltage	$R_S \leq 50\ k\Omega$			10	mV
Average Temperature Coefficient of Input Offset Voltage			6.0	30	$\mu V/°C$
Input Offset Current				70	nA
Average Temperature Coefficient of Input Offset Current	$25°C \leq T_A \leq 70°C$ $0°C \leq T_A \leq 25°C$		0.01 0.02	0.3 0.6	nA/°C nA/°C
Input Bias Current				300	nA
Large Signal Voltage Gain	$V_S = ±15V$, $V_{OUT} = ±10V$, $R_L \geq 2\ k\Omega$	15			V/mV
Output Voltage Swing	$V_S = ±15V$, $R_L = 10\ k\Omega$ $R_L = 2\ k\Omega$	±12 ±10	±14 ±13		V V
Input Voltage Range	$V_S = ±15V$	±12			V
Common Mode Rejection Ratio	$R_S \leq 50\ k\Omega$	70	90		dB
Supply Voltage Rejection Ratio	$R_S \leq 50\ k\Omega$	70	96		dB

Note 1: For operating at elevated temperatures, the device must be derated based on a 100°C maximum junction temperature and a thermal resistance of 150°C/W junction to ambient or 45°C/W junction to case.

Note 2: For supply voltages less than ±15V, the absolute maximum input voltage is equal to the supply voltage.

Note 3: Continuous short circuit is allowed for case temperatures to 70°C and ambient temperatures to 55°C.

Note 4: These specifications apply for $0°C \leq T_A \leq 70°C$, $±5V \leq V_S \leq ±15V$ and C1 = 30 pF unless otherwise specified.

Voltage Regulators

LM340 series 3-terminal positive regulators

general description

The LM340-XX series of three terminal regulators is available with several fixed output voltages making them useful in a wide range of applications. One of these is local on card regulation, eliminating the distribution problems associated with single point regulation. The voltages available allow these regulators to be used in logic systems, instrumentation, HiFi, and other solid state electronic equipment. Although designed primarily as fixed voltage regulators these devices can be used with external components to obtain adjustable voltages and currents.

The LM340-XX series is available in two power packages. Both the plastic TO-220 and metal TO-3 packages allow these regulators to deliver over 1.0A if adequate heat sinking is provided. Current limiting is included to limit the peak output current to a safe value. Safe area protection for the output transistor is provided to limit internal power dissipation. If internal power dissipation becomes too high for the heat sinking provided, the thermal shutdown circuit takes over preventing the IC from overheating.

Considerable effort was expended to make the LM340-XX series of regulators easy to use and minimize the number of external components. It is not necessary to bypass the output, although this does improve transient response. Input bypassing is needed only if the regulator is located far from the filter capacitor of the power supply.

features

- Output current in excess of 1A
- Internal thermal overload protection
- No external components required
- Output transistor safe area protection
- Internal short circuit current limit
- Available in plastic TO-220 and metal TO-3 packages

voltage range

LM340-05	5V	LM340-15	15V
LM340-06	6V	LM340-18	18V
LM340-08	8V	LM340-24	24V
LM340-12	12V		

schematic and connection diagrams

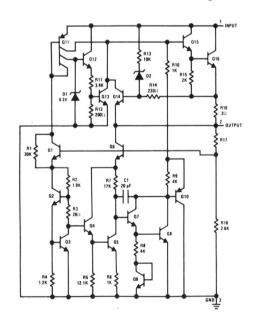

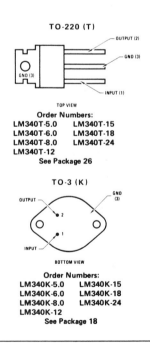

TO-220 (T)

TOP VIEW

Order Numbers:
LM340T-5.0 LM340T-15
LM340T-6.0 LM340T-18
LM340T-8.0 LM340T-24
LM340T-12
See Package 26

TO-3 (K)

BOTTOM VIEW

Order Numbers:
LM340K-5.0 LM340K-15
LM340K-6.0 LM340K-18
LM340K-8.0 LM340K-24
LM340K-12
See Package 18

absolute maximum ratings

Input Voltage (V_O = 5V through 18V)	35V
(V_O = 24V)	40V
Internal Power Dissipation (Note 1)	Internally Limited
Operating Temperature Range	0°C to 70°C
Maximum Junction Temperature	
TO-3 Package	150°C
TO-220 Package	150°C
Storage Temperature Range	−65°C to +150°C
Lead Temperature	
TO-3 Package (Soldering, 10 sec)	300°C
TO-220 Package (Soldering, 10 sec)	230°C

electrical characteristics

LM340-5 (V_{IN} = 10V, I_{OUT} = 500 mA, 0°C $\leq T_A \leq$ 70°C, unless otherwise specified)

PARAMETER	CONDITIONS	MIN	TYP	MAX	UNITS
Output Voltage	T_j = 25°C	4.8	5	5.2	V
Line Regulation	T_j = 25°C, 7V $\leq V_{IN} \leq$ 25V				
	I_{OUT} = 100 mA			50	mV
	I_{OUT} = 500 mA			100	mV
Load Regulation	T_j = 25°C, 5 mA $\leq I_{OUT} \leq$ 1.5A			100	mV
Output Voltage	7V $\leq V_{IN} \leq$ 20V, 5 mA $\leq I_{OUT} \leq$ 1.0A, $P_D \leq$ 15W	4.75		5.25	V
Quiescent Current	T_j = 25°C		7	10	mA
Quiescent Current Change	7V $\leq V_{IN} \leq$ 25V			1.3	mA
	5 mA $\leq I_{OUT} \leq$ 1.5A			.5	mA
Output Noise Voltage	T_A = 25°C, 10 Hz $\leq f \leq$ 100 kHz		40		µV
Long Term Stability				20	mV/1000 hr
Ripple Rejection	f = 120 Hz		60		dB
Dropout Voltage	T_j = 25°C, I_{OUT} = 1.0A		2		V

LM340-15 (V_{IN} = 23V, I_{OUT} = 500 mA, 0°C $\leq T_A \leq$ 70°C, unless otherwise specified)

PARAMETER	CONDITIONS	MIN	TYP	MAX	UNITS
Output Voltage	T_j = 25°C	14.4	15	15.6	V
Line Regulation	T_j = 25°C, 17.5V $\leq V_{IN} \leq$ 30V				
	I_{OUT} = 100 mA			150	mV
	I_{OUT} = 500 mA			300	mV
Load Regulation	T_j = 25°C, 5 mA $\leq I_{OUT} \leq$ 1.5A			300	mV
Output Voltage	17.5V $\leq V_{IN} \leq$ 30V, 5 mA $\leq I_{OUT} \leq$ 1.0A, $P_D \leq$ 15W	14.25		15.75	V
Quiescent Current	T_j = 25°C		7	10	mA
Quiescent Current Change	17.5V $\leq V_{IN} \leq$ 30V			1	mA
	5 mA $\leq I_{OUT} \leq$ 1.5A			.5	mA
Output Noise Voltage	T_A = 25°C, 10 Hz $\leq f \leq$ 100 kHz		90		µV
Long Term Stability				60	mV/1000 hr
Ripple Rejection	f = 120 Hz		50		dB
Dropout Voltage	T_j = 25°C, I_{OUT} = 1.0A		2		V

signetics

TIMER 555

LINEAR INTEGRATED CIRCUITS

DESCRIPTION

The NE/SE 555 monolithic timing circuit is a highly stable controller capable of producing accurate time delays, or oscillation. Additional terminals are provided for triggering or resetting if desired. In the time delay mode of operation, the time is precisely controlled by one external resistor and capacitor. For astable operation as an oscillator, the free running frequency and the duty cycle are both accurately controlled with two external resistors and one capacitor. The circuit may be triggered and reset on falling waveforms, and the output structure can source or sink up to 200mA or drive TTL circuits.

FEATURES

- TIMING FROM MICROSECONDS THROUGH HOURS
- OPERATES IN BOTH ASTABLE AND MONOSTABLE MODES
- ADJUSTABLE DUTY CYCLE
- HIGH CURRENT OUTPUT CAN SOURCE OR SINK 200mA
- OUTPUT CAN DRIVE TTL
- TEMPERATURE STABILITY OF **0.005% PER °C**
- NORMALLY ON AND NORMALLY OFF OUTPUT

APPLICATIONS

PRECISION TIMING
PULSE GENERATION
SEQUENTIAL TIMING
TIME DELAY GENERATION
PULSE WIDTH MODULATION
PULSE POSITION MODULATION
MISSING PULSE DETECTOR

PIN CONFIGURATIONS (Top View)

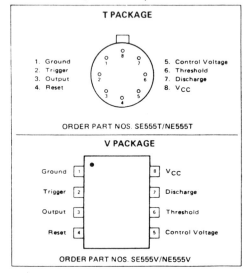

ABSOLUTE MAXIMUM RATINGS

Supply Voltage	+18V
Power Dissipation	600 mW
Operating Temperature Range	
NE555	0°C to +70°C
SE555	−55°C to +125°C
Storage Temperature Range	−65°C to +150°C
Lead Temperature (Soldering, 60 seconds)	+300°C

BLOCK DIAGRAM

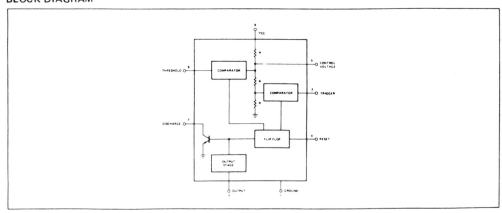

322

SIGNETICS TIMER ∎ 555

ELECTRICAL CHARACTERISTICS $T_A = 25°C$, $V_{CC} = +5V$ to $+15$ unless otherwise specified

PARAMETER	TEST CONDITIONS	SE 555 MIN	SE 555 TYP	SE 555 MAX	NE 555 MIN	NE 555 TYP	NE 555 MAX	UNITS
Supply Voltage		4.5		18	4.5		16	V
Supply Current	$V_{CC} = 5V$ $R_L = \infty$		3	5		3	6	mA
	$V_{CC} = 15V$ $R_L = \infty$		10	12		10	15	mA
	Low State, Note 1							
Timing Error (Monostable)	$R_A = 2K\Omega$ to $100K\Omega$							
Initial Accuracy	$C = 0.1 \mu F$ Note 2		0.5	2		1		%
Drift with Temperature			30	100		50		ppm/°C
Drift with Supply Voltage			0.05	0.2		0.1		%/Volt
Threshold Voltage			2/3			2/3		X V_{CC}
Trigger Voltage	$V_{CC} = 15V$	4.8	5	5.2		5		V
	$V_{CC} = 5V$	1.45	1.67	1.9		1.67		V
Trigger Current			0.5			0.5		μA
Reset Voltage		0.4	0.7	1.0	0.4	0.7	1.0	V
Reset Current			0.1			0.1		mA
Threshold Current	Note 3		0.1	.25		0.1	.25	μA
Control Voltage Level	$V_{CC} = 15V$	9.6	10	10.4	9.0	10	11	V
	$V_{CC} = 5V$	2.9	3.33	3.8	2.6	3.33	4	V
Output Voltage (low)	$V_{CC} = 15V$							
	$I_{SINK} = 10mA$		0.1	0.15		0.1	.25	V
	$I_{SINK} = 50mA$		0.4	0.5		0.4	.75	V
	$I_{SINK} = 100mA$		2.0	2.2		2.0	2.5	V
	$I_{SINK} = 200mA$		2.5			2.5		
	$V_{CC} = 5V$							
	$I_{SINK} = 8mA$		0.1	0.25				V
	$I_{SINK} = 5mA$					.25	.35	
Output Voltage (high)								
	$I_{SOURCE} = 200mA$		12.5			12.5		
	$V_{CC} = 15V$							
	$I_{SOURCE} = 100mA$							
	$V_{CC} = 15V$	13.0	13.3		12.75	13.3		V
	$V_{CC} = 5V$	3.0	3.3		2.75	3.3		V
Rise Time of Output			100			100		nsec
Fall Time of Output			100			100		nsec

NOTES
1. Supply Current when output high typically 1mA less.
2. Tested at $V_{CC} = 5V$ and $V_{CC} = 15V$
3. This will determine the maximum value of $R_A + R_B$. For 15V operation, the max total R = 20 megohm.

EQUIVALENT CIRCUIT

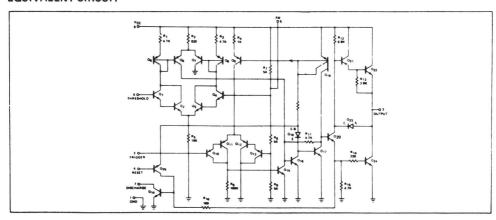

PHASE LOCKED LOOP

NE/SE565

NE/SE565-F,K,N

DESCRIPTION

The SE/NE565 Phase-Locked Loop (PLL) is a self-contained, adaptable filter and demodulator for the frequency range from 0.001Hz to 500kHz. The circuit comprises a voltage-controlled oscillator of exceptional stability and linearity, a phase comparator, an amplifier and a low-pass filter as shown in the block diagram. The center frequency of the PLL is determined by the free-running frequency of the VCO; this frequency can be adjusted externally with a resistor or a capacitor. The low-pass filter, which determines the capture characteristics of the loop, is formed by an internal resistor and an external capacitor.

FEATURES

- Highly stable center frequency (200ppm/°C typ.)
- Wide operating voltage range (±6 to ±12 volts)
- Highly linear demodulated output (0.2% typ.)
- Center frequency programming by means of a resistor or capacitor, voltage or current
- TTL and DTL compatible square-wave output; loop can be opened to insert digital frequency divider
- Highly linear triangle wave output
- Reference output for connection of comparator in frequency discriminator
- Bandwidth adjustable from < ±1% to > ±60%
- Frequency adjustable over 10 to 1 range with same capacitor

APPLICATIONS

- Frequency shift keying
- Modems
- Telemetry receivers
- Tone decoders
- SCA receivers
- Wideband FM discriminators
- Data synchronizers
- Tracking filters
- Signal restoration
- Frequency multiplication & division

PIN CONFIGURATIONS

F,N PACKAGE
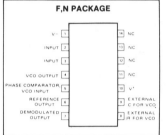

K PACKAGE
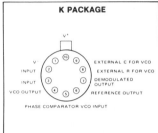

BLOCK DIAGRAM

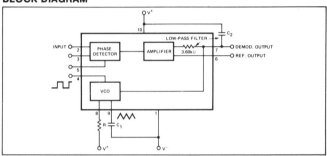

EQUIVALENT SCHEMATIC

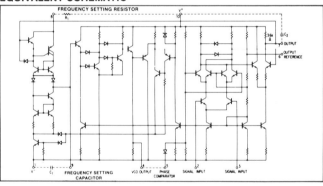

ABSOLUTE MAXIMUM RATINGS T_A = 25°C unless otherwise specified.

PARAMETER	RATING	UNIT
Maximum operating voltage	26	V
Input voltage	3	Vp-p
Storage temperature	−65 to +150	°C
Operating temperature range		
NE565	0 to +70	°C
SE565	−55 to +125	°C
Power dissipation	300	mW

Signetics

PHASE LOCKED LOOP

NE/SE565

NE/SE565-F,K,N

ELECTRICAL CHARACTERISTICS $T_A = 25°C$, $V_{CC} = \pm 6V$ unless otherwise specified.

PARAMETER	TEST CONDITIONS	SE565 Min	SE565 Typ	SE565 Max	NE565 Min	NE565 Typ	NE565 Max	UNIT
SUPPLY REQUIREMENTS								
Supply voltage		±6		±12	±6		±12	V
Supply current			8	12.5		8	12.5	mA
INPUT CHARACTERISTICS								
Input impedance[1]		7	10		5	10		kΩ
Input level required for tracking	$f_o = 50kHz$, ±10% frequency deviation	10	1		10	1		mVrms
VCO CHARACTERISTICS								
Center frequency								
Maximum value	$C_1 = 2.7pF$	300	500			500		kHz
Distribution[2]	Distribution taken about $f_o = 50kHz$, $R_1 = 5.0kΩ$, $C_1 = 1200pF$	-10	0	+10	-30	0	+30	%
Drift with temperature	$f_o = 50kHz$		200	525		300		ppm/°C
Drift with supply voltage	$f_o = 50kHz$, $V_{CC} = \pm 6$ to ± 7 volts		0.1	1.0		0.2	1.5	%/V
Triangle wave								
Output voltage level		1.9			1.9			V
Amplitude			2.4	3		2.4	3	Vp-p
Linearity			0.2			0.5		%
Square wave								
Logical "1" output voltage	$f_o = 50kHz$	+4.9	+5.2		+4.9	+5.2		V
Logical "0" output voltage	$f_o = 50kHz$		-0.2	+0.2		-0.2	+0.2	V
Duty cycle	$f_o = 50kHz$	45	50	55	40	50	60	%
Rise time			20	100		20		ns
Fall time			50	200		50		ns
Output current (sink)		0.6	1		0.6	1		mA
Output current (source)		5	10		5	10		mA
DEMODULATED OUTPUT CHARACTERISTICS								
Output voltage level	Measured at pin 7	4.25	4.5	4.75	4.0	4.5	5.0	V
Maximum voltage swing[3]			2			2		Vp-p
Output voltage swing	±10% frequency deviation	250	300		200	300		mVp-p
Total harmonic distortion			0.2	0.75		0.4	1.5	%
Output impedance[4]			3.6			3.6		kΩ
Offset voltage (V6-V7)			30	100		50	200	mV
Offset voltage vs temperature (drift)			50			100		μV/°C
AM rejection		30	40			40		dB

NOTES
1. Both input terminals (pins 2 and 3) must receive identical dc bias. This bias may range from 0 volts to -4 volts.
2. The external resistance for frequency adjustment (R1) must have a value between 2kΩ and 20kΩ.
3. Output voltage swings negative as input frequency increases.
4. Output not buffered.

TYPICAL PERFORMANCE CHARACTERISTICS

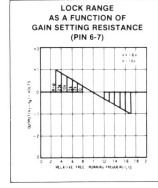

LOCK RANGE AS A FUNCTION OF GAIN SETTING RESISTANCE (PIN 6-7)

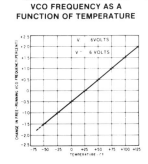

CHANGE IN FREE-RNNING VCO FREQUENCY AS A FUNCTION OF TEMPERATURE

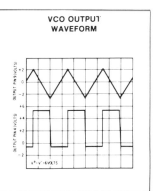

VCO OUTPUT WAVEFORM

MC1306P

1/2-WATT AUDIO AMPLIFIER

The MC1306P is a monolithic complementary power amplifier and preamplifier designed to deliver 1/2-Watt into a loudspeaker with a 3.0 mV(rms) typical input. Gain and bandwidth are externally adjustable. Typical applications include portable AM-FM radios, tape recorder, phonographs, and intercoms.

- 1/2-Watt Power Output (12 Vdc Supply, 8-Ohm Load)
- High Overall Gain — 3.0 mV(rms) Sensitivity for 1/2-Watt Output
- Low Zero-Signal Current Drain — 4.0 mAdc @ 9.0 V typ
- Low Distortion — 0.5% at 250 mW typ

1/2-WATT AUDIO AMPLIFIER

PLASTIC PACKAGE
CASE 626

MAXIMUM RATINGS (T_A = +25°C unless otherwise noted)

Rating	Symbol	Value	Unit
Power Supply Voltage	V^+	15	Vdc
Load Current	I_L	400	mAdc
Power Dissipation (Package Limitation) T_A = +25°C	P_D	625	mW
Derate above T_A = +25°C	$1/\theta_{JA}$	5.0	mW/°C
Operating Temperature Range	T_A	0 to +75	°C
Storage Temperature Range	T_{stg}	−65 to +150	°C

Maximum Ratings as defined in MIL-S-19500, Appendix A.

ELECTRICAL CHARACTERISTICS (V^+ = 9.0 V, R_L = 8.0 ohms, f = 1.0 kHz, (using test circuit of Figure 3), T_A = +25°C unless otherwise noted.)

Characteristic	Symbol	Min	Typ	Max	Unit
Open Loop Voltage Gain Pre-amplifier R_L = 1.0 k ohm Power-amplifier R_L = 16 ohms	A_{VOL}	− −	270 360	− −	V/V
Sensitivity (P_O = 500 mW)	S	−	3.0	−	mV(rms)
Output Impedance (Power-amplifier)	Z_o	−	0.5	−	Ohm
Signal to Noise Ratio (P_O = 150 mW, f = 300 Hz to 10 kHz)	S/N	−	55	−	dB
Total Harmonic Distortion (P_O = 250 mW)	THD	−	0.5	−	%
Quiescent Output Voltage	V_o	−	$V^+/2$	−	Vdc
Output Power (THD ≤10%, V^+ = 12 V)	P_o	500	570	−	mW
Current Drain (zero signal)	I_D	−	4.0	−	mA
Power Dissipation (zero signal)	P_D	−	36	−	mW

2N3905 • 2N3906
PNP SMALL SIGNAL GENERAL PURPOSE AMPLIFIERS AND SWITCHES
DIFFUSED SILICON PLANAR* EPITAXIAL TRANSISTORS

- P_D ... 625 mW @ $T_A = 25°C$
- V_{CEO} ... −40 V (MIN)
- h_{FE} ... 100-300 @ 10 mA (2N3906)
- NF ... 4.0 dB (MAX) WIDEBAND (2N3906)
- COMPLEMENTS ... 2N3903, 2N3904

See TO92-1 Package Outline

E B C
1 2 3

ABSOLUTE MAXIMUM RATINGS (Note 1)

†Maximum Temperatures
- Storage Temperature — −55°C to +150°C
- Operating Junction Temperature — 150°C
- Lead Temperature (60 seconds) — 260°C

†Maximum Power Dissipation (Notes 2 & 3)
- Total Dissipation at 25°C Ambient Temperature — 0.625 W
- at 70°C Ambient Temperature — 0.400 W
- at 25°C Case Temperature — 1.0 W

Maximum Voltages and Current
- V_{CBO} Collector to Base Voltage — −40 V
- V_{CEO} Collector to Emitter Voltage (Note 4) — −40 V
- V_{EBO} Emitter to Base Voltage — −5.0 V
- I_C Collector Current — 200 mA

ELECTRICAL CHARACTERISTICS (25°C Ambient Temperature unless otherwise noted)

SYMBOL	CHARACTERISTIC	2N3905 MIN.	2N3905 MAX.	2N3906 MIN.	2N3906 MAX.	UNITS	TEST CONDITIONS
h_{FE}	DC Pulse Current Gain (Note 5)	30		60			$I_C = 0.1$ mA, $V_{CE} = -1.0$ V
		40		80			$I_C = 1.0$ mA, $V_{CE} = -1.0$ V
		50	150	100	300		$I_C = 10$ mA, $V_{CE} = -1.0$ V
		30		60			$I_C = 50$ mA, $V_{CE} = -1.0$ V
		15		30			$I_C = 100$ mA, $V_{CE} = -1.0$ V
$V_{CE(sat)}$	Collector Saturation Voltage		−0.25		−0.25	V	$I_C = 10$ mA, $I_B = 1.0$ mA
	(Note 5)		−0.4		−0.4		$I_C = 50$ mA, $I_B = 5.0$ mA
$V_{BE(sat)}$	Base Saturation Voltage	−0.65	−0.85	−0.65	−0.85	V	$I_C = 10$ mA, $I_B = 1.0$ mA
	(Note 5)		−0.95		−0.95		$I_C = 50$ mA, $I_B = 5.0$ mA
I_{CEX}	Collector Cutoff Current		50		50	nA	$V_{CE} = -30$ V, $V_{EB} = -3.0$ V
I_{BL}	Base Cutoff Current		50		50	nA	$V_{CE} = -30$ V, $V_{EB} = -3.0$ V
BV_{CBO}	Collector to Base Breakdown Voltage	−40		−40		V	$I_C = 10$ μA, $I_E = 0$
BV_{CEO}	Collector to Emitter Breakdown Voltage (Note 5)	−40		−40		V	$I_C = 1.0$ mA, $I_B = 0$
BV_{EBO}	Base to Emitter Breakdown Voltage	−5.0		−5.0		V	$I_C = 0$, $I_E = 10$ μA
C_{ob}	Output Capacitance		4.5		4.5	pF	$I_E = 0$, $V_{CB} = -5.0$ V, $f = 100$ kHz
C_{ib}	Input Capacitance		10		10	pF	$I_C = 0$, $V_{EB} = -0.5$ V, $f = 100$ kHz
f_T	Current Gain Bandwidth Product	200		250		MHz	$I_C = 10$ mA, $V_{CE} = -20$ V, $f = 100$ MHz
t_d	Delay Time (see test circuit no. 333)		35		35	ns	$I_C \cong 10$ mA, $I_{B1} \cong 1.0$ mA, $V_{CC} = -3.0$ V
t_r	Rise Time (see test circuit no. 333)		35		35	ns	$I_C \cong 10$ mA, $I_{B1} \cong 1.0$ mA, $V_{CC} = -3.0$ V
t_s	Storage Time (see test circuit no. 239)		200		225	ns	$I_C \cong 10$ mA, $I_{B1} \cong 1.0$ mA, $I_{B2} \cong -1.0$ mA, $V_{CC} = -3.0$ V

Additional Electrical Characteristics on following page.

*Planar is a patented Fairchild process.
†Fairchild exceeds JEDEC registered value for this parameter.

NOTES:
1. These ratings are limiting values above which the serviceability of any individual semiconductor device may be impaired.
2. These are steady state limits. The factory should be consulted on applications involving pulsed or low duty cycle operations.
3. These ratings give a maximum junction temperature of 150°C and junction to ambient thermal resistance of 200°C/W (derating factor of 5.0 mW/°C); junction to case thermal resistance of 125°C/W (derating factor of 8.0 mW/°C).
4. Rating refers to a high current point where collector to emitter voltage is lowest.
5. Pulse conditions: length = 300 μs; duty cycle = 2%.
6. For product family characteristic curves, refer to Section 5 — SS23.

2N/PN2218A/19A • 2N/PN2221A/22A
NPN SMALL SIGNAL GENERAL PURPOSE AMPLIFIERS AND SWITCHES
DIFFUSED SILICON PLANAR* EPITAXIAL TRANSISTORS

- P_D ... 625 mW @ $T_A = 25°C$ (PN DEVICES)
- V_{CEO} ... 40 V (MIN) @ 10 mA
- h_{FE} ... 40-120 (2N/PN2218A, 2N/PN2221A) @ 150 mA
  ... 100-300 (2N/PN2219A, 2N/PN2222A) @ 150 mA
- t_{on} ... 35 ns (MAX) @ 150 mA, t_{off} ... 285 ns (MAX) @ 150 mA
- COMPLEMENTS ... 2N/PN2904A SERIES

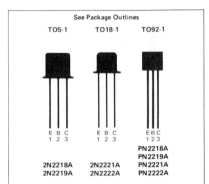

See Package Outlines: TO5-1, TO18-1, TO92-1

PN2218A / PN2219A / PN2221A / PN2222A
2N2218A / 2N2219A / 2N2221A / 2N2222A

ABSOLUTE MAXIMUM RATINGS (Note 1)

Maximum Temperatures	2N2218A/19A 2N2221A/22A	PN2218A/19A PN2221A/22A
Storage Temperature	−65°C to +200°C	−55°C to +150°C
Operating Junction Temperature	175°C	150°C
Maximum Power Dissipation	2N2218A 2N2221A	PN2218A/19A
(Notes 2 & 3)	2N2219A 2N2222A	PN2221A/22A
Total Dissipation		
at 25°C Case Temperature	3.0 W 1.8 W	1.0 W
at 25°C Ambient Temperature	0.8 W 0.5 W	0.625 W

Maximum Voltages and Currents
- V_{CBO} Collector to Base Voltage   75 V
- V_{CEO} Collector to Emitter Voltage (Note 4)   40 V
- V_{EBO} Emitter to Base Voltage   6.0 V
- I_C Collector Current   800 mA

ELECTRICAL CHARACTERISTICS (25°C Ambient Temperature unless otherwise noted)

SYMBOL	CHARACTERISTIC	2N/PN2218A 2N/PN2221A MIN.	MAX.	2N/PN2219A 2N/PN2222A MIN.	MAX.	UNITS	TEST CONDITIONS
h_{FE}	DC Current Gain	20		35			$I_C = 100 \mu A$, $V_{CE} = 10$ V
		25		50			$I_C = 1.0$ mA, $V_{CE} = 10$ V
	(Note 5)	35		75			$I_C = 10$ mA, $V_{CE} = 10$ V
	(Note 5)	40	120	100	300		$I_C = 150$ mA, $V_{CE} = 10$ V
	(Note 5)	25		40			$I_C = 500$ mA, $V_{CE} = 10$ V
	(Note 5)	15		35			$I_C = 10$ mA, $V_{CE} = 10$ V, $T_A = -55°C$
	(Note 5)	20		50			$I_C = 150$ mA, $V_{CE} = 1.0$ V
$V_{CE(sat)}$	Collector to Emitter Saturation Voltage (Note 5)		0.3		0.3	V	$I_C = 150$ mA, $I_B = 15$ mA
			1.0		1.0	V	$I_C = 500$ mA, $I_B = 50$ mA
$V_{BE(sat)}$	Base to Emitter Saturation Voltage (Note 5)	0.6	1.2	0.6	1.2	V	$I_C = 150$ mA, $I_B = 15$ mA
			2.0		2.0	V	$I_C = 500$ mA, $I_B = 50$ mA
h_{fe}	High Frequency Current Gain	2.5		3.0			$I_C = 20$ mA, $V_{CE} = 20$ V, $f = 100$ MHz
I_{CEX}	Collector Reverse Current		10		10	nA	$V_{EB} = 3.0$ V, $V_{CE} = 60$ V
I_{CBO}	Collector Reverse Current		10		10	nA	$I_E = 0$, $V_{CB} = 60$ V
			10		10	μA	$I_E = 0$, $V_{CB} = 60$ V, $T_A = 150°C$
I_{EBO}	Emitter Cutoff Current		10		10	nA	$I_C = 0$, $V_{EB} = 3.0$ V
C_{ob}	Output Capacitance		8.0		8.0	pF	$I_E = 0$, $V_{CB} = 10$ V, $f = 100$ kHz
C_{ib}	Input Capacitance		25		25	pF	$I_C = 0$, $V_{EB} = 0.5$ V, $f = 100$ kHz
$R_E (h_{ie})$	Real Part of Common Emitter High Frequency Input Impedance		60		60	Ω	$I_C = 20$ mA, $V_{CE} = 20$ V, $f = 300$ MHz
BV_{CBO}	Collector to Base Breakdown Voltage	75		75		V	$I_C = 10 \mu A$, $I_C = 0$

Additional Electrical Characteristics on following page.

*Planar is a patented Fairchild process.

NOTES:
1. These ratings are limiting values above which the serviceability of any individual semiconductor device may be impaired.
2. These are steady state limits. The factory should be consulted on applications involving pulsed or low duty cycle operation.
3. These ratings give a maximum junction temperature of 175°C, junction to case thermal resistance of 50°C/W (derating factor of 20 mW/°C), and junction to ambient thermal resistance of 188°C/W (derating factor of 5.33 mW/°C) for 2N2218A, 2219A. For the 2N2221A and 2N2222A, junction to case thermal resistance of 83.5°C/W (derating factor of 12 mW/°C) junction to ambient thermal resistance of 300°C/W (derating factor of 3.33 mW/°C). These ratings give a maximum junction temperature of 150°C, junction to case thermal resistance of 125°C/W (derating factor of 8.0 mW/°C); junction to ambient thermal resistance of 200°C/W (derating factor of 5.0 mW/°C) for PN2218A, PN2219A, PN2221A, PN2222A.
4. Rating refers to a high current point where collector to emitter voltage is lowest.
5. Pulse conditions: length = 300 μs; duty cycle = 1%.
6. For product family characteristic curves, refer to Section 5 – SS15.

2N/PN2218A/19A • 2N/PN2221A/22A

ELECTRICAL CHARACTERISTICS (25°C Ambient Temperature unless otherwise noted) (Cont'd)

SYMBOL	CHARACTERISTIC	2N/PN2218A 2N/PN2221A MIN.	MAX.	2N/PN2219A 2N/PN2222A MIN.	MAX.	UNITS	TEST CONDITIONS
BV_{CEO}	Collector to Emitter Breakdown Voltage (Note 5)	40		40		V	I_C = 10 mA, I_B = 0
BV_{EBO}	Emitter to Base Breakdown Voltage	6.0		6.0		V	I_C = 0, I_E = 10 μA
I_{BL}	Base Current		20		20	nA	V_{EB} = 3.0 V, V_{CE} = 60 V
t_d	Turn On Delay Time (see test circuit no. 231)		10		10	ns	I_{CS} = 150 mA, V_{CC} = 30 V, I_{B1} = 15 mA
t_r	Rise Time (see test circuit no. 231)		25		25	ns	I_{CS} = 150 mA, V_{CC} = 30 V, I_{B1} = 15 mA
t_s	Storage Time (see test circuit no. 232)		225		225	ns	V_{CC} = 30 V, I_C = 150 mA, I_{B1} = I_{B2} = 15 mA
t_f	Fall Time (see test circuit no. 232)		60		60	ns	V_{CC} = 30 V, I_{CS} = 150 mA, I_{B1} = 15 mA
T_A	Active Region Time Constant		2.5		2.5	ns	I_C = 150 mA, V_{CE} = 30 V
$r_{b'} C_c$	Collector to Base Time Constant		150		150	ps	I_C = 20 mA, V_{CE} = 20 V, f = 31.8 MHz
NF	Noise Figure				4.0	dB	I_C = 100 μA, V_{CE} = 10 V, R_G = 1.0 kΩ, BW = 1.0 Hz, f = 1.0 kHz
h_{ie}	Input Resistance	1.0	3.5	2.0	8.0	kΩ	I_C = 1.0 mA, V_{CB} = 10 V, f = 1.0 kHz
		0.2	1.0	0.25	1.25	kΩ	I_C = 10 mA, V_{CB} = 10 V, f = 1.0 kHz
h_{oe}	Output Conductance	3.0	15	5.0	35	μmho	I_C = 1.0 mA, V_{CE} = 10 V, f = 1.0 kHz
		10	100	25	200	μmho	I_C = 10 mA, V_{VB} = 10 V, f = 1.0 kHz
h_{re}	Voltage Feedback Ratio		500		800	$\times 10^{-6}$	I_C = 1.0 mA, V_{CB} = 10 V, f = 1.0 kHz
			250		400	$\times 10^{-6}$	I_C = 10 mA, V_{CB} = 10 V, f = 1.0 kHz
h_{fe}	Small Signal Current Gain	30	150	50	300		I_C = 1.0 mA, V_{CB} = 10 V, f = 1.0 kHz
		50	300	75	375		I_C = 10 mA, V_{CB} = 10 V, f = 1.0 kHz

Appendix B

SELECTED PROOFS AND DEVELOPMENTS

Proof of Miller's Theorem

An Amplifier's Pulsed Response

Closed Loop Output Resistance of an Operational Amplifier Circuit

Slew Rate

Infinite Gain Amplifier with Multiple Feedback Model

PROOF OF MILLER'S THEOREM (SIMPLIFIED)

The inverting amplifier A_1 in Figure B-1a has a finite voltage gain A but, to simplify the analysis, its input impedance Z_i is idealized or equal to infinity. The input voltage is v_S and the output voltage is v_O with Z_F a two-terminal feedback-connected impedance. The difference in potential across Z_F is

$$v_F = v_S - (-v_O) = v_S + v_O$$

The source current is equal to this potential difference divided by Z_F ($Z_i = \infty$).

$$i_S = v_F/Z_F = \frac{v_S + v_O}{Z_F}$$

The voltages v_O and v_S are related by the amplifier's gain A.

$$v_O = Av_S$$

Substituting for v_O in the equation for i_S produces

$$i_S = \frac{v_S + Av_S}{Z_F} = \frac{v_S(1+A)}{Z_F}$$

The ratio v_S/i_S is defined as the circuit's (not the amplifier's) input impedance Z_{in}.

$$Z_{\text{in}} \text{ (Miller)} = \frac{v_S}{i_S} = \frac{Z_F}{A+1}$$

The impedance $Z_F/(A+1)$ is the impedance seen by the source (with respect to ground), called the Miller input impedance.

The impedance Z_F can also be translated to an output impedance with respect to ground. This impedance is the ratio of the output voltage and the output current.

$$Z_{\text{out}} \text{ (Miller)} = \frac{v_O}{i_O} = \frac{-v_O}{-i_S}$$

Without a load, the output current is the same as the source and feedback current. In the

SELECTED PROOFS AND DEVELOPMENTS

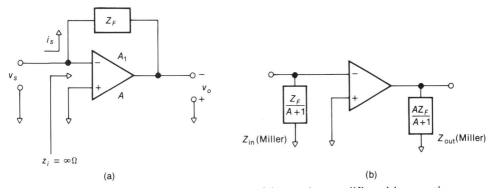

Figure B-1. Millerizing an amplifier circuit: (a) inverting amplifier with a negative feedback network (Z_F); (b) impedance translated circuit.

equation

$$i_S = \frac{v_S + v_O}{Z_F},$$

we substitute for v_S ($= v_O/A$) and obtain

$$i_S = \frac{v_O/A + v_O}{Z_F} = \frac{v_O\left(\frac{1+A}{A}\right)}{Z_F}$$

The ratio v_O/i_S is the Miller output impedance.

$$Z_{out}\,(\text{Miller}) = Z_F\left(\frac{A}{1+A}\right)$$

The impedance translated circuit is shown in Figure B-1b.

AN AMPLIFIER'S PULSED RESPONSE

The output voltage of an amplifier whose input is a voltage step is mathematically modeled by

$$v_o = V_{FV}(1 - e^{-t/RC})$$

where V_{FV} is the final value of the output voltage ($t \gg 0$) and RC is the product of the resistance and capacitance of the lag network,

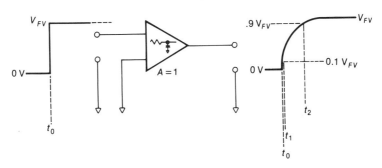

Figure B-2. An amplifier's pulsed response.

which establishes the upper cutoff frequency. The output voltage (Figure B-2) rises exponentially until the voltage V_{FV} is reached. The risetime T_R is defined as the time it takes the output voltage to go from 0.1 (10%) V_{FV} to 0.9 (90%) V_{FV}. At the 10% point, $v_o = 0.1\ V_{FV}$ and the time that this occurs is called t_1. Mathematically,

$$0.1\ V_{FV} = V_{FV}(1 - e^{-t_1/RC})$$
$$-0.9\ V = -e^{-t_1/RC}$$
$$e^{t_1/RC} = 1.1$$

Solving for t_1 produces

$$t_1 = RC \ln 1.1 = 0.1 RC$$

At the 90% point, $v_o = 0.9\ V_{FV}$ and the time that this occurs is called t_2.

$$0.9\ V_{FV} = V_{FV}(1 - e^{-t_2/RC})$$
$$-0.1 = -e^{-t_2/RC}$$
$$e^{t_2/RC} = 10$$

Solving for t_2 produces

$$t_2 = RC \ln 10 = 2.3 RC$$

Since $T_R = t_2 - t_1$, then

$$T_R = 2.2 RC$$

CLOSED LOOP OUTPUT RESISTANCE OF AN OPERATIONAL AMPLIFIER CIRCUIT

An amplifier circuit's output resistance r_{out} is the resistance seen between the amplifier circuit's output terminals under ac conditions. Since the reference for the output is typically ground or common, r_{out} is the resistance between the output terminal and ground. R_1 and R_2 of the circuit in Figure B-3 are the gain-setting resistors. The amplifier's finite input and output resistance is modeled by r_i and r_o. The amplifier has an open loop gain of value $-A$. The circuit output resistance is the parallel combination of $R_2 + R_1 \| r_i$ and some resistance we will call r_M. The resistance r_M will be the equivalent resistance to ground looking into the output circuit of the amplifier. It is called the Miller resistance and can be understood by considering the two circuits in Figure B-4. The voltage v_o' is the output voltage of the ideal amplifier and v_o is the output voltage of the circuit. Between the two voltages is r_o, which models the amplifier's finite output resistance. The Miller effect allows us to replace the resistance r_o, connected between the two floating terminals, with a

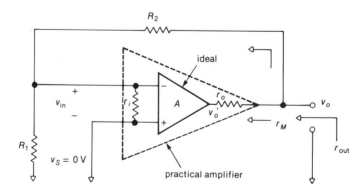

Figure B-3. Circuit for determining r_{out}.

SELECTED PROOFS AND DEVELOPMENTS

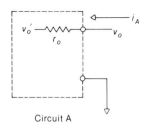
Circuit A

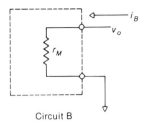
Circuit B

Figure B-4. Applying Miller's Theorem to find r_{out}.

resistance r_M, connected to ground. In order for the circuits A and B to be equivalent as far as v_o is concerned, we must have

$i_A = i_B$ where

$i_B = v_o/r_M$ and

$$i_A = \frac{v_o - v_o'}{r_o}$$

Thus, $\dfrac{v_o - v_o'}{r_o} = \dfrac{v_o}{r_M}$ or

$$r_M = \frac{r_o}{1 - v_o'/v_o}$$

The voltages v_o and v_o' are related by the following relationships.

$v_o' = -Av_{in}$ and

$$v_{in} = \frac{R_1 \| r_i}{R_2 + R_1 \| r_i} v_o = F v_o$$

F is the feedback factor (modified by r_i) and reflects the amount of output signal that is fed back to the input. Thus,

$v_o' = -AF v_o$ or

$$\frac{v_o'}{v_o} = -AF$$

Substituting for v_o'/v_o in the expression for r_M yields

$$r_M = \frac{r_o}{1 + AF}$$

The output resistance of the circuit is

$r_{out} = (r_M) \| (R_2 + R_1 \| r_i)$

The Miller resistance r_M is equal to the amplifier's output resistance divided by a very large number $(1 + AF)$.

$$r_M \ll (R_2 + R_1 \| r_i)$$

Thus, $r_{out} \cong \dfrac{r_o}{1 + AF} \cong \dfrac{r_o}{AF}$

If $r_o = 100\ \Omega$, $A = 100{,}000$, $F = 1/10$ ($A_{CL} = 10$), then $r_{out} = .01\ \Omega$. Other practical circuit considerations prevent the output resistance from ever being this low.

The output resistance of both the noninverting and inverting amplifier circuits is the same. The equivalent circuit for finding r_{out} is the same in both cases when the signal source is set to 0 V or shorted to common.

SLEW RATE (S_R)

Slew rate limit is the maximum rate of change ($\Delta v_o/\Delta t$) of the amplifier's output voltage. This parameter affects the gain-frequency performance of an operational amplifier under large signal conditions. A sinusoidal output signal will cease being small signal when its maximum rate of change exceeds the slew rate limit of the amplifier. A sinusoidal output voltage is mathematically modeled using the sine function.

$$v_o = V_p \sin 2\pi f t$$

V_p is the peak output voltage, and f and t are the frequency and time variables. The rate of

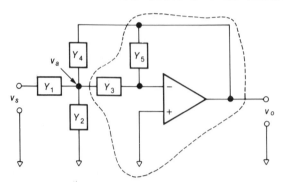

Figure B-5. Infinite gain, multiple feedback filter model.

change of v_o can be found using differential calculus.

$$\frac{d}{dt}v_o = \frac{\Delta v_o}{\Delta t} = 2\pi f V_p \cos 2\pi f t$$

The maximum rate of change for a sinusoidal signal occurs at the zero ($t = 0$) crossing.

$$\left.\frac{d}{dt}v_o\right|_{t=0} = 2\pi f V_p \text{ or,}$$

$$S_R = 2\pi f_{max} V_p$$

The frequency f_{max} is the highest distortion-free (<5%) frequency and is a function of the peak voltage and the S_R (slew rate) specification of the amplifier. The gain-frequency performance of the amplifier will be governed by small signal conditions if the output voltage signal frequency is less than f_{max}.

$$f \leq \frac{S_R}{2\pi V_p} \text{ (small signal)}$$

INFINITE GAIN AMPLIFIER WITH MULTIPLE FEEDBACK FILTER MODEL

The filter model in Figure B-5 contains five components that are either resistors or capacitors. At least two of the components must be capacitors for a second order filter. The devices are modeled by their admittances (Y). The A_1, Y_3, Y_5 section of the filter behaves as an inverting amplifier. If the voltage at node a is V_a, then v_a and v_o are related by

$$v_o = (-Y_3/Y_5) v_a$$

Applying KCL to node a yields the following relationship.

$$(v_a - v_S) Y_1 + v_a Y_2 + v_a Y_3 + (v_a - v_o) Y_4 = 0$$

$$v_a (Y_1 + Y_2 + Y_3 + Y_4) - v_o Y_4 = Y_1 v_S$$

Since $v_a = (-Y_5/Y_3) v_a$, then

$$-\frac{Y_5}{Y_3} v_o (Y_1 + Y_2 + Y_3 + Y_4) - v_o Y_4 = Y_1 v_S$$

$$-Y_5 v_o (Y_1 + Y_2 + Y_3 + Y_4) - v_o Y_3 Y_4$$
$$= Y_1 Y_3 v_S \text{ and}$$

$$v_o \{Y_5 (Y_1 + Y_2 + Y_3 + Y_4) + Y_3 Y_4\}$$
$$= -Y_1 Y_3 v_S$$

Solving for the transfer function produces

$$T(s) = v_o/v_s = \frac{-Y_1 Y_3}{Y_5 (Y_1 + Y_2 + Y_3 + Y_4) + Y_3 Y_4}$$

INDEX

Acceptor, 278
Ac diode resistance (r_d), 286
Ac equivalent circuit, 54
Ac ground, 56
Active devices, 23
Active filters, 257
Adjustable output regulator, 202, 205
All-pass filter, 244
α, 290
Amplification, 17
Amplitude, 211
AM radio dc circuits, 43
   local oscillator, 238
Antilog amplifier, 157
Assembly, 304
Astable mode, 230
Atom, 274
Attenuation slope, 249
Audio filters, 269

Band gap (ΔV_{BE}) reference, 201
Band-pass filters, 157, 244, 253, 262
Band-reject filters, 244, 254
Bandwidth, 19, 109, 149, 152
Barrier potential, 282
Base curve, 28
Base-driven model, 57, 68, 73
Base lag network, 98
Base resistance (r_b), 99
Batch processing, 295
Bel gain, 83

Bel voltage gain, 95
β, 27, 78, 290
Bias circuit design, 45
Bias circuit models, 35
Bias current (I_B), 161
Bipolar ICs, 294
Bipolar processing, 300
Bipolar transistors, 23, 288
Bode plot, 92, 104, 140, 247
Bonding, 305
Butterworth filter, 263
Bypass capacitor circuit, 108

Capture range, 237
Cascaded stages, 82
C_c, 98, 291
C_e, 98, 291
Center frequency, 248
Centered Q point, 118
Characteristic frequency, 248, 262
Chebyshev filter, 263
Clapp oscillator, 221
Classes of operation, 116
Class A amplifier, 118
Class AB amplifier, 124, 141
Class B amplifier, 124
Class C amplifier, 133
Closed loop gain, 148
CMRR, 165
Collector family of curves, 26
Collector feedback bias circuit, 39

335

336 INDEX

Collector-to-base feedback, 40
Collector lag network, 99
Colpitts oscillator, 221
Common base (CB) amplifier, 75
Common collector (CC) amplifier, 66, 69, 74, 80
Common emitter (CE) amplifier, 64, 68, 142
Comparators, 176
Complementary transistor pair, 129
Complex frequency, 246
Complex numbers, 92, 102, 247
Concept, 6
Conditions for oscillation, 213
Conduction, 274
Conductors, 275
Control system, 144
Corner frequency, 94, 103
Covalent bond, 276
Critical frequency, 94, 103
Crossover distortion, 126
Crystal oscillator, 221
Crystal pulling, 299
Current, 280
Current gain (β), 78, 290
Current limiting, 199
Current mirror, 127, 171
Current mode amplifiers, 170
Current regulator, 203
Current source, 157, 159
Current-to-voltage converter, 156
Cutoff frequency, 22, 94, 103
Cutoff point, 114
Cyclic frequency (f), 211, 246

Darlington transistor pair, 70
Data sheet parameters, 161
Dc diode resistance (R_D), 286
Dc equivalent circuit, 34
Depletion layer, 281
Depletion layer capacitance, 98, 286
Deposition, 294, 303
Derating factor, 123
Detector, 178
Dicing, 304
Die, 294
Differential amplifier, 85, 142, 154
Diffusion, 281, 294, 300
Diffusion capacitance, 98, 286

Digital-to-analog converter, 157
Digital interface, 181
Discrete, device, and distributed components, 22, 54, 265
Distortion, 20, 126, 212
Donor atom, 277
Dopant, 294
Dot orientation, 267
Drain faimly of curves, 27
Drift current, 281
Dual output supply, 205
Duty cycle, 218

Eddy currents, 265
Efficiency, 119
Electric field, 280, 283
Electron-hole pair, 279
Electrons, 275
Emitter bias circuit, 42
Emitter diode ac resistance (r_e), 58, 291
Emitter driven model, 57
Emitter follower, 74, 80
Energy, 6, 209
Energy band model, 278
Epitaxy, 294, 301
Error amplifier, 198
Excessive power dissipation protection, 199

f_c, 94, 103
Feedback, 142, 194
Feedback factor, 145
Filters, 188, 244
Filtering, 244
Fixed output regulators, 202, 204
FM demodulator, 237
Forward-biased diode, 284
Free-running multivibrator, 181
Frequency, 211
Frequency of oscillation, 213, 219
Function generators, 228, 234

Gain, 18
Gain-bandwidth-product (GBW), 140
Gyrator, 156

Half-wave rectifier, 157
Hartley oscillator, 221, 241
Heat sinks, 123
High current regulator, 204

INDEX

High-pass filters, 244, 251, 258, 269
Hole, 276
Hysteresis, 185, 265

IC components, 306
IC fabrication, 296
Ideal amplifier, 146
Ingot, 297
Input lead network, 107
Input resistance, 19, 67, 148, 152
Insulators, 275
Integrated circuit, 293
Integrating amplifier, 155
Integration, 295
Interfacing, 76
Intrinsic semiconductor, 276
Inverting amplifier, 150, 169, 172
Ion, 282
Ionization energy, 278

JFET bias circuits, 49
JFET common-source amplifier, 87
JFET transistor, 24

Kirchhoff's Current Law (KCL), 8
Kirchhoff's Voltage Law (KVL), 7

Lag network, 92, 97
Large signal, 118, 141
LC oscillators, 214, 221
Lead network, 102, 106
Line regulation, 191
Load line, 26
  ac, 114
  dc, 113
Load regulation, 192
Local oscillator, 241
Lock range, 237
Log amplifier, 157
Logarithms, 83
Loop gain, 145, 213
Lower cutoff frequency, 19, 91, 102, 265
Low-pass filters, 244, 250, 258, 269

Magnitude of a complex number, 93
Masks, 302
Metallization, 303
Midpoint bias, 40, 118
Miller capacitance, 13, 98

Miller's theorem, 12
Modified collector feedback bias circuit, 41
Modified lead network, 105
Monolithic regulators, 200
Monostable mode, 230
Motor speed control system, 238
Multiple-feetback filters, 260, 270

Negative conductance oscillator, 213
Negative feedback, 142
Negative regulator, 204
Negative resistance oscillator, 213
Negative saturation, 140
Noise, 20, 189
Nomographs, 232
Noninverting amplifiers, 146, 174
Norton amplifiers, 171
Norton's theorem, 10
n type semiconductor, 278, 296
Nucleus, 275
nV_{BE} bias circuit, 41

Offset current (I_{OS}), 164
Offset voltage (V_{OS}), 161
Open-loop gain, 148
Operating point, 26, 34
Operational amplifiers, 138
Order of a filter, 247
Oscillation, 209
Oscillators, 209
Output lead network, 107
Output power, 119
Output resistance, 19, 72, 148, 157
Ouput voltage swing, 166
Oxidation, 294, 302
Oxide removal, 294, 302

Packaging, 306
Passband gain, 249
Passive elements, 22
Pass transistor, 194, 198
Phase angle, 96
Phase of a complex number, 93
Phase detector, 235
Phase locked loop (PLL), 236
Phase relationships, 62
Photomasks, 302
Photoresist, 298, 302
Planar process, 281, 294, 298

pn junction theory, 281
Polar coordinate form, 93
Positive feedback, 212
Positive saturation, 140
Power amplifier ICs, 130
Power amplifiers, 113
Power dissipation, 122, 191
Power gain, 82
Power supply, 185
Principles, 6
Probing, 303
Programmability, 212
p type semiconductor, 278, 296
Pulsed response, 109

Q point, 22, 34, 113, 118
Quadrature oscillator, 227
Quality factor (Q), 249
Quiescent current, 189

Radian frequency (ω), 211, 246
Range, 212
r_b', 30, 99, 291
r_c', 30, 291
RC phase shift oscillator, 225
r_d, 286
R_D, 286
r_e', 58, 291
r_e' (large signal), 121
r_e' and temperature, 60
Recombination, 282
Rectangular coordinate form, 93
Rectangular wave oscillators, 218, 230
Rectifier circuits, 188
Reference detector, 178
Regulation, 185
Regulators, 185
Resonant, 214
Responses of a filter, 245
Reverse-biased diode, 284
Ripple rejection, 189
Risetime (T_R), 110
Rumble filter, 269

Sag time (T_S), 110
Sample-and-hold circuit, 157
Saturation point, 114

Schmitt trigger, 175, 232
Scratch filter, 269
Second-order filters, 260
Selective oxide removal, 294
Semiconductor materials, 275, 299
Semiconductor theory, 273
Series regulator circuits, 192, 194
Shunt regulator circuit, 193
Silicon atom, 275
Simulated inductor, 158, 258
Single-supply amplifier, 149
Slew rate (S_R), 141
Small signal, 60, 118, 140
Solid state, 294
Source resistance, 76
Square wave oscillator, 218
Stability, 144, 212
Stability factors, 47
Standard filter forms, 247
Summing amplifier, 153
Superposition theorem, 14
Supply current (I_S), 166
Swamped-emitter amplifier, 66, 79

Temperature coefficient, 191, 283
Testing, 306
Thermal resistance, 122
Thermal shutdown, 199
Thevenin's Theorem, 9
Thomson filter, 263
Timer, 230
T_R, 109
Transconductance, 24
Transfer characteristic, 139
Transfer function, 246
Transformer, 186, 264
Transient response, 20, 109
Transistor
    ac model, 58, 291
    dc model, 29
    junction temperature, 122
    theory, 288
    voltage gain, 62
Triangular wave, 232
Trickle bias, 126
T_S, 109
Tuned collector oscillator, 215

INDEX

Unijunction transistor (UJT), 213
Unipolar ICs, 294
Unipolar transistors, 24
Upper cutoff frequency, 19, 91, 100, 267

Valence bond, 276
Valence bond model, 274
Variable output lab supply, 206
VCCS, 160
Voltage-controlled oscillator (VCO), 232, 236

Voltage divider bias circuit, 35
Voltage reference, 198

Wafer, 294, 297, 300
Waveshape, 210
Wien-bridge oscillator, 215, 217
Window comparator, 180
Windows, 302

Zener regulator circuit, 189